The Ultimate
LNAT Collection

ISBN 978-1-912557-30-1

Published by *RAR Medical Services Limited* trading as **Infinity Books**

www.uniadmissions.co.uk
info@uniadmissions.co.uk
Tel: +44 (0) 208 068 0438

Oxbridge
LAW PROGRAMME

UNIADMISSIONS

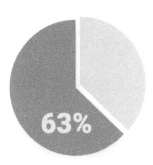

63%

UNIADMISSIONS 2019 Oxbridge Law Programme Success Rate

13%

The Average Oxford & Cambridge Law Success Rate

300+
Students successfully placed at Oxbridge in the last 3 years

50
Places available on our Oxbridge Law Programme in 2020

WHY DO OUR STUDENTS SEE SUCH HIGH SUCCESS RATES?

1

30 HOURS OF EXPERT TUITION.
UniAdmissions will guide you through a comprehensive, tried & tested syllabus that covers all aspects of the application - you are never alone.

2

UNPARALLED RESOURCES.
UniAdmissions' resources are the best available for the LNAT. You will get access to all of our resources, including the Online Academy, books and ongoing tutor support.

3

WEEKLY ENRICHMENT SEMINARS.
You'll get access to weekly enrichment seminars which will help you think like and become the ideal candidate that admissions tutors are looking for.

4

INTENSIVE COURSE PLACES.
By enrolling onto our Oxbridge Programme you will get reserved places for all of the Intensive Courses relevant to your application, such as the LNAT Intensive Course.

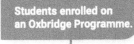

Students enrolled on an Oxbridge Programme.

Average success rate of the top 20 UK schools.

Average Oxford & Cambridge success rate.

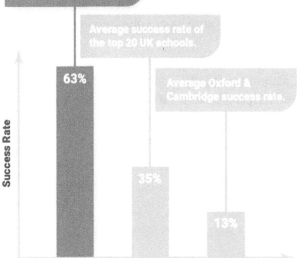

Success Rate

63%

35%

13%

UNIADMISSIONS Oxbridge Programme Average Success Rate

About the Authors

Will graduated from Downing College, Cambridge with a Law degree and has worked for *UniAdmissions* since 2015. His primary role is as a personal tutor, working with Oxbridge candidates to assist them through all stages of the application process. Since working for UniAdmissions, Will has successfully tutored a number of students preparing to sit the LNAT and is very familiar with all aspects of the paper.

He has an obsession with accuracy and in his spare time, he enjoys road cycling and keeping up with legal developments. He is keen to help others gain the same opportunities he has been afforded, and thus hopes to continue working with UniAdmissions to tutor law applicants from all backgrounds.

Rohan is the **Director of Operations** at *UniAdmissions* and is responsible for its technical and commercial arms. He graduated from Gonville and Caius College, Cambridge and is a fully qualified doctor. Over the last five years, he has tutored hundreds of successful Oxbridge and Medical applicants. He has also authored fourty books on admissions tests and interviews.

Rohan has taught physiology to undergraduates and interviewed medical school applicants for Cambridge. He has published research on bone physiology and writes education articles for the Independent and Huffington Post. In his spare time, Rohan enjoys playing the piano and table tennis.

The Ultimate
LNAT Collection

Dr William Antony
Dr Rohan Agarwal

UniAdmissions

How to use this Book

Congratulations on taking the first step to your LNAT preparation! First used in 2004, the LNAT is a difficult exam and you'll need to prepare thoroughly in order to make sure you get that dream university place.

The *Ultimate LNAT Collection* is the most comprehensive LNAT book available – it's the culmination of three top-selling LNAT books:

➢ *The Ultimate LNAT Guide*
➢ *LNAT Practice Papers: Volume 1*
➢ *LNAT Practice Papers: Volume 2*

Whilst it might be tempting to dive straight in with mock papers, this is not a sound strategy. Instead, you should approach the LNAT in the two steps shown below. Firstly, start off by understanding the structure, syllabus and theory behind the test. Once you're satisfied with this, move onto doing the 400 practice questions found in *The Ultimate LNAT Guide* (not timed!).

Then, once you feel ready for a challenge work through the four Mock Papers found in *LNAT Practice Papers* – these are a final boost to your preparation.

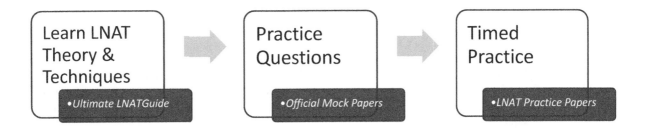

As you've probably realised by now, there are over 600 questions to tackle meaning that this isn't a test that you can prepare for in a single week. From our experience, the best students will prepare anywhere between four to eight weeks (although there are some notable exceptions!).

Remember that the route to a high score is your approach and practice. Don't fall into the trap that "you can't prepare for the LNAT"– this could not be further from the truth. With knowledge of the test, some useful time-saving techniques and plenty of practice you can dramatically boost your score.

Work hard, never give up and do yourself justice. Good luck!

The Ultimate LNAT Guide

The Basics

What is the LNAT?

The Law National Admissions Test (LNAT) is a 2 hour and 15-minute written exam for law students who are applying to certain UK universities. It is a computer-based exam that can be sat at different times with unique questions at each sitting. You register to take the test online and book a time slot at a test centre near you.

Test Structure

Section A:

There are 12 passages and 42 questions in the whole of Section A and you have 95 minutes to complete this section. It is a test of your critical thinking and comprehension skills, your ability to identify specific details in large passages and also understand the gist of the passages. It also tests your ability to understand arguments.

Section B:

One essay must be answered from a choice of 3, for which there are 40 minutes to complete. In particular, this tests your ability to construct coherent arguments and to argue persuasively.

Who has to sit the LNAT?

You have to sit the LNAT if you are applying for any of the following universities that ask for it in the current application cycle. You should check this list in May to see if the universities you are considering require it. The following is a list of the universities and courses requiring the LNAT for 2017 entry. As it is subject to change, it is included for guidance only.

University Name	Courses requiring LNAT
University of Birmingham	M100, MR11, MR12, M1N!, M2L6, M240
University of Bristol	M100, MR11, MR12
Durham University	M101, M102
University of Glasgow	M114, M1R7, M1R1, M121, M1R2, M122, M1R3, M1M9, 1RR, M1R4, M123, MN11, MN12, MV13, ML11, MQ13, MQ15, ML17, MV11, MV15, ML12, MR17
Kings College London	LM21, M100, M121, M122, M190
University of Nottingham	M100, M101, M1R1, M1R2
University of Oxford	M100, M190, M191, M192, M193, M194
SOAS, University of London	M100. *M102 (LLB Senior Status) does not require LNAT
University College London	M100, M101, M102, M141, M142, M144, M145, M146

Why is the LNAT used?

The LNAT was established in 2004 and is currently used by a consortium of nine UK universities. Law schools receive highly competitive applications with straight A grades at AS level and strong personal statements. They need to select the best from a pool of the very good applicants they already have and the LNAT fulfils this role.

The LNAT mark and essay are additional pieces of information that are used by admissions tutors when deciding between applicants. They are not used instead of A Levels and GCSEs but merely considered with them.

The multiple choice section examines your critical thinking and verbal reasoning skills, which are very important to do well at the law degree. The LNAT is an aptitude test and while it cannot be studied for, you can very usefully prepare for it. In our experience, it is possible to improve your LNAT score with only a small amount of work and with organised preparation, the results can be fantastic.

Different universities give different weightings to the LNAT. For example, the University of Bristol tends to give a candidate's LNAT a 25% weighting in the admissions decision (with the A Level at 40%, GCSEs at 20% and personal statement at 15%).

When do I sit the LNAT?

Registration for the LNAT normally opens from 1st August online.

You can decide when to sit the LNAT at the time you register for it online. You can choose your date and time slot. While there are available sittings throughout the year, you must sit the LNAT between 1st September through to 20th January for LNAT Universities. You can take the LNAT either before or after you send your UCAS application.

Given the early UCAS application deadline for Oxford University, you must sit the LNAT on or before 20th October in the admissions cycle.

No matter when you take the test, you will not be able to see your result before you send off your UCAS application. Results are issued to candidates in mid-February.

For Oxford:

Registration opens — 1 Aug 2018 → Testing begins — 1 Sept 2018 → Registration closes — 5 Oct 2018 → Testing finishes — 20 Oct 2018

For all other universities:

Registration opens — 1 Aug 2018 → Testing begins — 1 Sept 2018 → Registration closes — 15 Jan 2019 → Testing finishes — 20 Jan 2019

NB: Please check the dates as they are subject to change every year

How much does it cost?

In 2017, the cost for candidates was £50 at UK or EU test centres. If you are sitting the LNAT at a test centre outside the EU, the cost is £70.

Those candidates who struggle to fund the LNAT may be able to seek an LNAT bursary. This is done via the LNAT consortium's website and you are required to complete an application form online and attach one piece of documentary evidence. This must be done before paying for the test.

If you're a UK applicant, you will be able to claim an LNAT bursary in any of the following circumstances:
➢ Applicants who receive the 16-19 bursary in England, the top rate of EMA (Educational Maintenance Allowance) in Scotland, Wales or Northern Ireland or the Adult Learning Grant
➢ Where you reside with a family member who currently receives either:
  o Income support
  o Income-based Jobseekers' Allowance
  o Employment and Support Allowance
➢ Where you reside with a family member receiving child tax credits + you're named on the award + the household income as stated on the award is less than £30k.
➢ If you receive Pupil Premium payments or free school meals

If you're an EU applicant, you can claim for an LNAT bursary if you receive an equivalent state benefit in your country of residence.

If you're outside the EU, you are not eligible for an LNAT bursary.

Can I re-sit the LNAT?

You cannot resit the LNAT in the same application cycle. Whatever score you get is with you for the year. That's why it's so important to make sure you're well prepared and ready to perform at your very best on the test day.

If I reapply, do I have to resit the LNAT?

Yes. If you apply in a different admissions cycle then you need to retake the LNAT and the score from your new test will be sent to universities.

When do I get my results?

If you sit the LNAT by the 20th January deadline, you will receive your results by email in early February 2017. You will just receive your score of Section A and the average score for the cohort in the admissions cycle.

Section B is not marked by Pearson but is in fact marked by the universities' admissions tutors themselves. Accordingly, you will not get a mark back on Section B (but it is still taken into account in the admissions process)

Where do I sit the LNAT?

You can sit the LNAT at any Pearson test centre – these are the computer test centres where the driving theory test is taken. Once you book the test, you can choose the most convenient test centre to sit it at.

How is the LNAT scored?

Once you finish the test, your score is calculated by the computer. For all tests taken before 20th October, the universities that require the LNAT will receive the result directly from the test provider on the 21st October. On any day after 20th October, your test result will be sent directly to the LNAT universities within 24 hours of you taking the test.

Section A: Is scored out of 42 and the average score tends to vary from year to year, depending on the difficulty of the test. There are no mark deductions for incomplete or incorrect answers, so it's a good idea to answer every question even if it's a guess.

Section B: Your essay is sent directly to the universities and they will mark it themselves. They are testing your ability to construct a reasoned, persuasive and balanced argument, and to write in cogent English. There is a lot of essay writing at degree level, so Section B is a vital part of the LNAT.

How is the LNAT used?

The weighting which the universities place on the LNAT varies. The University of Bristol, for example, place a 25% weighting on it. The universities use the LNAT for a reason and place a lot of importance on both Section A and B. Therefore, the better you do at the LNAT, the better your chances of securing an offer. Given how competitive the law admissions process is, it is vital to produce a solid score on the LNAT. Each university will offer its own guidance on the LNAT on their admissions website.

How does my score compare?

This completely depends on how tough the test is and thus, the average scores. The average tends to vary each year but in the 2015-16 admissions cycle, the average mark for Section A was 23.3 out of 42.

Access Arrangements

If you need extra time, medication in the exam or any other special arrangement for the exam, you must apply with the Examination Access Requirements form, which is available from the LNAT website. It is important to apply before you book the test and it is necessary to attach evidence demonstrating your eligibility.

If you ordinarily receive extra time in public examinations, you would likely be granted it for the LNAT.

General Advice

Practice

Preparing for the LNAT will certainly improve your LNAT score. You won't be familiar with the type of questions in the LNAT and answering such questions in timed conditions. With practice, you'll become significantly quicker in reading the passages in Section A and your speed will increase greatly. Practice will also help you learn and hone techniques to improve your accuracy. This will make you calm and composed on test day, allowing you to perform at your best.

Start Early

It is easier to prepare if you practice little and often – this will be much more effective than just looking at practice questions on the night before the test. So start your preparation well in advance, ideally by August. This way, you will have a lot of time to work through practice questions to build up your speed and incorporate time-saving techniques into your approach. How to start? Well, by reading this you're obviously on track!

How to Work

It is necessary to focus on <u>both</u> Section A and Section B in your preparation. Even though you only get a score for Section A, universities place significant weight on Section B.

Section A:
It would be worth reading about the section and then going through a passage in your own time and understand what the questions are asking for. In particular, it would be useful to understand how an argument works by reading about the advice for Section A in this book. This will help you identify arguments in passages.
Even going through one passage and its set of questions per day would be worthwhile preparation whilst going through the worked solutions to any questions you answered incorrectly.

Closer to the test day, you would need to work in timed conditions, which is approximately 8 minutes per passage in order to become comfortable working under pressure. It is particularly important in the LNAT to be able to get the main idea or argument of the passage as a whole but to also focus on the details. Practice on Section A type questions will invariably make you faster.

Section B:
This is a very important part of the LNAT. A complaint of an admissions tutor when the LNAT first came out was that many candidates were unable to write a reasoned argument. So it would be worthwhile reading through the advice for Section B in this book and then to go about writing answers to topics you're unfamiliar with (as well as the examples given in this book). This will help you feel comfortable writing in exam conditions.

Crucially, Section B requires a broad range of knowledge and could involve any area of general knowledge. However, the main skill being assessed is your ability to construct a reasoned and persuasive argument, which is a vital skill for a law degree. Indeed, it would be worth getting a second person to take a look at a sample essay (such as an English/History A Level teacher). Even if you think you are fine writing essays in A Levels, writing an essay on an unfamiliar topic can sometimes unnerve people and the best way to overcome this is practice.

Repeat Tough Questions

When checking through answers, pay particular attention to questions you have got wrong. Look closely through the worked answers in this book until you're confident you understand the reasoning - then repeat the question later without help to check you can now do it. If you use other resources where only the answer is given, have another look at the question and consider showing it to a friend or teacher for their opinion.

Statistics show that without consolidating and reviewing your mistakes, you're likely to make the same mistakes again. Don't be a statistic. Look back over your mistakes and address the cause to make sure you don't make similar mistakes when it comes to the test. You should avoid guessing in early practice. Highlight any questions you struggled with so you can go back and improve.

Positive Marking

There is no negative marking in the LNAT – marking is only positive and you won't lose points for making a wrong answer. Therefore, if you aren't able to answer a question, you should guess.

For each question, there are 5 possible answers, thereby giving you a 20% chance of guessing correctly. If you need to guess, see if you can eliminate some of the options as this will improve your chances of making a successful guess.

Booking your Test

If you're applying to Oxford, it's necessary to take the test before the 20th October. It makes no difference what day you take it on as the results of all test results taken between the 1st September and 20th October are only sent to universities on the 21st October. That said, it would be best not to leave it too late in case of unforeseen circumstances, such as illness.

For all other universities, you just need to take the test before the 20th January. While it would be preferable not to leave it to the last minute, it is not necessary to do it in September or even before the 20th October. It would be much better to ensure that you're comfortable with the LNAT and have completed as much practice as possible. This would stand you in good stead for a solid LNAT result.

Mock Test

There are two full LNAT papers freely available at www.lnat.ac.uk and it would be worth going through them in timed conditions after you have worked through the questions in this book. There are also a further four full mock papers available in this book.

A word on timing...

"If you had all day to do your LNAT, you would get 100%. But you don't."

Whilst this isn't completely true, it illustrates a very important point. Once you've practiced and know how to answer the questions, the clock is your biggest enemy. This seemingly obvious statement has one very important consequence. **The way to improve your LNAT score is to improve your speed.** There is no magic bullet. But there are a great number of techniques that, with practice, will give you significant time gains, allowing you to answer more questions and score more marks.

Timing is tight throughout the LNAT – **mastering timing is the first key to success**. Some candidates choose to work as quickly as possible to save up time at the end to check back, but this is generally not the best way to do it. LNAT questions can have a lot of information in them – each time you start answering a question it takes time to get familiar with the instructions and information. By splitting the question into two sessions (the first run-through and the return-to-check) you double the amount of time you spend on familiarising yourself with the data, as you have to do it twice instead of only once. This costs valuable time. In addition, candidates who do check back may spend 2–3 minutes doing so and yet not make any actual changes. Whilst this can be reassuring, it is a false reassurance as it is unlikely to have a significant effect on your actual score. Therefore, it is usually best to pace yourself very steadily, aiming to spend the same amount of time on each question and finish the final question in a section just as time runs out. This reduces the time spent on re-familiarising with questions and maximises the time spent on the first attempt, gaining more marks.

It is essential that you don't get stuck with the hardest questions – no doubt there will be some. In the time spent answering only one of these, you may miss out on answering three easier questions. If a question is taking too long, choose a sensible answer and move on. Never see this as giving up or in any way failing, rather it is the smart way to approach a test with a tight time limit. With practice and discipline, you can get very good at this and learn to maximise your efficiency. It is not about being a hero and aiming for full marks – this is almost impossible and very much unnecessary. It is about maximising your efficiency and gaining the maximum possible number of marks within the time you have.

Top tip! Ensure that you take a watch that can show you the time in seconds into the exam. This will allow you have a much more accurate idea of the time you're spending on a question. In general, if you've spent more than 9 minutes on a section A passage – move on regardless of how close you think you are to finishing it.

Section A

Section A is the multiple choice section. In the exam, you will be presented with 12 passages and 42 questions, with approximately 3-4 questions per passage. There is a total of 95 minutes for this section and you cannot use any of the time for Section B in Section A – you only have a maximum of 95 minutes.

The aim of this section is to test the following skills:
➢ Comprehension
➢ Interpretation
➢ Deduction

This tests your ability to understand the different parts of a passage. It is important to understand what constitutes a good argument:
1. **Evidence:** Arguments which are heavily based on value judgements and subjective statements tend to be weaker than those based on facts, statistics and the available evidence.
2. **Logic**: A good argument should flow and the constituent parts should fit well into an overriding view or belief.
3. **Balance:** A good argument must concede that there are other views or beliefs (counter-argument). The key is to carefully dismantle these ideas and explain why they are wrong.

Sometimes, the question requires you to consider whether an argument is 'strong' or 'weak'. All arguments include reasons (premises) which aim to support a conclusion. Here, we are considering whether the reasons provide weak or strong support.

The parts of an argument:
An argument is an untimely attempt to persuade with the use of reasons. This can be distinguished from an assertion, which is simply a statement of fact or belief.

Assertion: It is raining outside.
Argument: I can hear the continuous sound of water splashing on the roof. Therefore, it must be raining outside.

The argument involves an attempt to persuade another that it is raining and it includes a reason as to why the speaker thinks it is raining, which is the splashing on the roof. The assertion, on the other hand, is not backed up with a reason – it is simply a statement.

An argument involves a <u>premise</u> and a <u>conclusion</u>.
A premise is simply a statement from which another can be inferred or follows as a conclusion.
A conclusion though is a summary of the arguments made.

For example:
> **Premise 1:** All dogs bark.
> **Premise 2:** My pet is a dog.
> **Conclusion:** My pet barks.

The conclusion here follows from both of the premises.

Explanation
Sometimes, it will be necessary to distinguish an argument from an explanation and you will need to be careful here as it can be difficult to distinguish sometimes. In essence, an argument will always involve an attempt to persuade the reader as to a point of view. Explanations, on the other hand, do not. Explanations may describe why something is the way it is or account for how something has occurred.

For example:

1. **Explanation:** We can hear the sound of water drops because the tap is leaking.
2. **Argument:** We can hear the sound of water drops. Therefore, we need to call the plumber.

Example 1 just accounts for *why* water drops can be heard – there is no attempt to persuade the reader that there are either water drops or that the tap is leaking. The tap leaking is just asserted as an explanation for the sound of the water drops.

In example 2, the author is advancing an argument as the author is making the case to call the plumber. The premise being the sound of water drops.

Premise vs. Conclusion

- A **Conclusion** is a summary of the arguments being made and is usually explicitly stated or heavily implied.
- A **Premise** is a statement from which another statement can be inferred or follows as a conclusion.

Hence, a conclusion is shown/implied/proven by a premise. Similarly, a premise shows/indicates/establishes a conclusion. Consider for example: *My mom, being a woman, is clever as all women are clever.*

Premise 1: My mom is a woman + **Premise 2:** Women are clever = **Conclusion:** My mom is clever.

This is fairly straightforward as it's a very short passage and the conclusion is explicitly stated. Sometimes the latter may not happen. Consider: *My mom is a woman and all women are clever.*

Here, whilst the conclusion is not explicitly being stated, both premises still stand and can be used to reach the same conclusion.

You may sometimes be asked to identify if any of the options cannot be "reliably concluded". This is effectively asking you to identify why an option **cannot** be the conclusion. There are many reasons why but the most common ones are:

1. Over-generalising: *My mom is clever therefore all women are clever.*
2. Being too specific: *All kids like candy thus my son also likes candy.*
3. Confusing Correlation vs. Causation: *Lung cancer is much more likely in patients who drink water. Hence, water causes lung cancer.*
4. Confusing Cause and Effect: *Lung cancer patients tend to smoke so it follows that having lung cancer must make people want to smoke.*

Note how conjunctives like hence, thus, therefore, and it follows, give you a clue as to when a conclusion is being stated. More examples of these include: "it follows that, implies that, whence, entails that".

Similarly, words like "because, as indicated by, in that, given that, due to the fact that" usually identify premises.

Assumptions

It is important to be able to identify assumptions in a passage as questions frequently ask to identify these.

An assumption is a reasonable assertion that can be made based on the available evidence.

A crucial difference between an assumption and a premise is that a premise is normally mentioned in the passage, whereas an assumption is not. A useful way to consider whether there is a particular assumption in the passage is to consider whether the conclusion relies on it to work – i.e. if the assumption is taken away, does that affect the conclusion? If it does, then it's an assumption.

> *Top tip!* Don't get confused between premises and assumptions. A **premise** is a statement that is explicitly stated in the passage. An **assumption** is an inference that is made from the passage.

Fact vs. Opinion

Sometimes you will be required to distinguish between a fact and an opinion. A fact is something that can be tested to be true or false. An opinion, on the other hand, cannot be tested to be true or false – it is someone's view on something and is a value judgement.

For example: "Tuition fees were reduced by the Welsh government in 2012. Many viewed this as a fair outcome."
Fact: Tuition fees were reduced by the Welsh government.
Opinion: It is a fair outcome.
> What one person sees as being 'fair' may not be 'fair' to another person – even if many people see a particular policy as fair. It is a normative statement that cannot be tested as true or false.

Correlation vs. Causation

Just because two incidents or events have occurred does not mean that one has caused the other. For example: "French people are known for having a glass of wine with dinner and they have a larger life expectancy than we do. Therefore, we should consume wine to be healthier."

This argument is flawed. There are 2 events: (i) French people known for having wine and (ii) French people having a larger life expectancy. There is no suggestion in the extract that (i) wine is causally related to (ii) or that having wine actually leads to a longer life. Accordingly, in itself, the premises do not adequately support the conclusion – there could be other reasons such as diet or exercise.

Approaching Section A

Responses

For each question, there are 5 options to choose from. Only one can be correct. Therefore, if you cannot find the correct one initially, you can use the process of elimination to find the correct one.
If you are stuck on a particular passage or question, do not spend too long on it as this can take time away from your other questions. It would be best to leave it until the end if you have time left.

The Passage

Take every fact in the passage as true and your answer must be based on the information in the passage only – so do not use your own knowledge, even if you feel that you personally know the topic. For example, if the question asks who the first person was to walk on the moon, then states "the three crew members of the first lunar mission were Edwin Aldrin, Neil Armstrong, and Michael Collins". The correct answer is "cannot tell" – even though you know it was Neil Armstrong and see his name, the passage itself does not tell you who left the landing craft first. Likewise, if there is a quotation or an extract from a book which is factually inaccurate, you should answer based on the information available to you rather than what you know to be true.

Read the Questions First

Different strategies work well for different people but indeed, having a look at the questions before going through the passage can help you focus on the important details in the passage in the first reading of it, thereby saving you time.

It would be best to try this strategy with some of the passages in this book to see if it works for you.

Timing

Even if you finish the questions before the 95 minutes runs out, you **cannot** use any of this extra time on Section B – you can only use this 95 minutes on Section A so you might as well go back through any questions that you found difficult or whether you were uncertain in any areas.

Knowledge

No knowledge is required for Section A and no knowledge of the law or current affairs is required.

Common Types of Questions

➢ What unstated assumption is being made?
➢ Which of the following is an assertion?
➢ What is the main idea in the passage?
➢ What is the main argument in the passage?
➢ Which of the following is an argument in favour of…?
➢ What is meant by?
➢ What conclusion is reached by the author?
➢ Which of the following weakens or strengthens the writer's argument?
➢ Which of the following is an assertion of fact?

Reading Non-Fiction

As well as critically analysing the passages in the book, a brilliant preparation for the LNAT is to engage in further non-fiction reading and to consider some of the following questions:
➢ What issues are being raised?
➢ What assumptions are made?
➢ What is the conclusion?
➢ Is there adequate support for the conclusion?
➢ Whose perspective is it coming from?
➢ How would you create a counter-argument?

Critically reading non-fiction, such as in a quality newspaper, will not only help improve your Section A performance but would also improve your knowledge bank for the Section B essay.

Top tip! Though it might initially sound counter-intuitive, it is often best to read the question *before* reading the passage. Then you'll have a much better idea of what you're looking for and are therefore more likely to find it quicker.

Section A Questions

Passage 1 – Controlled Drugs

There is a consensus among Parliamentarians that the current drug policy is simply not working. Approximately 1 in 12 adults in the UK have taken an illicit drug in the last year (amounting to 2.8 million people) and 1 in 5 young adults have taken an illicit drug. It is thus clear that the Government needs to do more. However, while it is clear that there needs to be a shift in policy, politicians cannot agree on what changes are needed.

Possessing a banned drug is a criminal offence but how can it be that all these individuals are potentially criminals? Is it moral to label these individuals as criminals? Around the world, there have been growing calls for the legalisation of drugs. In 2001, Portugal legislated to decriminalise the use of small amounts of drugs. Since then, drug consumption in Portugal has been below the European average and the percentage of young people aged 15-24 consuming drugs in Portugal has decreased. It is clear that the legalisation of drugs has not had the effect that opponents of the policy claimed it would have. Accordingly, decriminalising drugs may be a pointer in the right direction for the UK.

A key justification for criminalising the possession of drugs is that it would reduce the propensity of drug consumption (or deter people from consuming drugs). However, there is no strong evidence to support this notion. Once a person is in possession of a controlled drug, they have committed a criminal offence, yet this has not deterred the 2.8 million users. Further, a study by the European Union's Drugs Monitoring Agency found no correlation between harsher punishments for drug offences and lower drug consumption. This makes the argument for legalisation much more compelling.

Moreover, drug consumption in itself is a victimless crime in that it doesn't harm anyone apart from the drug user. Furthermore, the majority of users only consume drugs in small amounts which are unlikely to harm themselves. Any negative health effects that can be incurred are limited to the individual. This is in contrast to smoking, where 'passive smoking' can have a serious impact on others.

Opponents of legalisation have suggested that a drug addiction can lead to other crimes, such as theft and robbery, as the individual resorts to secondary crimes to fund their expensive addiction. Accordingly, they argue that taking controlled drugs can be criminogenic. However, this misses the point. The underlying reason for which individuals participate in such secondary crimes (e.g. robbery or theft) is the very high prices of controlled drugs, which are, in turn, a consequence of their prohibition. The very fact that they are illegal means that only criminal gangs end up supplying the controlled drugs, leading to the high prices. If the prohibition is removed, the increase in supply would reduce the price of the drugs and thus, reduce the 'need' to resort to crimes such as theft or robbery.

Legalisation is preferable to criminalisation but that is not to say that legalisation alone would suffice. Excess drug use should be seen as a public health issue, rather than a problem for the criminal law. While a drug addiction can lead to medical issues, so too can excess alcohol. Is it not incoherent for a society to allow any amount of alcohol consumption and yet totally prohibit the smallest consumption of controlled drugs? Accordingly, the freedom that individuals have to choose whether to consume alcohol should be accorded to them in regard to drugs.

1. What is the meaning of **criminogenic** in its context in this passage?
    A.   That consuming a controlled drug is a crime
    B.   That taking controlled drugs can lead to other crimes being committed
    C.   That taking controlled drugs is a victimless crime
    D.   That criminalisation is not the best response to reduce the consumption of drugs
    E.   Crimes such as theft or robbery

2. Which of the following is presented as being *paradoxical* by the author?
    A.   That smoking is not prohibited and yet drugs are prohibited
    B.   That alcohol is not prohibited and yet drugs are prohibited
    C.   That drug consumption is a victimless crime
    D.   That it is not drugs per se that lead to robbery or theft but the high prices of the drugs
    E.   That a justification for criminalising drugs is to reduce the consumption of drugs but there is no strong
         evidence to support that point

3. What is the main argument in the passage?
    A.   Drug use is a public health issue, rather than a problem for the criminal law
    B.   Drug consumption is victimless
    C.   Drug consumption is not criminogenic
    D.   That controlled drugs should be regulated
    E.   That controlled drugs should be legalised

4. What practical effect does the author believe would come about if the consumption of drugs were legalised?
    A.   Drug consumption would fall
    B.   Drug consumption would increase
    C.   Drug users would take part in fewer secondary crimes (such as robbery and theft)
    D.   It makes society fairer
    E.   Drug use would be seen as a public health issue

5. Which of the following would most weaken the author's main argument?
    A.   Drug consumption has a tendency to increase one's propensity for violence
    B.   Criminalisation is moral
    C.   Drugs have more negative health effects than alcohol
    D.   Drug dealers could turn to other crimes – such as people trafficking
    E.   It is not clear that there isn't a deterrent effect of criminalisation

Passage 2 – Sweeney Todd

Despite the fact that some associate musicals with cheesy joy, the genre is not limited to gleeful stories, as can be demonstrated by the macabre musical, 'Sweeney Todd'. The original story of the murderous barber appears in a Victorian penny dreadful, 'The String of Pearls: A Romance'. The penny dreadful material was adapted for the 19th century stage, and in the 20th century was adapted into two separate melodramas, before the story was taken up by Stephen Sondheim and Hugh Wheeler. The pair turned it into a new musical, which has since been performed across the globe and been adapted into a film starring Johnny Depp.

Sondheim and Wheeler's drama tells a disturbing narrative: the protagonist, falsely accused of a crime by a crooked judge, escapes from Australia to be told that his wife was raped by that same man of the court. In response, she has committed suicide, and her daughter - Todd's daughter - has been made the ward of the judge. The eponymous figure ultimately goes on a killing spree, vowing vengeance against the people who have wronged him but also declaring 'we all deserve to die', and acting on this belief by killing many of his clients; men who come to his barbershop. His new partner in crime, Mrs Lovett, comes up with the idea of turning the bodies of his victims into the filling of pies, as a way of sourcing affordable meat - after all, she claims, 'times is hard'.

Cannibalism, vengeance, murder, and corruption - these are all themes that demonstrate that this show does not conform to a happy-clappy preconception of its genre.

Sondheim and Wheeler's musical has been adapted into a number of formats over the years, including the film 'Sweeney Todd: The Demon Barber of Fleet Street' directed by Tim Burton. The nature of a film production necessitated a number of changes to the musical. Burton even acknowledged that while it was based on the musical, they were out to make a film and not a Broadway show. Accordingly, a three-hour musical was cut into a two-hour film, which brought a number of challenges: some of the songs and the romance between Todd's daughter and Anthony (a sailor) had to be removed.

There was initially concern though as the film actors, while critically acclaimed in their profession, were not professional singers. However, that turned out to be a non-issue as the film's soundtrack received glowing reviews, in particular, Depp's voice which received positive critical appraisals.

6. Which of the following statements are best supported by the above passage?
   A. Sondheim is a brilliant musician and lyricist
   B. Most musicals deal with morbid themes
   C. Wheeler is an avid penny dreadful fan
   D. Generalisations can be misleading
   E. Film adaptations lead to fundamental changes in the storyline

7. All the adjectives below are explicitly supported by the passage as ways of describing the crimes described within it, except:
   A. Comic
   B. Culinary
   C. Vengeful
   D. Sexual
   E. Disturbing

8. Which of the following statements best sums up Todd's belief?
   A. Bad people should die so good can live and prosper
   B. Good people should die because the bad have basically taken over
   C. All men should die
   D. All humans merit death
   E. Death is unavoidable

9. Which of the following statements is best supported in the above passage?
   A. There are four themes in 'Sweeney Todd'
   B. Legal corruption is the predominate theme of 'Sweeney Todd'
   C. Several 'Sweeney Todd' themes are morbid
   D. There is nothing positive in 'Sweeney Todd'
   E. Sadness is the focus of Sweeny Todd

10. Which of the following is true?
    A. Mrs Lovett and Sweeney Todd are in a romantic relationship
    B. All of the songs from the musical were removed or adjusted
    C. The storyline of the film adaptation was fundamentally different to the musical
    D. The film did not receive positive critical acclaim
    E. The film actors did not have professional musical experience

Passage 3 – Youth Unemployment

Youth Unemployment -that is: those young people who are in search of work but are unable to get into work -is disturbingly high. The current youth unemployment figure for the UK is at an unsettling 12%. This is much higher than that of other developed countries such as Germany and Switzerland and the societal implications of this are greater than what the politicians acknowledge. The longer a young person is unemployed, the less likely they are to find a job at all. This has destructive effects on the country: it increases the government's spending on welfare pay-outs, reduces the economy's capacity and increases the likelihood of crime. The personal impact of youth unemployment is equally devastating; lower quality of life, low self-esteem, and lack of confidence and even depression, which can lead to a never-ending cycle of unemployment. It is, thus, clear that youth unemployment is a dangerous virus that demands immediate government attention.

There are a number of reasons for the high youth unemployment rate, such as the sluggish state of the economy and the global financial crash in 2007/08. When the economy is not doing well, businesses tend to lay off workers in response to a lack of sales. This happened in 2008 when the economy slumped and unemployment drastically increased. Since then, the economy has only recovered lethargically.

However, this alone does not account for the entirety of the youth unemployment rate. Since the economic slump of 2008, total unemployment has reduced to 5.4%, while youth unemployment is at a much higher 12%. Why is there such a big difference? Is it just an inherent feature of society? Do businesses not want young people? A number of young people report that there aren't enough jobs for them. Yet at the same time, businesses say they are desperate to find skilled young people. They just can't find young people with the right skills to suit their needs. For example, Dulux, the paint manufacturer, has pointed out that there simply aren't enough skilled painters and decorators. In London, two-thirds of construction firms have had to turn down work as they don't have enough practical and skilled workers. And herein lies the problem – many young people do not have the skills that businesses are looking for.

That is not to say that it is the fault of those who are unemployed. The root of the problem is the lack of courses that are geared to the kind of skills that businesses want and the existing structural inadequacies within our education system. The head of Ofsted recently pointed out that the lack of high-quality vocational courses in England is a concern. Vocational courses have traditionally been seen as a 'second-rate' option in the country, with the academic A Levels being the 'gold' standard. This view must change – not everyone is destined for academia and vocationally trained individuals have an important role in our society. Would you rather have a well-read English graduate or a vocationally trained engineer fix your central heating? Thus, the government must make high-quality vocational provision a priority. Vocational education tends to be incorrectly seen as second-rate by students and this must change. Putting an emphasis on vocational courses will address the skills shortage in the UK, make the UK more productive, and crucially improve the prospects of our young generation. In addition, the education sector and businesses should engage with each other more closely to ensure that skill deficits are addressed in the national curriculum.

A report from the Institute for Public Policy Research (IPPR) suggests that youth unemployment tends to be lower in countries where there is a vocational route into employment and not just an academic one. This shines a lot of light on the situation in the UK.

11. Which of the following is **not** a potential personal impact of youth unemployment?
    A. Lower quality of life
    B. Increases the government's spending on welfare pay-outs
    C. Lack of confidence
    D. Low self-esteem
    E. Depression

12. Which of the following is the underlying reason for the high youth unemployment rate?
    A. The global financial crash of 2007/08
    B. Not enough jobs for young people
    C. Lack of skills
    D. The head of Ofsted
    E. The lack of high-quality vocational courses

13. Which of the following is implied but **not** stated in the passage?
    A. There is a mismatch between the skills that young people have and the skills that employers are looking for
    B. Young people don't have the skills that businesses are looking for
    C. Teachers should encourage young people to undertake vocational courses
    D. Businesses should provide training to improve the skills of young people
    E. Unemployment is bad

14. Which of the following is the author's main argument in the passage?
    A. An increased emphasis should be placed on vocational courses
    B. An increase in the skills of young people needs to be brought about
    C. Better jobs for young people are needed
    D. That unemployment has caused a lack of skills
    E. That there aren't enough skilled young people

15. According to the author, what can businesses do to reduce youth unemployment?
    A. Create more jobs
    B. Increase young people's skills
    C. Engage with the education sector
    D. Train more young people
    E. Create vocational courses

Passage 4 – The English Reformation

In the early 1500s, King Henry VIII set the English Church on a different course forever. Henry was undoubtedly a devout catholic when he took the throne. Indeed, he was a staunch defender of Catholicism in the face of threats from religious reformers, such as Luther. Impressed by Henry VIII's defence, the Pope gave him the title 'Defender of the Faith'. So how did Henry come to separate from the Roman Church?

Although historians are not universally in agreement, many put Henry VIII as the key driver behind separating the Church of England from Rome. Henry was disappointed in his marriage with Catherine of Aragon as, in spite of multiple pregnancies, they only had one daughter together. Henry though was desperate to conceive a son. He had a monumental ego and was, thus, concerned about his legacy. In order to secure his dynasty and ensure that the Tudor reign remained strong, he needed a legitimate son. Accordingly, he was eager to secure a divorce with his current wife and marry Anne Boleyn with the aim of having a legitimate son with her. The English church was under the authority of the Roman Catholic Church (of whom the Pope was the leader) and in order to separate from Catherine, Henry needed to obtain an annulment from the Pope. Despite the mammoth efforts of Henry's right-hand man, he was unable to secure an annulment of the marriage from Rome, which would have been the straightforward option. It became clear that Rome was not going to budge on this and from then, Henry began to pursue a separation from the Roman Church.

Historians also point to another reason for Henry's desire to break away from Rome. He liked the idea of being the only head of the church and the supreme leader. His ego influenced many of his key decisions, such as engaging in wars abroad, and this decision was no different.

A number of historians suggest that Thomas Cromwell was the man behind the separation. Indeed, Cromwell played a significant role in engineering it. With control of the King's parliamentary affairs, he persuaded Parliament to enact a supplication pronouncing Henry as 'the only head' of the church, establishing the doctrine of royal supremacy. This was in clear conflict with Papal authority and began the process of breaking away from the Roman Church. But while it is clear that Cromwell had a vital role in the break from Rome, the obvious must still be repeated – were it not for Henry's desire of a break, there would not have been such a break.

Through a series of Acts of Parliament over two years, the break from Rome was secured and ties between the English church and Rome were severed. One such Act of Parliament in 1934, the Act of Supremacy, declared the King as 'the only Supreme head in earth of the Church of England.' This drastic change put the English church on a new course and while there were no major day-to-day changes initially, it planted the seed for the differences we see today between the Roman Catholic Church and the Church of England.

16. What was the ultimate cause of the Church of England's breakaway from Rome?
    A. Henry VIII's ego
    B. Rome wouldn't grant him a divorce
    C. Henry wanted a son
    D. Royal Supremacy
    E. Religious reasons

17. What does 'dynasty' mean in the Passage?
    A. Family
    B. Henry's control of the Kingdom
    C. Succession of people from Henry's family to the throne
    D. Exertion of dominion by the Tudors
    E. The power of Henry VIII

18. Why did Henry want a son?
    A. To secure Royal Supremacy
    B. He wanted to divorce Catherine
    C. To secure the Tudor reign
    D. Males were preferred in the 16th Century
    E. None of the above

19. Which of the following is an unstated assumption?
    A. Henry had an ego
    B. There was no opposition to the reformation
    C. The public was supportive of the break from Rome
    D. Henry needed Cromwell to make the break from Rome
    E. Henry believed that he couldn't get a divorce through Rome

20 What is the Royal Supremacy?
    A. The breaking away from Rome
    B. The idea of the King being the supreme authority
    C. The King becoming the leader of the church
    D. The authority of the Pope over the Church
    E. The Act of Supremacy 1934

Passage 5 – Charities and Public Schools

What constitutes a charity is a matter of public significance, but also an important issue in determining the taxable income a charity receives. In the popular sense, charities are seen as institutions which primarily help the poor, however, a question has been raised as to why public schools should be considered as charities considering the fees required to attend them.

In order to be classified and registered as a charity, it is necessary for an institution to demonstrate that its purposes are for the public benefit. Once accorded charity status, the institution gains a number of fiscal benefits from the government. For example, no corporation tax is paid on most types of income. In contrast, corporation tax, which currently stands at 20% of all profits, is paid by all other private businesses. The law should not allow a 'free-for-all' where any profit-making company can be a charity by just doing a minuscule charitable act, as this would have a negative impact on the public purse. Nonetheless, charitable status is highly sought out by many organisations for these reasons and has become highly controversial in the case of public schools.

Public schools charge a fee for admission, in contrast with state schools, which are funded by the Government. Accordingly, as public schools are private institutions, becoming charitable will help their finances. Whether this should be possible hinges on what acting for the public benefit means and requires in the context of education.

In 2011, the Independent Schools Council (ISC), representing public schools, sought a judicial review of the Charity Commission's guidance on what is required for a public school to demonstrate a 'public benefit'. The ISC argued that they did provide a public benefit, but they did face opposition. The Education Review Group, who helped draft the Commission's disputed guidance, also intervened in the case, advancing arguments in the trial. Ultimately, the tribunal held that the Commission's guidance was wrong as a matter of law and required them to change it. The trial judge decided that in order to operate for the public benefit, a sufficient section of society must directly benefit from the education provided, which he said must include children whose parents would be unable to afford the fees without assistance.

In the trial, a number of arguments were advanced on either side of the issue. One such argument was that independent schools are a net cost to society as they remove able pupils from state schools and present barriers to social mobility. However, the tribunal did not consider such an argument as it related to a 'political' issue, rather than a judicial one.

Further, private education provision can provide a multitude of benefits to society. Indeed, it educates the children whose parents pay for the provision. While this may not seem inherently charitable because parents are paying for the education, there are public benefits too. Firstly, the provision of education in itself is a benefit – having an educated population benefits not only the individuals through enabling them to enjoy a higher living standard, but also the general economy. More taxes will be paid and there will be less crime. That is not to say that we should ignore the gap between public schools and state schools. Indeed, state schools that are struggling should be willing to receive help from public schools and public schools should, in accordance with their public duty, offer such help.

21. Which of the following is definitely true based on the passage?
    A. Any institution that provides a public benefit gains fiscal benefits from the government
    B. Every organisation would rather be a charity than a private company
    C. If an organisation is not a charity, it does not provide a public benefit
    D. Charities may not have to pay corporation tax
    E. The law should not allow a free-for-all

22. Which of the following is required if an organisation is to become a charity?
    A. To help the poor
    B. Nothing
    C. To exist for the public good
    D. To demonstrate that the fiscal benefits gained would be for the public benefit
    E. To not make private profits

23. Who is most likely to have advanced the argument that public schools are a net cost to society based on the passage?
    A. The Independent Schools Council
    B. The tribunal judges
    C. State schools
    D. The Education Review Group
    E. The government

24. Which of the following is an opinion as opposed to a fact?
    A. The tribunal did not consider the argument that there was a net cost to society
    B. No corporation tax is paid on most types of income
    C. Public schools charge a fee for admission
    D. The law should not allow a free-for-all
    E. A charity has to demonstrate that it operates for the public benefit

25. Which of the following would have adequately supported the argument that there is a public benefit from public schools before the tribunal?
    A. Public schools educate the children whose parents pay for it
    B. Public schools provide scholarships to others who can't afford the fees
    C. Public schools can make better use of money, as opposed to it being paid through tax
    D. Public schools are better than state schools
    E. State schools can learn from public schools

Passage 6 – Amazon vs. Hachette

The public does not normally witness corporate trade negotiations or disputes. They are generally held behind closed doors and in private for the mutual benefit of the companies in the dispute. However, there was an exception in the dispute between the international publisher, Hachette, and Amazon in 2014. Both of them are powerful organisations with market power, however, this episode has shown that one is more powerful than the other.

It is first necessary to go into the background of this dispute. The sale of a book involves three main protagonists. Arguably the most important, the author writes the book. The publisher prints and distributes the book. The retailers then act as the point of sale to consumers. In the US, there are five very large publishers who have enjoyed significant market dominance. When distributing their books, publishers want them in the biggest retailers and crucially, the biggest of them all by far is Amazon. It is estimated by some that 50% of all book sales (both printed and electronic) across the US go through Amazon. It is the most dominant bookseller and, therefore, it is imperative for publishers to get their books on Amazon. In order to do this though, each publisher needs to enter into a legal contract with Amazon, which is normally a private arrangement.

In 2014, Hachette and Amazon were in negotiations to renew their contract for the pricing and distribution of Hachette's books. While the exact issues in the negotiations remain private, it became clear that the negotiations weren't going well. Amazon stopped selling a number of Hachette's books and delayed deliveries of many by weeks. Famous books such as those by *JK Rowling* were delayed. Was this just business? Or did Amazon go too far? It infuriated both Hachette and the authors of the books that Hachette publish. It showed the length that Amazon would go to in order to get what they want. Hachette's authors, who normally stayed out of publisher-retailer contracts, weighed in and criticised Amazon. Amazon had used their enormous market power to restrict the sales of the books from Hachette to try to get their way and many authors argued that Amazon had abused their market power. However, it was only a minority of authors (and mainly the successful ones) that spoke out.

In reality, Hachette and their authors are not the innocent victims in all of this. Hachette, with their market power in publishing, conspired with the other major US publishers and Apple to fix the prices of eBooks (i.e. to keep them artificially high) in 2012. When the US Department of Justice sued, the publishers (including Hachette) made a settlement for $164m. So it's a bit rich for Hachette to complain about Amazon's aggressive price strategy.

In regard to individual authors, the major publishers haven't always been friendly either. It is a monstrous task for an up and coming author to get even a small book deal with a publisher. Publishers generally have a narrow view as to what a suitable book is and are primarily focused on what they think the monetary returns will be. Amazon, though, has taken a new step. They have introduced a suite of services that allows authors to self-publish their work through Amazon. It allows individuals to publish both an eBook and a print book. Amazon, with their vast resources, are also able to offer a 'print-on-demand' service whereby Amazon prints each book to order. This bypasses the need for traditional publishers or any need for a large pot of cash to fund a print run. Surely, this genius innovation should be applauded. It allows many more small-time authors to self-publish their works and disrupts the unfairness that the big publishers created. Yes, Amazon has excessive market power, but at least they're using it to the benefit of small authors, unlike the traditional publishers. So let them engage in whatever tactics they want to with Hachette.

26. What is the author's view as to the balance of power between Amazon and Hachette?
    A. Hachette is more powerful than Amazon
    B. Amazon is more powerful than Hachette
    C. They are both powerful
    D. The author doesn't have a view as to which
    E. They have both exerted market power

27. The author stated that it is 'a bit rich for Hachette to complain about Amazon's aggressive strategy'. What is he suggesting about Hachette's complaint?
    A. It attacks Amazon's views
    B. It's ironic
    C. It's unfair
    D. It's sarcastic
    E. It's awkward

28. What is the main conclusion of the author's article?
    A. That Amazon has a lot of market power
    B. That Hachette has a lot of market power
    C. Amazon are disruptive
    D. Amazon has done more for small-time authors than Hachette has done
    E. Amazon's actions against Hachette are just business

29. What did the author imply by the use of the word 'monstrous' in the passage?
    A. That the publishers are monsters
    B. That it's a big task to get a book deal
    C. That it is unacceptably too tough to get a book deal
    D. That authors have to work hard
    E. That authors have to act like monsters

30. Which of the following would, if true, most undermine the author's argument in the final paragraph?
    A. Amazon charge high fees to authors for their 'self-publishing' services
    B. Amazon is not the first company to offer services in self-publishing
    C. Hachette offers the same 'self-publishing' services
    D. Amazon has caused a loss to some authors from their aggressive negotiations with Hachette
    E. Amazon are taking away business from the publishers

31. Which of the following would the author most likely disagree with?
    A. It would be easier for authors to use Amazon's services than that of a traditional publisher
    B. Hachette and Amazon have a lot of market power
    C. Amazon has abused their market position
    D. Hachette has abused their market position
    E. Authors have been given a hard deal by the main publishers

Passage 7 – Online Courts

There are two main ways to resolve a private dispute concerning a question of law. Firstly, you can come to a private settlement with the person you disagree with – this can either (i) just be between the two sides of the case or (ii) involve a third party as a mediator. Alternatively, if that doesn't work, you can sue the other person and seek redress from the courts. It is well known, though, that the court system in the UK is expensive, inefficient, and not suited to the needs of the ordinary person.

Let's say, for example, that you are having an electrician complete some wiring in your house at a price of £300. You pay the electrician and he leaves, but it then transpires that he completed the task erroneously and you want your money back. How do you go about getting it back? Would you go to court? The court fee alone is £35 for each side in the case though and any legal advice from a solicitor would cost approximately £200 per hour. It must be noted that there is a well-established principle that the losing side in a civil case pays the winning side's legal costs. However, in the event that you don't win the case, you could potentially lose even more money from having to pay the winning side's (the electrician's) legal costs.

The cost is disproportionately high here in relation to the value of the claim and herein lies a flaw in the justice system. The high cost limits access to justice. The existing court process takes too long, involves too much paperwork and unnecessarily involves the use of expensive lawyers. Lord Dyson, a leading judge in the Court of Appeal (the second highest court in the UK), echoed these comments. Crucially, there are ways to make the system more efficient and the best way is to introduce an online court.

Firstly, we must accept the truth: lawyers aren't always needed for small-time disputes. We can look to eBay for inspiration. eBay is a well-renowned online auction site where private individuals can sell goods to other individuals. It is not always smooth sailing, however, and frequently disputes arise (for example, when a damaged defective item is sold). When disagreements arise, the seller and buyer are encouraged to negotiate online. If negotiation fails, eBay offers an online resolution service whereby an eBay official decides on the case and makes a binding decision. No lawyers and not even any face-to-face interaction with the eBay official. For simple matters, it would be unnecessary and inefficient to hire a lawyer to complete the task. Crucially, this means much lower costs than in a courtroom.

This system should be used by the justice system for small claims. There should be an online mediation system to allow each side to negotiate in an online discussion area.

Anyone watching the ITV hit television show, 'Judge Rinder', would realise that small-time claims don't require a lawyer. The setting for the show resembles a courtroom where small-time disputes are heard before a 'judge' who mediates between each side. The 'judge' then makes a decision that binds each side. There are no lawyers and each person represents him or herself. While there are a wealth of differences between the TV show and actual court proceedings, it does show that lawyers are not always needed to resolve disputes.

Accordingly, the Government should go one step further and establish an online court to resolve small-time legal disputes. This would involve real judges from the judiciary deciding cases online. They would review the documentary evidence submitted online by the parties and, if necessary, conduct a hearing via video link. While it would cost a lot to set up the online system, it would, in the long run, result in significant cost efficiencies both to the government and to the users of the online court.

32. Which of the following most undermines the author's argument in the second and third paragraphs?
    A. For small claims, people don't need to go to court to resolve disputes – there are alternative methods (such as private mediators)
    B. Such small cases are normally successfully settled outside of court
    C. The UK's justice system is cheaper than that of many other countries in the world
    D. The losing party pays the winning party's legal costs
    E. Solicitors fees for court cases are greater than their fees for out-of-court settlement

33. Assume that you and the electrician go to court. You each take 1 hour of legal advice from a solicitor. How much are the total legal fees for the losing side in the case?
    A. £35
    B. £200
    C. £235
    D. £300
    E. £470

34. Which of the following is an assertion of opinion?
    A. '60 million disagreements between traders and buyers are settled online'
    B. 'the seller and buyer are encouraged to negotiate online'
    C. 'this means much lower costs than in a courtroom'
    D. 'This system should be applied by the justice system'
    E. None of the above

35. Which of the following is implied but not stated about the TV show 'Judge Rinder'?
    A. Judge Rinder decides real cases
    B. The two sides are not represented by lawyers
    C. It is fake
    D. The TV show is not set in a real courtroom
    E. It is innovative

36. Which of the following was not argued by the author?
    A. That an online court should be introduced
    B. The online court should be modelled on the 'Judge Rinder' TV Show
    C. The online court should be modelled on eBay's dispute resolution system
    D. That existing legal costs are too high
    E. Lawyers are not always needed for legal disputes

37. The author argues that the justice system is inefficient. Which of the following best describes why his argument is weak?
    A. It involves a generalisation – the author only referred to small claims but this does not necessarily mean that the whole justice system is inefficient
    B. People don't have to use the justice system to reach a settlement
    C. Only one judge's approval was cited
    D. The author did not say how much the online court would cost
    E. eBay and the justice system are not comparable

Passage 8 – Cars

We live in a world of technological change and it seems that nothing is immune from it. Our phones, computers, kitchens, gardens and cars are all undergoing significant change. If businesses want to keep the custom of consumers, they must engage in technological change and find new, innovative ways to improve their products.

Ever since the introduction of the car over 100 years ago, one thing has remained constant: a human being has always driven the car. While the look, feel, and efficiencies of cars have improved enormously, cars have always required a human being to drive them. Indeed, the law requires human beings to be in control of cars. However, all of this is going to change.

Car companies – and some traditionally non-car companies – have been developing 'driverless cars' at a monumental rate. Famously, TopGear presenter Jeremy Clarkson tried an autonomous BMW around their race track in 2011. While driving on a race track is not comparable to driving on busy roads, there have been significant developments since. Google, for instance, have been testing autonomous cars on open roads in California. A key feature of autonomous vehicles is that they are capable of sensing their environment without human input.

Driverless cars are expected by industry experts to be the norm within 20 years. According to the Society of Motor Manufacturers and Traders, the market for autonomous car technology is expected to contribute £51 billion to the UK economy and over 300,000 jobs. A lot needs to happen in the meantime, though. Car companies need to rigorously test their cars on open roads and the public need to be convinced of their utility. Testing on open roads will allow companies to develop the accuracy and safety of the technology used. Indeed, such testing is essential to develop autonomous cars – how else can we be sure that they will be safe in the real world?

The current law requires a human being to be in control of a car. However, governments have issued special dispensations to car companies wanting to test autonomous cars on public roads. The UK government allow autonomous vehicles to be tested as long as a driver is ready to take over in the case of a system fault. The government has also announced that 40 miles of road in Coventry is to be equipped with technology to aid autonomous vehicles. The significance of the government's support is that it will accelerate the development of autonomous cars and will encourage worldwide car companies to set up permanent research facilities in the UK to test autonomous cars. If the UK can become a world leader in autonomous vehicles, the industry may well contribute a lot more than the expected £50 billion to the UK economy.

However, not everyone is convinced of the success of the driverless car. The CEO of Porsche, a luxury car company, recently dismissed the use of driverless cars, saying that his cars are meant to be driven. That may well be correct for the luxury car market – people want to drive the cars they spend £100,000+ on. However, it does not follow that it's the same for the rest of the car market. People will see that the benefits of autonomous cars outweigh the use of traditional cars. Firstly, autonomous cars are expected to be safer than traditional cars. For instance, a computer system can react much faster than a human to a dangerous situation. Secondly, the driver becomes a passenger and can do something else with his time – the age old saying that time is money still rings true today. This alone will encourage drivers to buy driverless cars. While people may find driverless cars strange initially, they will get used to them. The first desktop computer seemed strange but it is now virtually ubiquitous. So peculiarity should actually be an incentive to development.

38. What is the **main** point the author is making by using the first desktop computer analogy [final paragraph]?
    A. Desktop computers are strange
    B. The public should adapt to the driverless car
    C. People find new things strange
    D. People eventually adapt to new things
    E. The public will adapt to the driverless car

39. What is the underlying **assumption** that the author made in using the first desktop computer analogy?
    A. That people will adapt to new things
    B. People find new things strange
    C. Desktop computers are strange
    D. The public should adapt to the driverless car
    E. Peculiarity should actually be an incentive to development

40. What does the Top Gear autonomous BMW test suggest about the potential for autonomous cars on roads?
    A. It shows that autonomous cars can work on public roads
    B. There would not be accidents from the use of autonomous cars
    C. Autonomous cars will be well received by the public
    D. It would appeal to celebrities, such as Jeremy Clarkson
    E. None of the above

41. What is an underlying theme of the article?
    A. To describe the advent of the driverless car
    B. The autonomous car industry will contribute £50 billion to the UK economy
    C. The public will want to drive an autonomous car
    D. Development of driverless cars has been at a fast rate
    E. There are arguments for and against the adoption of driverless cars

42. Which of the following most undermines the author's argument in the final paragraph?
    A. The fact that driverless cars will be expensive
    B. Cars are meant to be driven
    C. The public would not be convinced by the safety of it
    D. A survey showed that many people are doubtful of the uptake of the autonomous car
    E. The CEO of Porsche is correct

Passage 9 – Beauchamp and Childress

Euthanasia, derived from the Greek word for 'easy death', involves the purposeful killing of a sick patient where the actual death is caused by a third party. Assisted suicide is the purposeful killing of one's self, made possible with the support of another person. So the individual takes the final action to end their life (e.g. injecting lethal drugs) but there is assistance from another person (e.g. a doctor providing the lethal drugs). Both assisted suicide and euthanasia involve at least a second person in the death whereas suicide just involves the deceased person. For this reason, assisted suicide and euthanasia are more controversial.

Beauchamp and Childress highlight four principles (Autonomy, Beneficence, Non-maleficence and Justice) that they believe should have a role in the decision-making process of medical ethics. These can conflict with one another and also, each principle can potentially conflict with itself.
Autonomy is the ability to self-govern one's own actions. The obvious point is that in order to respect autonomy, people should have the right to decide for themselves if they would like to die, for whatever reason. Otherwise, it would not accord sufficient respect to their position as autonomous human beings capable of reason.

However, when a person is mentally ill to the extent that their capacity to understand their own situation and determine their own actions is diminished, the ability to be autonomous is effectively lost. Therefore, no decisions relating to their own death should be taken by them. But still, this view is only relevant if the person is suffering from some form of mental incapacity.

However, the Society for the Protection of the Unborn Children suggests that a "patient's freedom entails a responsibility to act ethically" thereby implying a condition (or a limit) is attached to a person's autonomous abilities. Thus, they believe that refusing a patient's request for assisted suicide or euthanasia would not restrict the person's autonomy as, in making such a request, the person is not adhering to their responsibility.

However, this definition of autonomy is not universally accepted and incorporates an artificial condition. Further, it follows from the Society's argument that they would argue that suicide should become illegal.

The overruling idea is that patients should be able to make decisions regarding assisted suicide and euthanasia for themselves.

On the other hand, the principle of **non-maleficence** is that there is a commitment by medical professionals to do no harm. This may be seen as a very clear reason not to cause or assist in a death and militates against allowing euthanasia and assisted suicide. However, the principle should also be interpreted to mean: do not cause pain. Not yielding to a patient's own wish of death may prolong their suffering and, thus, result in further pain. Hence, "harm" can be interpreted in two different ways here to support opposing views.

These principles can be useful in deciding whether to allow an assisted suicide. When deciding on particular cases, certain arguments can be ruled out or be more relevant and thus, have more weight placed on them to come to an overall conclusion.

For example, the court in the case involving Diane Pretty in 2002 did not allow her husband to assist in her death. The court arguably gave priority to the principle of non-maleficence over personal autonomy. Yet still, in the case of Miss B, she won the case for her treatment to be withdrawn (a ventilator to be turned off) with the intention of letting her die without it - this is because it was different from Pretty's case. Here, the second person would not actually be 'acting' or causing harm. The hospital simply didn't act (in not providing treatment). Hence, it appears that the principle of non-maleficence had a significant weighting in both these cases as a dangerous precedent might have been set if a second party was allowed to participate in a killing because of the consequences for the exploitation of the vulnerable.

43. What does the Society for the Protection of Unborn Children assume about an intentional killing?
    A. It is not a function of autonomy
    B. It is unethical
    C. It is a crime
    D. It goes against autonomy
    E. It involves harm to the patient

44. Which of the following constitutes an explanation of why there is a gap in the author's argument that it follows from the Society's argument that suicide should become illegal?
    A. Suicide is not necessarily unethical
    B. Assisted suicide and euthanasia can't be assimilated with suicide
    C. Suicide is not illegal
    D. The Society did not argue that suicide should be illegal
    E. Suicide is not wrong

45. According to the passage, what is the implicit argument that the Society is making?
    A. Assisted suicide and euthanasia should not be allowed
    B. Assisted suicide and euthanasia are unethical
    C. Assisted suicide and euthanasia are illegal
    D. Suicide is illegal
    E. Suicide should be illegal

46. Which principle has priority according to the author?
    A. Autonomy
    B. Non-maleficence
    C. Euthanasia
    D. Assisted Suicide
    E. No principle has priority over another

47. According to the author, why did the court allow treatment to be withdrawn in the case of Miss B?
    A. The principle of non-maleficence was satisfied
    B. In order to respect the patient's autonomy
    C. It was not euthanasia
    D. The patient was not vulnerable
    E. The patient made the choice out of her own free will

Passage 10 – The IMF

The International Monetary Fund (IMF) is an international organisation, with 188 member countries whose main goal is to encourage monetary cooperation and secure financial stability across the world. It was formed in 1944 in response to the issues raised by the Great Depression in 1930. The IMF's most well-known feature is the ability to make loans to countries in need. All the member countries contribute money to a pool of funds, which can be distributed to any country experiencing difficulties. Indeed, this has saved many countries from financial ruin but the IMF's policy of attaching conditions has been severely criticised by many economists.

The IMF, among other things, operates a formal policy known as 'Surveillance', which involves reviewing each member's economy and their economic policies. The organisation then provides an assessment of each government's policies and gives suggestions as to what policies governments should take. Surveillance is also designed to warn countries of risks to their economies. In spite of this, no one at the IMF predicted the global financial crash which affected the entire world in 2007. What this made clear is that the IMF is unable to determine what the true risks in the world are. This surely would have dented confidence in the IMF and given their incompetence on this front, it would be unsurprising if countries don't rely so heavily on their advice.

When countries have financial difficulties – for example, when they have run out of money or have run up huge debts to international creditors – the IMF can step in and provide a loan. These are not the run-of-the-mill bank loans which consumers get from banks. Crucially, the IMF is the last port of call and can prevent a country from going into bankruptcy. For instance, Ireland was hit very badly by the recession, unemployment dramatically increased, the government ran out of money and they had difficulty in borrowing money. The IMF stepped in (with others) and provided a bailout loan to Ireland. This proved to be what Ireland needed and they have since made a solid economic recovery, which just shows what the IMF can do.

However, it has not always been plain sailing. Another debt crisis unfolded in Greece in 2010, where the Greek government also ran out of money. The situation was more serious than in Ireland though, as the Greek government had spent even more money than what they had, borrowed a lot of money and then they couldn't afford to meet the repayments on their loans. The IMF (and the EU) did come along to bail them out with a loan but they imposed strict conditions, such as a requirement that the Greek government reduced spending significantly. These conditions differentiate an IMF loan from an ordinary bank loan. Banks don't dictate what borrowers can spend in their lives but the IMF dictates what proportion of a loan countries can spend. Many economists criticised the IMF for requiring this as it meant that the Greek government could not stimulate their flailing economy. Indeed, the economy did get worse and the IMF must take some responsibility for it.

While the IMF has an important role to play in the world, they must not attach such strict conditions to the loans they give. Doing so can potentially make a bad situation even worse, as it did in Greece.

48. The Passage does not suggest:
    A. That the IMF could have done better
    B. That the IMF has caused some problems for countries
    C. The IMF should be replaced
    D. The IMF should be improved
    E. The IMF has been incompetent

49. Which of the following does the author portray as a positive intervention by the IMF?
    A. Giving a loan to Greece
    B. Giving advice to countries
    C. Making loans to countries in need
    D. Attaching conditions to loans
    E. Giving a loan to Ireland

50. According to the Passage, what was the root cause of the Greek debt crisis?
    A. The IMF's strict conditions on its loan to Greece
    B. The global financial crisis
    C. The government spent more money than what it had
    D. The IMF requiring the Greek government to reduce spending significantly
    E. The IMF's loan

51. Which of the following would the author be least likely to agree with?
    A. The IMF needs to change its procedures
    B. The IMF has only been beneficial in the world
    C. The IMF should have a replacement
    D. The IMF made a positive difference in Ireland
    E. The IMF made a negative difference in Greece

52. Which of the following is implicit in the author's assertion that the IMF is incompetent?
    A. The IMF should have been able to give a warning about the financial crisis
    B. The IMF should be replaced
    C. The IMF is useless
    D. The IMF failed to warn countries about the global financial crisis
    E. The IMF did not help in the Greek debt crisis

Passage 11 – The European Convention

The Human Rights Act in the UK has caused significant controversy and there have been calls from figures within the government to scrap it. While it has accorded many basic rights and freedoms to individuals, there have been instances where it has led to controversy. This has dented public confidence in it.

To understand the position, it is necessary to look at the background story. After the atrocities of World War II, European countries got together and signed the European Convention on Human Rights (ECHR) in 1950 and it came into force in 1953. There are 46 signatories to the Convention. This is an international treaty which contains a number of rights and freedoms which every person is entitled to in their countries (these rights are colloquially known as the 'Convention rights'). It included articles guaranteeing liberties such as the right to a fair trial, right to life and a prohibition on discrimination. From then on, all the States that signed up to the treaty were bound to uphold the rights of their citizens. The treaty also established a Court based in Strasbourg, known as the 'European Court of Human Rights', which can determine whether a State has breached a person's human rights and can hear cases from aggrieved citizens. If they find that a State has breached a person's human rights, they can award compensation (an award which the particular government must satisfy). So since 1953, if the UK government made a law that infringed a person's human rights, that person could take the UK government to the European Court and seek redress. However, international law operates differently to national law and UK citizens could not make a claim in the UK courts for a breach of the ECHR. They could only make a claim against the government in the European Court. This meant that people had to incur significant and unnecessary expenses if they were to vindicate their human rights. This changed at the beginning of the millennium, though.

In 1998, the UK Parliament passed the Human Rights Act, which came into force in 2000. This incorporated the Convention rights into domestic law. Crucially, this means that where the government breached the European Convention on Human Rights, UK citizens could sue in domestic courts (rather than just the European Court in Strasbourg). The domestic courts can quash government decisions and instruments that are inconsistent with the Convention rights and the Act it makes human rights a part of the law of our country.

There have been concerns though that the European Court has given too much protection to dangerous individuals. In the case of Abu Qatada, the UK government wanted to deport a terrorist suspect to Jordan to face criminal prosecution for terrorism offences. However, the European Court of Human Rights had a different say on the matter and effectively blocked the move, saying that it would be a breach of his right to a fair trial. It determined that there was a real risk of evidence obtained from torture being used in Qatada's trial in Jordan. The UK wanted to deport him nonetheless and appealed the ruling. It ended up taking over 10 years, numerous appeals, and a treaty between Jordan and the UK, by which it was agreed that evidence from torture would not be used, to send Qatada back to Jordan. In another instance, the European Court held that prisoners in the UK, who are currently not allowed to vote, should get the right to vote. Both these matters were given enormous press coverage, compared to other cases involving human rights. In particular, newspapers focused on the fact that most of the judges in the European court are 'foreign' and 'unelected'. The fact that Abu Qatada claimed £700,000 in legal aid from the government to pay for his numerous appeals further infuriated the public, but arguably that was an issue relating to the domestic government and not the Convention.

These are just two isolated cases, though. In the vast majority of instances, there have been more positive cases, which have not been reported as widely. For example, the Court has held that the security services can't shoot to kill without good reason. After 9/11, it was held that it is inconsistent with human rights to imprison a terror suspect (or, indeed, any suspect) indefinitely without a criminal charge. In 2015, the Court of Appeal held that the Convention requires the police to investigate rape allegations. Can anyone disagree with these judgements? They are just a few examples of the numerous instances in which the European Convention on Human Rights has helped give rights to the ordinary person in the UK while balancing concerns for national security. Indeed, the European Court may not be perfect – but what system (or person) is perfect?

53. Which of the following is an unstated assumption in the second paragraph?
    A. European countries were disgusted at World War II
    B. That the UK was a participant to the European Convention on Human Rights
    C. That the UK introduced the Human Rights Act
    D. A person whose human rights were infringed could sue to the government abroad
    E. The European Court of Human Rights can award compensation

54. Since which year could UK citizens seek protection of their Convention rights through domestic law?
    A. 1950
    B. 1953
    C. 1998
    D. 2000
    E. The year in which the Human Rights Act was introduced

55. Which of the following is an argument made by the author?
    A. The European Court of Human Rights were wrong in the Abu Qatada case
    B. The European Convention on Human Rights has caused endless problems for the UK
    C. The European Convention on Human Rights is not beneficial to the UK
    D. The European Convention on Human Rights has benefited UK citizens
    E. If a State breaches the Convention, they can be sued

56. According to the author, why did the UK not deport Qatada back to Jordan for over 10 years?
    A. The European Court of Human Rights didn't allow it
    B. They were unsure as to whether the European Convention on Human Rights prevented it
    C. The possibility of torture in Jordan
    D. There was a risk that evidence from torture would be used against Abu Qatada in a trial in Jordan
    E. The UK did not want to give Abu Qatada a fair trial

57. What is the author's main point about the 'more positive' cases in the final paragraph?
    A. Positive cases do not receive as much attention as the negative cases
    B. The Human Rights Act does not lead to bad outcomes
    C. The European Convention is not perfect
    D. The benefits of the European Convention outweigh the negatives
    E. The European Court is unelected

58. Which of the following would be a suitable main conclusion to the article?
    A. The UK should keep the European Convention on Human Rights
    B. The UK should not keep the European Convention on Human Rights
    C. The European Convention on Human Rights should be altered
    D. The European Court of Human Rights needs to be improved
    E. There are drawbacks to the European Convention on Human Rights

Passage 12 – Business Objectives

Most institutions in the country are businesses – shops, factories, energy companies, airlines, and train companies, to name a few types. They are the bedrock of society, employ most people in it and it is, thus, crucial that we examine their values.

The overriding objective of businesses is to make the most profit (i.e. maximise on revenue and minimise on costs). The notion was first popularly expounded by Adam Smith in his book, 'The Wealth of Nations' in 1776. Furthermore, his view was that if an individual considers merely their own interests to create and sell goods or services for the most profit, the invisible hand of the market will lead that activity to maximise the welfare of society. For example, in order to maximise profits, sellers will only produce and sell goods that society wants. If they try to sell things people don't want, no one would buy it. This is how the free market works. Indeed, the focus on profit is the basis on which companies operate and encourages them to innovate and produce goods that consumers want, such as iPhones and computers. So there are clear benefits to the profit maximisation theory.

This is a more effective society than, for example, a communist society where the government decides what to produce – as the government has no accurate way of deciding what consumers need and want. Arguably, the poverty that communist regimes such as the Soviet Union created have instilled this notion further.

However, were companies left to their own devices to engage in profit maximisation, what would stop them from exploiting workers? What would stop them from dumping toxic chemicals into public rivers? Engaging in such practices would reduce their costs of production, which would increase their profits. However, this would be very damaging to the environment. Accordingly, other objectives should be relevant. Businesses can also do other bad things to make a profit as well. For example, selling products to people who don't want or need them.

Corporate social responsibility entails other possible objectives for businesses, such as a consideration of the interests of stakeholders. A stakeholder is, in essence, anyone who is significantly affected by a company decision, such as employees or the local community. One business decision can have huge impacts on stakeholders. For example, a decision to transfer a call centre from the UK to India would likely increase profits, as wage costs for Indian workers can be much lower than that of British workers. This increase in profits would benefit the shareholders, however, it negatively harms other stakeholders. It would make many employees redundant. Here, there is arguably a direct conflict between profit maximisation and employees' interest. Nonetheless, moving call centres abroad does not always work. Given the different cultures and accents, companies have received complaints from frustrated customers. This, in fact, led BT to bring back a number of call centres to the UK.

However, the objective of profit maximisation has not always led to maximum welfare for society. Arguably, as banks sought to maximise their profits, they lent money to individuals who could not afford to pay it back. Eventually, many borrowers stopped meeting their repayments and lost banks enormous amounts. This led to a need for banks to be bailed out by the government and Lehman Brothers; one of the largest US banks that collapsed. Arguably, though, this was more due to idiocy rather than profit maximisation alone – in the end, the banks lost billions.

59. What literary technique did the author use by referring to an 'invisible hand'?
    A. A simile
    B. A metaphor
    C. An organisation that ensures that welfare for society is maximised
    D. Irony
    E. Analogy

60. Which of the following groups is a stakeholder in a business that was implicitly mentioned in the passage?
    A. Employees
    B. People
    C. Shareholders
    D. Pressure Groups
    E. Business rivals

61. Why does the author discuss the example of call centres?
    A. To argue that businesses are bad
    B. To show that call centres should not be moved abroad
    C. To show that moving call centres abroad can harm UK jobs
    D. To show the consequences of profit maximisation
    E. Because the author is against call centres moving abroad

62. Which of the following best encapsulates the author's position?
    A. Profit maximisation should be abandoned
    B. Adam Smith is wrong
    C. Profit maximisation works
    D. Profit maximisation should not be the only business objective
    E. Employees should have a greater role in businesses

63. What is Adam Smith's view according to the Passage?
    A. Profit maximisation arose because of the invisible hand
    B. The invisible hand had an important role in establishing profit maximisation
    C. Profit maximisation leads to the invisible hand
    D. Profit maximisation maximises welfare for society
    E. Profit maximisation is the best option for society

Passage 13

Global oil prices have fallen over the past year to rock bottom – from over $100 per barrel in 2013 to $30 at the dawn of 2016. At the possible peril of oil producers, there appears to be no let-up in the oil price. The ultimate reason for this is a simple matter of demand and supply. The so-called 'shale boom' in the US, where fracking was used to extract oil in an alternative way, meant that there was a lot more oil in the world – i.e. the supply of oil was boosted. This reduced the need to import oil into the US and other sellers (like Saudi Arabia and Nigeria) had to sell their oil elsewhere. However, there was no one else to sell it to and thus, to get rid of it, the price had to be reduced.

A major problem for oil companies is that the oil price reduction has drastically reduced their revenues by over 70%. Crucially, the revenue reduction has harmed the profitability of many oil companies. BP have even run a loss of £4.5 billion in 2015 – a dangerously high figure which has led them to cut 7,000 jobs. Indeed, job cuts are the staple response to low profits. In particular, many skilled North Sea oil workers have been made redundant, thereby increasing unemployment in Northern Scotland (particularly in Aberdeen). This has had devastating impacts on families, some of whom may require state benefits until new work is found. There would also be knock-on effects in the local community. As significantly fewer people have an income from work, much less money will be spent in local shops. Instead, 7,000 people paying taxes to the government will end up seeking unemployment benefits from the government. Unfortunately, there seems to be little other option for businesses like BP where their woes were brought on by the oil price. They're losing money and must cut costs, including employees, in order to survive. Otherwise, the loss-making Oil Company will eventually go bust. What now needs to happen is that the government must provide adequate support to those made redundant and sponsor retraining schemes where required.

The low oil price has caused other problems as well. Many oil businesses are highly 'leveraged'. In other words, they have borrowed significant amounts of money from banks. Such money was borrowed in order to fund new projects and new oil rigs, but they are less likely to be profitable now. If, as is the case with many at the moment, an oil company is making a loss, they will no longer be able to pay back their loans to the banks. Given this new likelihood, many loans previously seen as 'safe' are seen as significantly risky and could cause monumental losses for banks exposed to oil companies. This requires incredibly careful attention from financial regulators as this may have wider effects on the economy.

It doesn't stop there. Oil companies engage many other businesses (or contractors) to build projects or to help with maintenance. Unlike consumer businesses (like Tesco or Sainsbury) who deal with millions of customers, these contractors only deal with a few customers (such as oil companies). Accordingly, those contractors who only supplied the oil industry will find lower business and might potentially end up bust themselves (which means that they will have to make their employees redundant too).

All this will involve work for law firms. While they may get less work from advising on exploration projects and on buying oil rigs, they will, at least, get more work from administrators advising on insolvency, from banks for the restructuring of debts, and from oil companies looking to merge with one another to save on costs.

The low oil price has already had immediate impacts on consumers. Fuel (both diesel and petrol) comes from oil. Therefore, the lower oil price should influence the fuel prices as it will cost much less to produce the fuel. This would benefit consumers. Again, fuel is a significant factor in airline ticket prices so there should be a corresponding impact there too. Finally, as fuel gets cheaper, transportation costs will fall. As such costs have a significant impact on imported goods (such as electronics and computers from China), there should be price drops there too.

64. According to the author, who has the most to gain from the low oil prices?
    A. Oil contractors
    B. Business rivals
    C. The government
    D. Law firms
    E. Consumers

65. What is the immediate reason given in the passage for the unemployment resulting from the cuts made by BP?
    A. Low oil prices
    B. Demand and supply
    C. Many workers have been made redundant
    D. A loss of £4.5 billion
    E. Low revenues

66. What unstated assumption is made when the author states that the 7,000 unemployed individuals will end up 'seeking unemployment benefits' from the government?
    A. Those individuals have become unemployed
    B. The government has unemployed benefits to give
    C. Those individuals will not find work elsewhere initially
    D. That the low oil price has caused the unemployment
    E. That the low profits has caused the unemployment

67. What was the main reason for the reduction in the price of oil?
    A. Lack of demand
    B. People aren't interested in buying as much oil anymore
    C. The 'shale boom' in the US
    D. Saudi Arabia couldn't sell their oil in the US, but couldn't sell it elsewhere either
    E. The peril of oil producers

68. According to the author, why do some oil companies have little option but to make the redundancies that they have made?
    A. Low oil prices
    B. Workers are no longer useful
    C. Redundancies are necessary to boost profits
    D. US fracking
    E. Survival

Passage 14 – Nuclear Energy

Energy is ubiquitous in all of our lives, such that life without it would be very different. Where our energy comes from and the consequences of using it has come centre stage in the last half century. Non-renewable energy comes from energy sources that will run out in our lifetimes. Examples include fossil fuels: such as coal, oil, and natural gas. Accordingly, the current over-reliance on these sources is not ideal as they will run out, maybe not in our lives, but certainly in our children's and grandchildren's lives. Therefore, we must take action to develop renewable energies in order to give future generations a sustainable future.

A further and more significant consequence of fossil fuels is that they pollute the atmosphere. Their use involves the emission of carbon into the atmosphere, which most scientists believe contributes to global warming over time. Many also believe that upsetting the carbon balance in the atmosphere will lead to more unforeseen weather patterns. Given that 80% of the UK's energy came from fossil fuels in 2013, it is necessary to further develop cleaner alternatives.

Nuclear power, already an important source of energy, has been a government priority recently. In fact, in 2015, the government have even offered very significant subsidies for nuclear power with ministers proclaiming that nuclear power is the future for the UK's energy needs, being the most beneficial source of energy. The big attraction of nuclear power is that it is a clean form of energy – it does not emit carbon into the atmosphere. It just emits harmless water vapour. Crucially, the increased use of nuclear power can help the UK meet its obligations under international treaties to reduce carbon emissions. However, the use of nuclear power raises some significant concerns.

Harnessing nuclear energy involves a complex process and numerous steps to ensure that it is done in a safe manner. Nuclear materials emit dangerous radioactive material, which has the potential to kill or cause serious injury. The danger of radiation can be seen from the disaster at Chernobyl, where an explosion took place at a nuclear power plant. This led to 31 immediate deaths and an estimated 4,000 deaths due to cancer, resulting from the radiation. The surrounding area, of approximately 1,600 square miles, has been declared uninhabitable for 20,000 years. While there are significant safeguards to prevent such an accident happening again, such that we can consider nuclear power effectively safe for the most part, the risk of an accident cannot be ruled out. In 2011, another serious radiation release occurred in Japan.

Producing nuclear energy leads to 'nuclear' waste which is officially separated into three categories: low-level waste, medium-level waste, and high-level waste. High-level waste includes the by-products of the reactions inside the nuclear reactor and this stays highly radioactive, and, thus, dangerous, for many thousands of years. Accordingly, safe disposal is critical. Suitable storage is at an enormous cost and requires effective government planning. Yet it does not seem that the government always consider the long term implications of nuclear waste. In 2007, for example, the High Court quashed the government's decision to build new nuclear power stations because of inadequate and misleading consideration given to waste disposal. This is worrying. If the government were to account for the implications and the true cost of dealing with nuclear waste, they would see that it's not necessarily the 'economical' future for the UK's energy needs.

69. Which of the following is not true based on the passage?
    A. All the non-renewable energy sources emit carbon into the atmosphere
    B. Carbon emissions are harmful to the atmosphere
    C. There is currently an over-reliance on fossil fuels
    D. The UK is under an obligation to reduce carbon emissions
    E. Nuclear waste can be dangerous

70. Which of the following does the author argue in the passage?
    A. Nuclear power has been a government priority recently
    B. Nuclear power stations will kill or cause serious injury
    C. Nuclear power should be made into a clean form of energy
    D. Society should develop clean renewable energies
    E. Nuclear power is economical

71. Which of the following would the author most likely agree with?
    A. Nuclear power is worse than fossil fuels
    B. Nuclear power is the most beneficial to society
    C. There should be greater focus on renewable energy sources (such as solar and tidal power) that do not create dangerous waste
    D. The government should consider the cost of nuclear waste
    E. Nuclear power is unsafe

72. What did the author refer to as 'worrying'?
    A. That the High Court quashed a government decision
    B. That the government wanted to build new nuclear power stations
    C. The high cost of dealing with nuclear waste
    D. The fact of the government giving inadequate consideration to nuclear waste disposal
    E. Nuclear waste disposal

73. Which of the following most weakens the government's argument that nuclear power is most beneficial to society?
    A. Other forms of energy are cheaper
    B. Other types of energy are important
    C. There are cheaper and cleaner alternatives
    D. The disaster in Chernobyl caused mass damage
    E. The High Court's decision

Passage 15 – Equal Pay

Figures from the Office for National Statistics show that there is a 14% difference in average pay between men and women in full-time jobs. Yet since the Equal Pay Act in 1970, it has been unlawful for an employer to pay differential wages to men and women doing the same work. So why, 45 years since then, are women earning less than men?

Essentially, it is against the law for an employer to directly discriminate on the grounds of gender. If a woman can show that they're being paid less than a man for doing exactly the same value of work for the same employer, they can sue their employer in an employment tribunal. However, the effectiveness of this has been inhibited by high tribunal fees, which act as a disincentive to women to bring forward their claims. Some have also reported fears of career progression being hindered by bringing a claim. Accordingly, it would be more effective for a governmental body to identify and bring prosecutions against employers who discriminate, rather than leaving it to individuals.

Nonetheless, direct wage discrimination does not entirely account for the gender gap in average pay in the whole of the UK.

The average salary for a male in an executive role is £40,625 whereas the average for a female in an executive role is £30,125. It does not conclusively follow that there is direct discrimination by individual employers on wages – executive roles differ from organisation to organisation and so do their pay structures.

A particular issue highlighted by the Fawcett Society is that 47% of women are in low paid jobs whereas the figure for men is just 17%. A reason for this is that women tend to be in lower paid sectors – for example, 80% of care and leisure workers are women while only 10% of those in better paid skilled trades are women. This cannot necessarily be attributed to discrimination. The reasons for entering lower paid professions are multi-factorial, and can include discrimination in recruitment, especially in traditionally male-dominated sectors. Indeed, traditional stereotypes may well permeate the career decisions of both girls and boys – traditionally, boys have tended to take up STEM subjects, particularly Physics, whereas girls have tended to take more of the Arts and Humanities subjects.

Even within sectors, women tend to occupy lower positions than men, which inevitably leads to lower salaries. Out of all the FTSE 250 directors, only 19.6% are female. Figures from HMRC further highlight that women are underrepresented in the top jobs as only 27% of women are higher rate tax payers. It is, thus, clear that women do not get as far in the world of work as men do, which is deeply concerning. Whatever the reason for the pay differential is, it needs to change as women are equally as talented as men.

While current laws may well be making a difference, they can't, in their current form, change attitudes. What can the government do about all of this? Some have suggested a mandatory minimum of women in certain roles. While this will undoubtedly help the figures, it may just be sticking a plaster on the wound. Yes, companies will improve the number of women represented in the workforce but it may not lead to the culture change that is required. It may further cause resentment and doubt as to whether those who reached those positions did so on merit.

The government has, alternatively, proposed to require any organisation with more than 250 employees to report the pay gap between men and women in their workforce. They will be legally required to show the number of men and women in each pay range, which will highlight the pay gaps within the organisation. The government then propose to create a league table of the best and worst employers. This would likely lead to significant public pressure on employers to improve the pay gaps within their organisation. Crucially, this would be more likely to stimulate the culture change required – from recruitment to promotions - while also helping the enforcement of existing laws by notifying employees of when there may be potential discrimination. If companies do not address the issue after this, it may, in fact, become necessary for the government to impose a mandatory minimum number of women in higher positions.

74. What is the author's preferred option for tackling the gender pay gap?
    A. Strengthen discrimination law
    B. Implement a reporting obligation
    C. A mandatory minimum number of women in higher positions
    D. Reduce tribunal fees
    E. The government should bring prosecutions against discriminatory employers

75. Which of the following is an argument made by the author?
    A. Women are equally talented as men
    B. Lower tribunal fees are necessary
    C. Current laws are not making a difference
    D. Maternity leave should be increased
    E. The government should prosecute employers who directly discriminate

76. Why does the author prefer a reporting obligation as opposed to a mandatory minimum?
    A. A reporting obligation will help change the culture
    B. A reporting obligation may be more effective
    C. It will reduce the gender pay gap
    D. It does not require people to pay high tribunal fees
    E. It allows the companies to take action themselves first

77. Which of the following are reasons for the differential in *average* pay between genders?
    A. Discrimination
    B. There are more women in lower paid sectors
    C. Men occupy higher positions in the same sector
    D. A, B, and C
    E. Resentment

78. What is the author's concern with the current discrimination laws?
    A. They do not make a difference
    B. The government needs to enforce it better
    C. Direct discrimination is only one part of the problem
    D. They only work in cases of discrimination by one employer
    E. The high costs of bringing a claim (high tribunal fees) under current discrimination laws

Passage 16 – Trade

We enjoy TVs, phones, computers, electronics, and clothes, but we do so at our eventual risk.

International trade is an enormous advantage to the entire world – it allows people in the world to enjoy goods made in other countries. However, it does benefit some countries more than others. The UK has for some time been a net importer of products from the rest of the world. Importing involves buying goods from abroad, whereas exporting involves selling goods abroad. Put simply, we buy more from other countries than we sell to them.

For consumers in the UK, importing may seem beneficial – we can benefit from televisions, phones, computers, food, and other cheaper goods from other countries. In particular, the development of low-cost manufacturing in China and India in the last 30 years has meant that it's much cheaper to produce goods there rather in the UK. It's not so much an innovation but more an expansion, due to the ability to pay workers much lower wages. While the UK is a net exporter of services (in particular financial services), it is a net importer of physical goods and when considering both goods and services together, the UK is a net importer.

Importing goods mean that British money is going abroad to foreign companies as opposed to British companies, but this is not a problem in itself. As long as Britain exports to the rest of the world as much as it imports, which economists call a 'trade balance', the position is satisfactory.

Oddly, even in spite of a weakening currency, the country is nowhere near to having a trade balance. In 2014, we imported £34bn more than what we exported. This means that £34bn leaves the UK each year. If that was instead spent on UK goods (rather than foreign goods), it would boost British businesses, British jobs, and reduce our reliance on debt.

Indeed, debt is a direct consequence of being a net importer. How do we, as a country, fund these extra £34bn purchases from abroad? Debt. The UK borrows a lot of money and these debts are held by foreign investors, which is fine in the short run but eventually, that money will have to be paid back and with interest.

Being a net importer (which is the same as not exporting enough) is also a symptom of a wider problem in the UK. Due to the process of de-industrialisation in the 1980s, the UK has produced significantly fewer manufactured and semi-manufactured goods. Given the lack of investment in those industries, it costs a lot more to manufacture those goods in the UK when compared to, for example, Germany. It is also a symptom of a poorer skills base among UK workers. Simply, when we are not as skilled as those in other countries, it costs more to produce things here and, thus, we have to sell our products at higher prices. In this global marketplace, people can simply get products more cheaply from other countries. This, in turn, reduces our exports.

79. What should a weakening currency ordinarily achieve according to the author?
    A. Lead to more imports
    B. Not lead to a trade balance between imports and exports
    C. Lead to a trade balance between imports and exports
    D. More trade
    E. £34bn of net imports

80. Which of the following best explains what a 'trade balance' is?
    A. When there is a balance in trade
    B. When there are no imports
    C. When exports are greater than imports
    D. When the level of exports is the same as the level of imports
    E. When there is a satisfactory level of trade

81. Why does the UK not have a trade balance?
    A. Foreign debt from abroad coming into the UK
    B. The UK's imports are much greater than their exports
    C. Cheap Chinese goods
    D. International trade
    E. De-industrialisation and a poor skills base among UK workers

82. Which of the following is the author's purpose in writing the article?
    A. To explain the UK's import and export levels
    B. To argue that the UK should not import as much
    C. To argue that the Chinese low-cost goods are bad for the UK
    D. To show that the UK have net imports of £34bn
    E. To show that the UK owes money to foreigners

83. Based on the last paragraph, why would an overseas individual buy a manufactured good from Germany as opposed to the UK?
    A. A preference for German goods
    B. Germany produces more goods than the UK
    C. The individual is German
    D. There is a lack of investment in the UK manufacturing sector
    E. The prices of German goods are cheaper than UK goods

Passage 17 – Star Wars

Star Wars has stormed back onto the box office, just 10 years after the last one. In 2012, Disney famously bought the rights to the Star Wars franchise from the original founder, George Lucas. With the intention of producing another trilogy, Disney released the first film, "Star Wars: The Force Awakens" in December 2015 to the delight of fans.

A few suggested that the film was based too much on the old style, but they were few and far between. The majority of cinema goers pronounced the film a success on social media and it immediately received positive critical acclaim. You know you've created a successful film when hard-to-please and meticulous critics give it a good rating. The actors themselves received heaps of praise as two up and coming British actors came into stardom.

The film didn't just achieve success on this front. As a result of being a box office hit, it has reached $2 billion in box office sales, one of only three films to have ever done so. According to analysts, the success doesn't stop there, though. Some suggest that merchandise sales will outstrip the takings from box office sales.

An important part of most successful films is the consumer products that go with them and Star Wars is no exception. Disney exploited the pre-existing goodwill that the first six Star Wars films built up and licensed and created a vast array of products – from toy action figures of the film's characters to remote controlled droids. Significantly, Disney did not create all of these products themselves. They have licensed the Star Wars brand to other companies, who are allowed to use the Star Wars brand on their own goods. This has meant that other companies can use their expertise to produce Star Wars related goods which Disney could not produce. Significantly, in return for using the Star Wars brand, such companies pay Disney a license fee. For example, Electronic Arts, the game software developer was granted a license to produce Star Wars video games. In return, they gave Disney $225 million.

Disney picked an unorthodox time of year to release the Star Wars film but this was another inspired decision, both in terms of film screenings and consumer products. Given the Christmas period, where parents would have been wondering what gifts to buy their children, Disney gave them a clear answer. In releasing the Star Wars products a month before the film, Disney gave fans and, crucially, parents the opportunity to consider buying them for Christmas. Unlike most children's films, many of today's parents were teenagers at the time of the original Star Wars trilogy (in 1977 to 1983) and, thus, would have had a greater inclination towards purchasing the Star Wars goods. It's clear that this merchandising strategy played out well as Star Wars merchandise sales boosted profits in their consumer products section by 23%.

Paradoxically, it seems, Disney's share price has gone down since the release of the film. It appears though that this is due to concerns about a different part of the Disney Company, such as its ESPN network.

84. Why did Disney's share price go down after the release of the latest Star Wars film?
    A. Some people did not like the film
    B. The film did not get as much revenue as expected by investors
    C. Concerns about the next Star Wars film
    D. Investors are concerned about other areas of the Disney company
    E. Disney's performance has been poor

85. Which of the following is an opinion rather than an assertion of fact?
    A. The film received positive critical acclaim
    B. Disney mobilised a commercial enterprise
    C. Many of today's parents were teenagers at the time of the first trilogy
    D. The new Star Wars trilogy is a resounding success
    E. Parents were wondering what gifts to buy their children

86. What prediction does the author specifically make about Star Wars merchandise?
    A. Other companies will produce Star Wars branded products
    B. That merchandise revenues will exceed box office revenues
    C. Parents will be interested in buying Star Wars merchandise
    D. It will make the Star Wars film a success
    E. The Star Wars merchandise will get good reviews

87. Which of the following would be a view held by the author?
    A. The Star Wars film was based too much on the old films
    B. Disney's share price should not have gone down
    C. Any successful film would have significant merchandising from toys
    D. The success of the merchandising contributed to the success of the Star Wars film
    E. The success of the Star Wars products created by other companies should not be credited to Disney

88. Which of the following is definitely false based on the passage?
    A. Disney licensed all their merchandise to other companies
    B. The film was successful
    C. The founder of Star Wars played a role in the new film
    D. The film was not a British film
    E. The film was not a success in the eyes of a few people

Passage 18 – Marriage

In 2014, the Marriage (Same Sex Couples) Act 2013 brought a revolution in society where, for the first time, individuals of the same sex could enter into a marriage. Previously, marriage was only exclusively available for opposite-sex couples (a man and a woman), a fundamental part of marriage for hundreds of years. In 2004, same-sex couples were allowed to enter into a civil partnership, being a marriage in all but name. Some see this as insignificant. Others argue that, by giving same-sex couples a different label, their relationship would be seen as different. Whether this is a legitimate difference was and is a matter of debate. In any event, Parliament took the bold step in 2013 by legislating to allow same-sex marriage. The institution of marriage has now evolved to allow both opposite-sex and same-sex couples to marry.

Some opponents of same-sex marriage argue though that marriage is, by definition, a relationship between a man and a woman. Therefore, it should remain as just between a man and woman. However, this reasoning is circular. Such reasoning follows as such: a marriage is between a man and woman, therefore, it's between a man and woman. It adds nothing to the argument at stake and does not address the question of what a marriage should be. It is an extraordinarily shallow argument.

In considering what marriage is, many opponents base their view on marriage being a religious construct. Marriage ceremonies frequently, but not exclusively, take place in churches and other places of worship. In Christianity, the institution of marriage is a sacrament. Crucially, the main denominations of Christianity do not recognise same-sex marriage. Accordingly, on religious grounds, it is opposed.

Significantly, the Marriage (Same Sex Couples) Act guarantees the availability of a civil (non-religious) marriage and allows a religious marriage if the particular place of worship wishes to allow it.

Some opponents take the position that marriage can only be between a man and a woman because of biological or 'natural' reasons. This is because, in order to procreate (i.e. to produce a child) naturally, there must be a man and a woman. However, as the courts have highlighted in a different context, the essential feature of being in an opposite-sex relationship does not necessarily involve children – a number of opposite-sex couples are married but without children. Same-sex relationships can have exactly the same qualities of intimacy, stability, and interdependence that opposite-sex couples can have. Further, a marriage cannot be nullified by a lack of children.

Proponents of same-sex marriage indeed base their argument on the view that marriage is a social construct. Accordingly, as society evolves to accept that there is not a relevant distinction between opposite-sex and same-sex couples, same-sex couples should be accorded the same rights as opposite-sex couples – i.e. to marry should they so wish.

Whether the modern institution of marriage is a religious construct or a social construct is a matter of debate. Some proponents of same-sex marriage have pointed out that marriage is no longer a religious construct, but a matter for society as a whole. Opponents though insist on the continuing significance of religion. These two views are not as inconsistent as they first seem. It is indeed true that marriage started out as solely a religious construct, with its origins being in the 12th Century.

Marriage was still a social construct though – the vast majority of society was religious and religion had a ubiquitous role in daily life. On this basis, it had significant social relevance. That's not to say that religion is not important today – it certainly still is. That said, it is not as universal now and ever since 1837, marriages need not have taken place through a religious ceremony.

What has been a constant though since the 12th Century and now is that the institution of marriage has reflected society and has continued to evolve in order to do so.

89. What is the author's belief as to the availability of same-sex marriage?
    A. It should be allowed
    B. It should not be allowed
    C. The concerns raised of same-sex marriage are valid
    D. That same-sex marriage is a religious construct
    E. None of the above

90. What is the author's concern in the second paragraph?
    A. Marriage is not just between a man and woman
    B. Marriage should not just be between a man and a woman
    C. It is illogical to suggest that marriage is just between a man and a woman
    D. The reasoning employed by the opponents' argument here was circular
    E. Marriage should be open to same-sex and opposite-sex couples

91. Which of the following best encapsulates the author's view as to marriage?
    A. It is not a religious concept
    B. Religion should not be relevant in marriage
    C. Marriage has always been a social construct
    D. Religion should not inform marriage
    E. Not as many people are religious in the 21st Century

92. What did the author imply about the state of marriage before 1837?
    A. Society was entirely religious
    B. It was only possible to get a marriage through a religious ceremony
    C. Marriage started out as a religious construct since the 12th Century
    D. It was different
    E. More individuals took it seriously

93. Based on the whole passage, what does the author imply about marriage today in the final sentence?
    A. Marriage is a social construct
    B. The acceptance of same-sex marriage is a reflection of society
    C. Religion is not relevant anymore
    D. Marriage has evolved to reduce the relevance of religion
    E. Same-sex marriage should be allowed

Passage 19 - Sugar Tax

According to an analysis in the Global Burden of Disease Study, 67% of men and 57% of women are obese or overweight. According to the study, the UK has one of the highest levels of obesity in Western Europe, which carries severe consequences for the population, health services, and the economy.

One proposal under consideration by the government is to impose a tax on sugary products, in proportion to the amount of added sugar in a particular beverage. This is particularly relevant as most adults and children consume too much sugar.

The government currently recommends that sugars should be limited to 5% of the total energy an individual consumes in a given day. This means a recommended maximum of 30g of sugar per day. Significantly, anyone who consumes a full can of ordinary coke is instantly above the recommended maximum. The government hope that their recommendations influence people's behaviour but given the high obesity rates in the UK, it is clear that it has not solved the problem. Accordingly, stronger action is clearly required but a sugar tax has led to controversy.

Being overweight and obese causes significant health issues such as diabetes, heart disease, and cancer. It not only reduces the quality of life but also one's life expectancy. It, thus, has major implications. Even though there are labels on products displaying the amount of sugar, people tend not to look at them and even when people do, they're not aware of the significance of it. Arguably, as individuals do appreciate an immediate or short-term negative health consequence of sugary food and drinks, many may not see it as a serious issue until they actually develop health problems. Not only are there serious personal consequences but the total social cost of obesity is at a mammoth £47 billion.

Imposing a sugar tax should both dis-incentivise excessive sugar consumption and make sure that those who are likely to add to the social cost of obesity contribute more to society.

It is sometimes attested that a sugar tax or any limit on sugar infringes on people's freedom, as the government is intervening in people's lives to reduce what they can consume. If an individual wants to consume sugar, they should be free to do so, with an absence of government interference. However, this argument can be countered by the fact that consumers can still consume sugary foods should they wish, they will just pay an extra tax which reflects the costs to society of obesity. Surely it is better that they pay for the health costs they cause than for the whole of society to pay, which is what currently happens. Accordingly, it would be fair to impose the sugar tax.

It is, finally, arguable whether a choice to consume sugary drinks involves the exercise of free will. Our decisions to consume say, a can of fizzy pop, are heavily influenced by advertising campaigns and the social acceptance of it. However, this does not necessarily lead to informed consent to the consequences of obesity and being overweight. A large swathe of society is ignorant of the health consequences so it appears that the public does not truly understands the risks of consuming too much sugar.

However, imposing a tax is not the end of the government's problems and as the chief nutritionist at Public Health England pointed out: obesity does not have a 'single bullet solution'. Other actions such as education are vitally important but whatever the cost of these policies, their benefits must surely outweigh the cost of picking up the pieces. It's better to put up a fence at the top of the cliff rather than an ambulance at the bottom.

94. Which of the following best accords with the main idea in the article?
    A. Obese individuals are to blame for their health consequences
    B. Sugar should be taxed
    C. Sugar should not be taxed
    D. There are other policies which should be done
    E. To consider the pros and cons of a sugar tax

95. Which of the following would, if true, most weaken the author's argument?
    A. A sugar tax is immoral
    B. A sugar tax affects the poor
    C. A sugar tax would lead to public protest
    D. A sugar tax would increase prices
    E. People would not reduce consumption in response to an increase in prices in sugary products

96. Which of the following would undermine the 'fairness' argument made by the author [sixth paragraph]?
    A. Some people would disagree with the sugar tax
    B. The sugar tax would not reduce the consumption of sugary products
    C. People are unaware of the health consequences of sugary products
    D. Sugar doesn't necessarily cause health problems in people, particularly when they consume small quantities
    E. Not everyone is ignorant of the health consequences

97. Which of the following is not true based on the passage?
    A. Around the world, more men are obese than women
    B. Social cost of obesity is estimated at £47 billion
    C. Being overweight causes health problems
    D. All cases of obesity can be blamed on excess sugar consumption
    E. Many individuals are ignorant of the health consequences of obesity

98. Which of the following is an <u>unstated</u> assumption in the passage?
    A. Sugar consumption is a cause of obesity and heart disease
    B. The social cost of obesity is £47 billion
    C. A full can of coke exceeds government sugar recommendations
    D. The government proposed the sugar tax
    E. A sugar tax will not solve all of the problems

99. Which of the following best illustrates the final sentence of the passage:
    A. The government should not provide health assistance to those who suffer from obesity
    B. The government should focus entirely on stopping people from consuming sugar
    C. Preventing a high consumption of sugar is more effective than trying to cure the health consequences of it
    D. Education is better than taxes
    E. Someone is falling off a cliff

Passage 20 – Tour de France

The Tour de France was established in 1903 to increase sales of the newspaper L'Auto, yet it went on to become the world's biggest cycle race and arguably the toughest sporting challenge on the planet. The modern version is a 23-day event, consisting of 21 days of racing where the competitors cycle at least 100 miles on each day. On half of the stages, the cyclists go up a number of mountains, including those in the Alps. Such a feat is absolutely astonishing both for seasoned fans and particularly for the uninitiated. It is, thus, easy to forget that the competitors themselves are human. Crowds line the streets in support of their favourite cyclists or just to see the spectacle. It is something that the whole of France and the world of cycling gets behind. So its humble beginnings would naturally surprise most people.

While the exact route of the Tour de France changes from year to year, the finish line is always at the heart of Paris on the world famous Champs-Elysées. The person with the fastest time wins the competition and is awarded the coveted Yellow Jersey.

What makes this sport even more intriguing is its structure. Approximately 200 cyclists take part and each rider is part of a team. In each team, there are nine riders. The team provides mechanical support, race support, medical assistance, massages after the race, and food specifically prepared for the demands of the race. The team's race director determines the strategy for each day of racing and for the competition as a whole. Gaining a place on a team is, some say, even harder than actually doing the race itself. The team must be convinced that you will be able to complete the course, that you're dedicated to the success of the team and if you're not going to be the team leader, that you'll sacrifice yourself for the leader.

It is not the organisers nor an official qualification round that determines who enters the competition; it is the teams. The teams have the freedom to choose who races for them. Cycling looks like a deceptively easy sport to understand – after all, what is there to a bunch of men racing around France for 3 weeks?

On flat stages, most of the cyclists tend to stay together in a group (or 'peloton' in French) for most of the day, which has a major influence on the team's racing strategy. Further, each team of cyclists has one individual designated as the team leader, which the others must support.

A significant factor in cycling is the slipstreaming effect. This effect is not just in cycling – it occurs whenever two objects move in space with one closely followed by the other. It applies to motor racing as well. It also explains why birds fly in a V-formation. When one cyclist follows another very closely, the cyclist behind saves a significant amount of energy. Conversely, that means that the cyclist in front uses up more energy. For these reasons, the cyclists tend to stay together in the races – each cyclist can take it lighter by being behind someone.

That leads on to a crucial point. The team chooses one rider to be the leader – it is simply determined on who, in their opinion, is the best cyclist at the time of choosing. The rest of the riders in the team are expected to support the lead cyclist in the race. They do this by, among other things, making sure the leader is behind them. As this happens with all teams, it is, in essence, each 'lead' rider against each other, each of whom is supported by their teammates. While this is the convention, it isn't always crystal clear how the best is chosen.

For example, when Team Sky had two very strong cyclists in 2012, a supporting rider demonstrated superior vigour on many of the stages. The leader of the team won the whole race but it was not without its controversy.

Given the extra amount of energy used up by the supporting cyclist in order to help his team leader, if they were going head-to-head, it would not be possible to say that the result would have been the same. That would have also led to a far more exciting spectacle as well. Interestingly enough, while the team leader won the race, in the end, the 'supporting rider' came 2nd, beating all the other team's lead cyclists – something unheard of in the historic sport. And yet, unlike every team leader (including his own), he didn't have the support that team leaders have.

100. Which of the following is an opinion?
    A. [The Tour de France] 'has become the world's biggest cycle race'
    B. It is the 'toughest sporting challenge on the planet'
    C. The competitors are human
    D. The route is 3000km
    E. A condition for becoming part of the team is that one would be willing to make sacrifices

101. Why does the lead rider in a team benefit from being the lead rider?
    A. The leader has the support of the team management
    B. The leader can beat the riders from the other teams
    C. The leader is the best of the rest in his team
    D. He is more likely to win the race
    E. He is dominant over his teammates

102. Why do the cyclists stay in a group in the races on flat stages?
    A. It is the best team strategy
    B. The slipstreaming effect
    C. It is the team's racing strategy
    D. To support the team leaders
    E. They are supported by riders at the front

103. Which of the following is implied but not stated by the author in the final paragraph?
    A. The correct result was reached in the 2012 race
    B. Team Sky's choice of leader was not correct for them
    C. A different result may have been reached in an equal race
    D. The slipstreaming effect meant that the supporting rider was a better rider
    E. The supporting cyclist may have been stronger in the 2012 race

104. Based on the passage, why would the supporting cyclist in the 2012 Tour de France have used extra energy?
    A. He was inefficient
    B. He was a better cyclist
    C. He was helping the team leader win, with the slipstreaming effect
    D. He was not as good a cyclist
    E. He was racing against the team leader

Passage 21 – Criminal Justice

The principle of equality before the law is a fundamental part of our justice system and the rule of law. While this has many facets, one is that people should not be treated more severely on account of their race or colour. Yet the treatment of ethnic minority individuals has been a never-ending concern due to the disproportionate numbers of ethnic minorities in prison. It has now reached the top of government and the Prime Minister has ordered a government review into discrimination against black and ethnic minority people in the criminal justice system.

Statistics from the Home Office indeed show that 25% of the prison population are ethnic minorities, and yet they only account for 14% of the UK population. Ethnic groups also receive longer custodial sentences than for white offenders. These are troubling statistics. It is, therefore, imperative to find out what the root cause of this is, and the government's review will hopefully go in depth enough to shed more light onto the matter.

Once a person is found guilty of a criminal offence, the courts then consider what their 'sentence' or punishment should be. Judges have discretion in choosing what sentences to give to offenders, even among those convicted of the same offence, and are required by law to consider a number of factors in making up their minds. For example, aggravating factors such as violence, previous offences, and pre-meditation, can increase the sentence the judge gives. Likewise, cooperation with the police may reduce the sentence. A guilty plea at the start of the trial leads to an automatic reduction of a third of the sentence. These factors explain why those convicted of the same offence can receive different sentences.

However, a study in 1992 by Roger Hood found that 7% of the over-representation of ethnic minorities in prison came from a higher use of custody than would have been predicted from legally relevant factors, which indicates discrimination in the exercise of judicial discretion. Indeed, other possible reasons also abound.

Cultural differences play a significant role. For example, admitting guilt in a criminal case (officially termed a 'guilty plea') can lead to a significant reduction in the sentence given (up to a third in some cases). Ethnic minority offenders, though, are significantly less likely to plead guilty than white offenders and are less likely to opt for legal advice. This may be a significant reason for the longer sentences. On the face of it, this does not appear to be discrimination. Given that judges are legally required to give the sentencing 'discount' for a guilty plea, it may be necessary to consider whether this discount should remain.

Finally, a black individual is six times more likely to be stopped and searched by police than a white person. The racial disparity in stop and searches are particularly disturbing given that the chances of arrest are not different for a black or white person. This may be for a number of reasons. It could be due to a police force's priority in high-crime rate neighbourhoods. However, a number of researchers have found that racial stereotypes exist, consciously and subconsciously, among police officers which could lead to a greater targeting of ethnic communities. Increased targeting, whether subconscious or conscious, would naturally lead to an increase in the numbers of ethnic minority offenders arrested. Accordingly, the significance of the disparity in stop and searches among racial groups requires urgent attention in the government's review. The government have already been encouraging police forces to reduce the disparity but it appears that more needs to be done. These figures in themselves do not conclusively demonstrate discrimination but may be indicative of it.

105. Which of the following is an opinion as opposed to an assertion of fact?
    A. The disparity requires urgent attention
    B. There is an over-representation of ethnic minorities in prisons
    C. Ethnic minority offenders are less likely to plead guilty
    D. Judges are required to give a sentencing discount for a guilty plea
    E. A black person is more likely to be stopped and searched than a white person

106. What is the implication of the second paragraph?
    A. More ethnic minorities are criminals
    B. There are a disproportionate number of ethnic minorities in prisons
    C. The government needs to find out the cause of it
    D. The statistics are not ideal
    E. There is racism in the criminal justice system

107. Which of the following is true, based on the passage?
    A. The higher sentences given to ethnic minorities is a breach of the rule of law
    B. Judges have a discretion to giving a guilty plea discount
    C. Ethnic minorities are more likely to have longer sentences than white individuals
    D. Judges consciously discriminate against ethnic minority defendants
    E. Socio-economic factors are not related to the longer sentences of ethnic minorities

108. What is the main reason for those convicted of the same offence being given different sentences?
    A. Discrimination
    B. Judicial discretion
    C. A different level of harm is caused
    D. More stop and searches of ethnic minorities
    E. The legal requirement to consider aggravating factors

109. How could the final paragraph explain the disproportionate number of ethnic minorities in the prison population?
    A. More ethnic minorities commit crimes
    B. The police are discriminating against ethnic minorities
    C. The increased targeting means that ethnic minority offenders are more likely to be caught than white offenders
    D. The chances of arrest do not differ between white and black individuals
    E. The government should act to reduce the number of ethnic minorities in prison

110. What is implied by the author in the final paragraph?
    A. There may be discrimination in the police
    B. The government needs to take further action
    C. There needs to be further investigation into the reasons for the disparity
    D. The reason for the disparity is multi-factorial
    E. Judicial discretion is not a factor for the disproportionately high % of ethnic minorities in prison

Passage 22 – Tax Incentives

Tax inversions have been on a trend throughout 2015. A well-known instance is when Pfizer, a large pharmaceutical company, made a deal to buy Allergan, a much smaller drug company which produces Botox. A tax inversion is where a company purchases a smaller company in a different jurisdiction so that it can relocate its headquarters to benefit from a lower corporate tax rate. A number of American companies have taken this route, to the annoyance of the US Government.

This also leads to another issue. A number of countries have offered competitive tax regimes or tax incentives to encourage foreign companies to set up there. A tax incentive involves the reduction or exemption of a tax liability. In particular, it appears that the UK are taking this approach in their tax policy – in 2015, the UK government announced a reduction in the corporate tax rate from 20% to 18% by 2020.

In the US in contrast, the corporate tax rate is at 35% for most companies and uniquely, it is on both the profits earned in the US and abroad. Both these features make the US rate one of the highest in the world.

Concerns have been raised about companies moving around to take advantage of competitive taxes in other countries. Such moves may well improve the profitability of companies and their returns to shareholders. However, the tax inverters still maintain their activities in the country that they left. Lower contributions from corporations mean that individual citizens have to pick up the tab and that the government has to reduce spending in society.

A further issue is that it may lead to unfairness to smaller companies, who have a limited ability to do a tax inversion, which in turn gives them a competitive disadvantage. On the other hand, though, supporters of tax inversions argue that the benefit of tax competition means that multinational companies can pass savings on to customers and invest more. The CEO of Pfizer claimed that their merger with Allergen will allow them to invest more in the US.

Proponents of low tax rates argue that regardless of geography, complex and high taxes are inefficient and stifle investment. Accordingly, taxes should be lowered and simplified in order to stimulate business activity and growth. When businesses grow, so too can jobs, wages, and prosperity generally. An incidental point is that the drive to attract business can encourage more efficient tax regimes. These can apply to both large and small business, and take forms such as the UK's wholesale reduction of corporation tax to 18%.

Famous businessman, Warren Buffett, has pointed out that on the contrary, investors want good roads, an educated and healthy workforce, and the rule of law - all of which help prosperity, but inextricably mean tax is needed.

111. Which of the following is true based on the passage?
    A. There were no tax inversions before 2015
    B. Tax inversions are bad
    C. Tax inversions always lead to more investment
    D. The Pfizer-Allergen deal will benefit society
    E. Allergen is in a different country to Pfizer

112. Which of the following is a defining feature of a tax inversion?
    A. A company buying another company
    B. Moving to another country
    C. A company making more profit
    D. A company moving its headquarters to a lower-tax regime
    E. A company evading tax

113. What is the main motivation for American companies buying companies in other countries?
    A. To expand reach to other countries
    B. To benefit customers
    C. To have lower taxes
    D. To leave America
    E. To benefit from simpler tax regimes

114. Which of the following is stated as a benefit to society of a company doing a tax inversion?
    A. Lower tax rate
    B. An efficient tax system and lower tax rate
    C. More jobs
    D. Lower taxes to both big and small businesses
    E. More businesses being attracted to the society

115. Which of the following is an argument that is implied but not stated in Warren Buffett's statement?
    A. The US should keep increasing its tax rate
    B. The UK's tax rate is too low
    C. Taxes are good for businesses
    D. Taxes should be raised
    E. Businesses should pay their taxes

Passage 23 – Black Death

In the Middle Ages, most people tended to form small communities on a feudal manor, which consisted of a village, church, and a castle presided over by a lord. Peasants typically held some land in the manor and in return, had to provide some services to the lord. While they could occupy a property, they were not allowed to leave the manor without permission.

The high mortality of the Black Death shocked medieval British society in more ways than could have been predicted at the time. Although historians disagree on the precise mortality rate, a number of estimates put the decline in population size at 50%. Perversely, this might have actually been to the benefit of the peasants that remained.

The decrease in population reduced the supply of labour and, thus, wages rose and peasants were able to demand higher wages. This allowed them to enjoy a higher standard of living. Due to the massive death rate, many vacant land holdings became available when people died without an heir. This transformed the position of peasants, as they could now move to different manors. Due to the labour shortage, the lords wanted the peasants to stay and were willing to increase wages and improve conditions. For example, some labourers were given a hot meal as opposed to a cold one. However, this increase in wage costs meant that lords could not farm as much of their demesne and had to change the way in which it was used.

This change in bargaining position shifted the economic balance of power and marked the breakdown of the feudal system. An example can be shown in Great Waltham and High Ester, where 7 in 51 marriages before the Black Death took place without merchet being paid. After the Black Death, though, this increased to 20 out of 46. Accordingly, it is clear that the power of the lords in the manorial system had weakened and they no longer held ultimate control over their peasants.

Peasants at the bottom of the social hierarchy were marked by a greater individualism. Their labour to the lord was no longer determined by their ties to the land or by custom but by market forces – i.e. which lord was offering the best pay. Given the greater availability of land, those who were doing well, such as merchants and clothiers, could also buy it and improve their status. This is remarkably reminiscent of a capitalist society as opposed to a feudal one.

Interestingly, women were valued a lot more and their wages increased. The Black Death's high mortality meant that much of the workforce was dead and needed to be replaced. However, there was not much of a transformation in society as the general attitudes towards women remained.

It is not a ubiquitous view though that the Black Death caused this change in the balance of power between peasants and the lords.

The Great Famine had already started the reduction of the population that led to increasing wages. Accordingly, some see the Black Death as an accelerator of social change, rather than an initial cause. Others also question the very basis of the view that the Black Death 'caused' the increase in entrepreneurialism among the lower orders in society. In order for a move from feudalism to capitalism to occur, there must surely be a change in attitudes (such as an increase in risk taking) – otherwise, even the death of half the population would not change things. Accordingly, some view the Black Death as merely giving an opportunity for the shift in the balance of power.

116. Which of the following fits in logically with the third paragraph?
    A. Peasants were previously tied to their manor
    B. Peasants transformed society
    C. Peasants were poor
    D. The power of the lords increased
    E. Peasants became more powerful than the lords

117. Why did women get higher wages?
    A. Women were valued a lot more
    B. Attitudes in society changed
    C. There were more women than men
    D. There was a shortage of labour
    E. The end of feudalism and a move to a capitalist society

118. Which of the following is true based on the passage?
    A. Peasants had permission to move to different manors
    B. The lords were happy for peasants to leave
    C. The Black Death caused the social change that was described
    D. The Black Death triggered the end of the manorial system
    E. Peasants were able to demand higher wages

119. Which of the following is an assumption made about the Great Famine?
    A. It reduced the population
    B. It had an impact on wages
    C. It occurred before the Black Death
    D. It was more severe than the Black Death
    E. It was a bigger cause of the shift in the balance of power

120. What is a common feature among the views in the final paragraph?
    A. The Great Famine had a role in shifting the balance of power
    B. The Black Death had a role in the shift of the balance of power
    C. The Black Death did not impact the shift in the balance of power
    D. Black Death was just an accelerator of social change
    E. Society was going to change even without the Black Death

121. Which of the following, if true, most weakens the author's argument in the fourth paragraph?
    A. In most manors, there was not a reduction in the number those paying merchet
    B. Most peasants were still paying merchet
    C. The lords still owned the land which was being used
    D. The lords still decided the wages to pay peasants
    E. The shift in the balance of power was small

Passage 24 – Cohabitation

Out of the 18.6 million families in the UK, 12.5m consist of married couples, and this has traditionally been the most common family type in the UK. However, cohabitation is on the rise, with statistics displaying an increase of 30% in cohabitation since 2004, thereby making cohabitation the fastest growing type of family in the UK. Accordingly, this should raise concerns as to whether the current legal regime for families is still suitable for society.

A difference between marriage and cohabitation is that, as a legal construct, many legal rights and responsibilities flow from the fact of being married. No additional legal rights flow from the fact of being a cohabitant.

In essence, a cohabitation is where two adults live together as a couple in an intimate and committed relationship. Strictly speaking, there is no legal definition of cohabitation – it is simply whenever people are living together in a relationship but are not married. It's more a descriptor denoting a relationship, which may be like a marriage, but is not formally one.

For example, a key difference is when one partner dies without creating a will. In the instance of a married couple, most of the property would pass to the spouse. Alternatively, the partner in a cohabitation would not automatically receive such property, unless it was stated in the will. Indeed, it is easy for a cohabiting couple to get round this by simply creating a will but thinking about the consequences of death is not particularly common among couples.

Furthermore, the process of separating is markedly different for a married couple compared to a cohabiting couple. An unmarried couple can just go their own ways and separate informally. On the other hand, separation from a marriage requires the consent of the court (although that is a mere formality) but crucially, the courts can, and frequently do, transfer property between spouses on divorce. For example, in the high-profile divorce between Heather Mills and Paul McCartney, Mills was awarded £24.3 million out of McCartney's estate. No such award would have been made had they been cohabitants.

While some have argued for such legal rights, both on death and separation, to be accorded to cohabitants, there are serious issues with such assertions. The most serious issue is that of consent. In regard to cohabitation, it is not possible to say that the parties consented to giving each other additional rights and responsibilities in respect of the other. Accordingly, it would infringe the parties' freedom and autonomy. Or would it? Campaigners retort that 58% of cohabiting couples believe their rights are the same as that of married couples anyway.

However, given that the rest of the couples surveyed do not consent to having new legal duties, surely the law should not impose that on them against their wishes. The law also becomes significantly more uncertain in giving rights and responsibilities to cohabitants: at what point do people living together become cohabitants? A legal definition would be required, which would descend into arbitrariness. In particular, it must be decided how long it would take for a couple to legally become a cohabitee. Further, to determine whether one is a cohabitee, it would likely be necessary to undertake an inquiry of a couple's relationship, which would hardly be an easy and inexpensive task.

While the status quo may not entirely reflect the public's understanding, improving the latter as opposed to forcing individuals to undertake greater legal duties and responsibilities would surely be the ideal solution.

122. How does a marriage and cohabitation differ?
   A   The individuals in cohabitation are not committed to each other for life
   B   The legal rights and obligations are lower for a cohabitation than a marriage
   C   The courts have given more rights to cohabitees recently as opposed to married couples
   D   In cohabitation, the partners can separate
   E   Married couples tend to have children

123. How can a cohabiting couple put themselves on the same legal basis as a married couple in the event of a death of one of the parties?
   A   They cannot under the current law
   B   They need to get married
   C   They need to separate
   D   Write a will
   E   Write a contractual agreement

124. What would the argument of the 'Campaigners' in the [sixth] paragraph most likely be?
   A   For cohabitants to get all the same legal rights and responsibilities as married couples
   B   Having legal rights and responsibilities would infringe the cohabiting couple's freedom
   C   Having legal rights and responsibilities would not infringe the cohabiting couple's freedom
   D   That 58% of cohabiting couples believe their rights and responsibilities are the same as married couples
   E   The parties' freedoms should not be infringed

125. Which of the following policies would the author most likely agree with?
   A   Making cohabitation a legal concept
   B   Give cohabiting couples rights and responsibilities in regard to each other
   C   Banning cohabitation
   D   Require cohabiting couples to enter into a marriage
   E   A campaign to raise awareness of the differences between cohabitation and marriage

126. Which of the following does the author disagree with?
   A   Cohabitation
   B   Legal rights and responsibilities
   C   That 58% of cohabiting couples consent to the same rights as in marriage
   D   Imposing additional legal rights and responsibilities on cohabiting couples
   E   The £24.3 million settlement for Heather Mills

Passage 25 – F1

Formula 1, the pinnacle of motorsport, should epitomise the best racing the world has to offer and yet television viewing figures are decreasing and the outcomes of races are more predictable than ever. Worryingly, a survey in 2015 of over 200,000 fans highlighted significant disquiet over the way the sport is going. One of the top three descriptors of the sport was 'boring' and 89% of fans said that F1 needs to be more competitive. This should not be a surprise given that one team has dominated since 2014 but hopefully, this will be a wakeup call. At the same time, costs in F1 are spiralling out of control, which make it difficult for smaller teams to catch up.

F1 has tried to project an image of being environmentally friendly by requiring all teams' engines to have a capacity of 1.6 litres, much smaller than in previous regulations, and electrical and kinetic energy recovery systems in order to harvest energy that would have otherwise been wasted. Indeed, an object of this was to attract more commercial partners. In particular, if F1 regulations are more relevant to road cars, manufacturers would be more likely to enter, or remain in, Formula 1 as it would have a greater relevance to their commercial operations. A corollary of this is that costs of engines have doubled, coming into the region of £18 million for a year's supply. This is double that of the previous engines and is the biggest single expense for the teams. Such costs are significantly inhibiting for smaller teams.

It is not just the cost of engines that is large – the entire budgets of teams stretch to 100s of millions. Red Bull is estimated to have the largest budget at over €460 million with Mercedes close behind. While these large operations can add to the prestige of the teams, it doesn't necessarily add to the excitement of the sport. Arguably the latter is more important and serious issues in F1 remain.

Since the new regulations were introduced, one team, in particular, the Mercedes F1 team have by all accounts dominated the sport with very few drivers from other teams winning. While Mercedes have undoubtedly produced the best car and thus, are deserving of their success, it has led to accusations that the spectacle has become boring. It is mostly clear which team is going to win before the race has even started. This issue always abounds when one team dominates and may make a Grand Prix more of a procession than a race.

Mercedes have received criticism for their continued winning streak. Indeed, a famous driver from a rival team, Sebastian Vettel, has suggested that this is why the sport is boring. However, dominance is not a new phenomenon in F1 though, it has happened before – Vettel himself enjoyed a number of years in a dominant Red Bull with little competition from the other drivers. In 2000-2004, Ferrari dominated F1. It is thus clear that dominance is a key feature in Formula 1.

Accordingly, Mercedes should not be criticised for doing what every other team is doing, which is building the best F1 car they can. Mercedes built the best of the rest which is why they're winners. However, the tendency of F1 to produce a dominating team should not be attributed to the Mercedes car, but to the nature of some of the rules and regulations of the sport. The rules must be created in such a way as to promote competition. While the drive for environmentally friendly engines did not aim to achieve this, it did not help and may have even hampered competition.

Rule changes should include making it easier for cars to overtake, reducing engine costs and even reducing the total costs of teams. Such changes would reduce the large disparity between teams and make for closer racing. Whether the F1 rulers can make such changes depends on their will but it is in their interest too. Ultimately, change is required as should television numbers continue to drop, F1 will no longer be a draw on sponsors' and TV companies' money. In the long run, at least, F1's success is inextricably linked to the fans and the powers that be in F1 should recognise that.

127. What does the word 'corollary' mean in the context in which it is used?
    A. A benefit
    B. A consequence
    C. An incidence
    D. Cost
    E. Price

128. Which of the following is an opinion as opposed to a fact?
    A. The entire budgets of teams stretch to 100s of millions
    B. One of the top three descriptors of the sport by fans was 'boring'
    C. Engine costs have doubled
    D. F1 is the pinnacle of motorsport
    E. Mercedes should not be criticised

129. Based on the whole passage, how might the drive for environmentally friendly cars have hampered competition?
    A. It caused the dominance of Mercedes
    B. It led to slower engines
    C. It increased engine costs, which hit smaller teams harder
    D. It reduced the excitement of racing
    E. It has led to a procession rather than a race

130. To what does the author attribute the success of Mercedes?
    A. Mercedes building the best car
    B. The rules and regulations of F1
    C. The engine changes
    D. The drive for environmentally friendly engines
    E. Spending the most money

131. Why is there dominance in F1 according to the author?
    A. The racing is not close enough
    B. Mercedes' winning streak
    C. The new environmentally friendly engines
    D. Large spending
    E. The rules and regulations of F1

Passage 26 – Reparations

The transatlantic slave trade lasted approximately four centuries, and Britain played a well-known role in it. Over this period, English ships made slaving voyages to Africa, where Africans were captured and transported to the Americas to be sold as goods. David Richardson, a prominent historian, has calculated that over 3 million enslaved Africans were transported to the Americas. The Royal African Company gave the slave trade a formal royal charter and it has been argued that the slave trade was the richest part of Britain's trade in the 18th century. Karl Marx has argued that the slave trade created the financial conditions for Britain's industrial revolution but the extent of economic benefit to Britain has been heavily disputed. Slavery was formally abolished through The Slavery Abolition Act 1833 across the British Empire.

Recently there have been calls for the UK to pay reparations to compensate for the atrocity of the slave trade centuries ago. On Prime Minister David Cameron's first state visit to Jamaica in 2015, he faced calls from politicians (including the Jamaican leader) for Britain to pay reparations for its involvement in the slave trade. According to calculations by researchers, total reparations could be between $5.9 trillion and $14 trillion.

It is a fundamental moral right of anyone to seek redress for damage caused to them. However, the damage was done centuries ago so many question whether modern Brits should be subject to reparations when no actual damage was done by them. They committed no wrong.

Indeed, one line of argument is that we embrace our history in other aspects, such as the success of Britain in World War II. The British government has also apologised for tragic events such as the Hillsborough tragedy. Accordingly, when embracing our history, we should surely embrace our full history. Indeed, Germany has paid over €89 billion in reparations to compensate Jewish victims of Nazi war crimes in World War II in the 60 years after it. However, opponents of reparations suggest that this is not analogous to giving reparations for the slave trade.

If it was decided that Britain should pay reparations, critics argue that it is not clear who should be paid. Did countries such as Jamaica suffer the entirety of the damage? From another perspective, the damage was done to individuals specifically as opposed to just the countries as a whole.

In a further argument outlined by Julia Hartley-Brewer, she points out that the majority of slaves were, in fact, sold by African rulers to Europeans in exchange for goods, such as weapons. Accordingly, should the current ruling classes of Africa benefit from reparations? However, even though others had a role in the slave trade, it should not displace any responsibility of one of the main antagonists.

Some argue that it's clear that given Britain's involvement, they owe a clear and substantial debt to the descendants of slaves who were harmed. However, opponents of reparations dispute the idea that the damage of slavery continues to the present day.

132. Why might the German reparations not be analogous to reparations for the slave trade?
    A. The Nazi war crimes were more recent
    B. The Nazi war crimes were worse
    C. Victims of the Nazi war crimes themselves were being paid
    D. More died from Nazi war crimes than through the slave trade
    E. Modern Brits are not at fault for the slave trade

133. Which of the following best *illustrates* the implicit argument made when considering who Britain should pay reparations to?
    A. A company should not claim damages when its employee is injured by another person
    B. When an employee is injured, its descendants should be able to claim damages
    C. When an employee is injured, the community should benefit
    D. When a company suffers damage, its employees should benefit
    E. When an employee suffers harm, they should be compensated by the community

134. What was the author's purpose in writing this article?
    A. To argue for reparations for the slave trade
    B. To argue against reparations for the slave trade
    C. To point out and consider arguments for and against giving reparations for the slave trade
    D. To argue for the position of modern British people
    E. To argue that Britain was not wholly responsible for the slave trade

135. What is the author's view as to Julia Hartley-Brewer's point [penultimate paragraph]?
    A. African rulers were to blame for the slave trade and not Britain
    B. The slave trade was wrong
    C. It should not affect the UK's responsibility
    D. The UK was the main antagonist
    E. That it is incorrect

Passage 27 - Animal Experimentation

Animal experimentation has become significantly controversial in recent years. With the advocacy of groups such as the National Anti-Vivisection Society, public discourse on animal testing has taken on a new light. Extremist animal rights groups have taken to violent demonstrations and the harassment of employees engaged in animal testing. Indeed, the arguments for animal rights are important but extremist actions tend to shut off the oxygen for such debate. Many animal rights groups tend to believe that any animal testing is unacceptable and cruel, without being willing to consider a balanced view, though. That's not to say that their arguments are not useful, but that there must be consideration of the benefits of animal testing.

Reasonable scientists do not aim to make animals suffer if unnecessary. In fact, the law does not allow that. The current legal position, as laid out in the Animals (Scientific Procedures) Act, is more restrictive than one may first envisage. Firstly, animal testing is not allowed unless a license is granted by the government to the researcher carrying out the testing, to the particular project as a whole and for the place at which the work is carried out. When deciding whether to grant a license, there must be a clear potential benefit to either people, animals or the environment. In essence, the potential costs and benefits are weighed up. Animal rights groups even disagree with this, though. Ultimately, animal testing is not undertaken unless necessary and harm is minimised at all points in the process. Crucially, the legal position makes sure that there is a net benefit to society from the use of animals in research. Furthermore, strides in technology reduce the need to use animal experimentation.

The benefits of using animal testing cannot be understated – there are approximately 120,000 sufferers of dementia in the UK, to which animal testing is being used in the development of treatments to mitigate against the symptoms of it.

Of course, that's not to say that a blanket allowance should be given – licenses should only be given where there are no other possible treatments. Developments in technology and computer modelling have meant that the need to engage in animal testing may well reduce.

Public opinion also appears to be on the side of animal researchers -a 2005 poll by MORI showed that 89% of the public accept the need to use animals, so long as animals do not suffer unnecessarily.

In the instance of tinnitus, which affects approximately 10% of the UK population, the use of animals in research has the potential to improve the quality of lives of millions of people. Testing has involved a wire being inserted into the brains of mice and sounds being played to gauge their neurological response to sound. The mice are sedated with anaesthetic and given pain killers. Crucially, animal research does not always necessarily involve harm to animals, which makes the uncompromising position of animal rights groups untenable.

136. Which of the following is true based on the passage?
  A.   Reasonable scientists do not make animals suffer
  B.   Animal testing is always necessary
  C.   The government considers the costs and benefits of engaging in animal testing
  D.   Not enough animal testing is done
  E.   Scientists are reasonable

137. What is the author's view as to the use of animal testing?
  A.   Animal testing should be allowed
  B.   Animal testing should not be allowed
  C.   Cannot tell
  D.   Animal testing is not necessary for society
  E.   Animal rights groups should be banned

138. Which of the following is an opinion?
  A.   Reasonable scientists do not aim to make animals suffer unnecessarily
  B.   Animal testing is not allowed unless a license is granted
  C.   Tinnitus affects 10% of the UK population
  D.   89% of the public agree with the need to use animals in testing
  E.   There are 120,000 sufferers of dementia in the UK

139. What is the author's objection to animal rights groups?
  A.   They advocate against animal testing, which is beneficial
  B.   They take an absolutist position
  C.   They do not understand the benefits of animal testing
  D.   They are out of step with public opinion
  E.   They do not help those with serious diseases, such as Parkinson's

140. Which of the following most undermines the author's viewpoint?
  A.   The estimated benefit of animal research is far higher than the actual benefit derived
  B.   Animals can feel harm
  C.   The public is misinformed as to the ability of animals to suffer harm
  D.   There are other methods of research that could be usefully used
  E.   Patients may not know that they're being treated with animal-tested products

Passage 28 – Copyright

Copyright is a significant form of intellectual property right. It is a legal right that is accorded to the creator of, among other things, films, soundtracks, and original literary, artistic, dramatic and musical works. It lasts for 70 years after the death of the author and any person who copies, sells or distributes a copyrighted work without authorisation can be sued by the author. The author can claim damages for a loss of earnings and seek a court injunction to stop the individual infringing the author's rights. Breach of this would be a contempt of court, which carries, even more, consequences for the individual.

However, the relevance of copyright in the modern world has been questioned more than ever. It is meant to protect song artists, but illegal downloading is all too prevalent in the world. Indeed, within a day of Kanye West's latest album release, there were reports of over 500,000 unlawful downloads, with an estimated cost of $10 million to West. A particular concern is whether the law is now out of touch, particularly when improvements in technology have meant that enforcement of copyright is significantly more arduous. For example, it is harder to close down websites, and this allows them to operate above the law. In the meantime, record labels have been developing technology to help inhibit infringement. However, the very nature of digital products, such as songs, means that preventing the illegal use of them will continue to be laborious.

Accordingly, copyright is now more relevant than ever. Removing copyright protection would leave artists' incomes even more in peril of the morals of the population. It would legalise an immoral act and mean an even greater amount of free downloading. The fact of copyright protection indicates to society that it is wrong to download music without paying for it, or without authorisation. Removing that protection will increase unauthorised use.

Indeed, a prime justification of copyright protection is the natural rights theory, first advanced by Locke. The premise here is that every person 'has property in his own person'. Accordingly, when using your labour with things from nature, those things become yours. Using labour confers natural rights over the resultant output of that labour. Therefore, allowing copyright protection is vital to recognising those rights.

Locke's approach though is only superficially satisfying in regard to copyright protection. Many critics point out that copyrighted works, such as books and songs are not the complete creation of the author. The author's work may well have been influenced by other creators and works that had an impact on the author. Further, academics Alpin and Davies suggest that 'labour' alone is too imprecise to determine the boundaries of intangible goods.

Nonetheless, an alternate theory put forward by Hettinger may provide the answer. Based on the utilitarian theory, as founded by Bentham, Hettinger suggests that laws should promote the creation of valuable intellectual works, such as music, books, and films. Accordingly, this requires artists to be granted copyright protection in what they produce in order to incentivise the production of such works. The provision of copyright allows authors greater control over what they can do with their work – they can decide on its sale and distribution and at a price of their choosing. If anyone tries to sell their work without their consent, that person can be taken to court. However, a concern with Hettinger's view is that it is not clear whether empirical evidence exists to back this up or even whether it would be possible to acquire such legitimate evidence.

141. What have improvements in technology done?
    A. Allowed illegal downloading to be stopped
    B. Discouraged illegal downloading
    C. Made illegal downloading free
    D. Made illegal downloading unfair
    E. Allowed illegal websites to evade authorities

142. What effect does the writer believe that removing copyright protection will have?
    A. Artists' incomes would be at the peril of the morals of the population
    B. Legalise an immoral act
    C. Legalise the free downloading of music
    D. Increase the number of free downloads
    E. Hamper efforts of record labels to increase protection

143. Which of the following is a conclusion which can be derived from Locke's theory of natural rights?
    A. Every person has natural rights
    B. Every person has a right to the things that they create
    C. It would not be right to give one's own creation for free
    D. Copyright protection should be increased
    E. People's creations should not be used without their consent

144. According to Hettinger, why should copyright exist?
    A. To allow authors a fair price for their work
    B. To ensure that a person's creation receives a just reward
    C. To accord with utilitarianism
    D. To incentivise the production of intellectual works
    E. To stop people free-riding

145. Which of the following is implied but not stated in Hettinger's argument?
    A. Having control over one's intellectual creation encourages creation
    B. Author's getting paid for their works encourages creation of their works
    C. There is no evidence backing it up
    D. There is evidence backing it up
    E. Locke's justification should not apply at all

146. Which of the following is common among Locke and Hettinger's justifications?
    A. They both critique the status quo
    B. They both advocate against the use of the law to provide copyright
    C. They both suggest that there should be copyright protection
    D. Authors always want to get paid
    E. Not paying an author for their work is always wrong

Passage 29 – Tax Avoidance

The tax affairs of large corporations have taken centre stage over the last few years, and mostly, but not wholly, through their own fault. While tax is a controversial matter in itself, for corporations, it has become even more significant. Tax reduces a business's paper profits and so businesses indeed have an incentive to pursue lower taxes. There are two main ways to do this directly: tax evasion and tax avoidance. Tax evasion is the illegal reduction of one's tax liability – this can be through misreporting a company's sales or simply not paying the correct liability. Tax avoidance, on the other hand, is the legal reduction of one's tax liability and is basically where a company's tax liability is lower than it should be – sometimes this is quite subjective but it can be obvious at other times. Tax avoidance can occur through setting up offshore subsidiaries, transfer pricing and benefiting from vague tax laws.

Tax rates are deceptively clear. Each company has to pay 20% of all UK profits as corporation tax to HM Revenue & Customs. If a company does not pay the 20% tax on its declared profits, it is breaking the law. Crucially, if no UK profits are made, the company is liable for no tax. However, this is an issue given that the profits that a company can legally declare may not be a sign of their true profits. This is where tax avoidance occurs.

However, Starbucks took headlines in 2012 when Reuters pointed out that since setting up in the UK in 1998, they had paid a meagre £8.6 million corporation tax on £3 billion of sales. In a number of years, they paid zero tax. This revelation led to boycotts and protests at Starbucks stores. Indeed, they were acting within the law, but as one protester pointed out: "They've shown utter contempt for our tax system."

It is unacceptable for such companies to be paying such little tax. In particular, companies that benefit from public infrastructure, such as roads and railway, and from public services, all of which require tax. It's, thus, quite ironic that Starbucks relied on the police to protect their stores from protests and yet they contributed a pittance to the public purse.

However, a significant issue is where the responsibility for such an unacceptable situation lies. Such companies are already acting lawfully and abiding by the rules. Some even question whether companies such as Starbucks are acting immorally, given that they're 'just' following the rules. The government do indeed have responsibility for implementing rules. While it may be quite difficult in this globalised world to establish laws to reduce tax avoidance, it is still the responsibility of world governments to coordinate efficient tax policies and prevent or reduce tax avoidance.

That's not to say that multi-national companies are absolved from responsibility, though. They have benefited enormously from basing their operations in the UK – with £3 billion of revenue in the example of Starbucks. It allowed shareholders to gain wealth while not contributing to society. It is quite simply unfair. Small businesses are less able to avoid tax because they can't shift their profits to subsidiaries abroad or hire expert tax lawyers. Individual citizens also can't avoid tax. So it is quite clear that politicians and companies need to restore balance in the tax system, however hard it may be.

147. What is the difference between tax evasion and tax avoidance?
    A. The public only condemns tax avoidance
    B. Google and Starbucks have only committed tax avoidance
    C. Tax evasion is morally bad
    D. Tax avoidance is moral
    E. Avoidance is legal but evasion is illegal

148. How much profit did Starbucks declare in the years when it paid zero tax?
    A. £8.6 million
    B. £3 billion
    C. 20%
    D. Zero
    E. Cannot tell

149. Why was there a boycott of Starbucks stores in 2012?
    A. Starbucks engaged in tax evasion
    B. Starbucks engaged in tax avoidance
    C. Consumers did not like Starbucks
    D. Too many companies were acting in defiance of the tax rules
    E. Starbucks paid the wrong form of tax

150. What is the implied argument in the fourth paragraph?
    A. Companies should not prioritise profit over tax
    B. Companies should pay for the infrastructure and police
    C. Companies should pay tax
    D. The police should not have protected Starbucks stores
    E. The tax rate should reduce

151. Which of the following is a matter of opinion?
    A. The tax rate
    B. Tax evasion
    C. Tax avoidance
    D. The amount of profit declared
    E. Starbucks had £3 billion of sales in the UK

152. Who holds responsibility, according to the author, for the tax avoidance in the UK?
    A. The tax system
    B. Small businesses
    C. Tax avoiding companies
    D. Government
    E. The government and tax avoiding companies

153. Why is there unfairness in the tax system generally according to the author?
    A. Tax avoidance
    B. Globalisation
    C. Starbucks avoiding tax
    D. Multinational companies avoiding tax while individuals have to pay their way
    E. The corporation tax rate is only 20%

Passage 30 – Susan B. Anthony

At the election of President and Vice President of the United States, and members of Congress in November 1872, Susan B. Anthony, and several other women, offered their votes to the inspectors of election, claiming the right to vote, as among the privileges and immunities secured to them as citizens by the fourteenth amendment to the Constitution of the United States. The inspectors, Jones, Hall, and Marsh, by a majority, decided in favour of receiving the offered votes, against the dissent of Hall, and they were received and deposited in the ballot box. For this act, the women, fourteen in number, were arrested and held on bail and indictments were found against them under the 19th Section of the Act of Congress of May 30th, 1870, (16 St. at L. 144.), independently charging them with the offence of knowingly voting without having a lawful right to vote. The three inspectors were also arrested, but only two of them were held to bail, Hall having been discharged by the Commissioner on whose warrant they were arrested. All three, however, were jointly indicted under the same statute—for having knowingly and wilfully received the votes of persons not entitled to vote.

Of the women voters, the case of Miss Anthony alone was brought to trial, a nolle prosequi having been entered upon the other indictments. Before the trial, Miss Anthony gave lectures in all of the twenty-nine districts in Monroe County, the location of the trial, where she argued that she had a lawful right to vote. US Supreme Justice, Judge Ward Hunt, was persuaded that Miss Anthony might have prejudiced potential jurors and moved the trial to Canandaigua, Ontario County. Miss Anthony continued to give lectures in Ontario County, but the trial was not altered again and set to go ahead.

Upon the trial of Miss Anthony before the U.S. Circuit Court for the Northern District of New York, at Canandaigua, in June, 1873, it was proved that before offering her vote she was advised by her counsel that she had a right to vote; and that she entertained no doubt, at the time of voting, that she was entitled to vote.

154. According to the above passage, how many people in total were arrested due to the group of women voting?
    A. Fourteen
    B. Three
    C. Seventeen
    D. Sixteen
    E. Fifteen

155. Based on the passage only, who was definitely brought to trial over the incident?
    A. Susan B. Anthony
    B. Susan B. Anthony and the inspectors
    C. Susan B. Anthony, Jones, and Marsh
    D. Jones, Marsh, and Hall
    E. None of the above

156. Which of the following best describes initial opinions of the election officers?
    A. United by each member's personal support of the women's votes
    B. Divided in response to the women's actions
    C. Apathetic about the women's actions
    D. United by general disapproval of the women's actions
    E. Unequivocal about the legality of the women's actions

157. Which defence for Susan B. Anthony is mentioned above?
    A. She did not realise she was not allowed to vote
    B. That all people born in the USA should be able to vote for their president
    C. That gender should not prevent her vote
    D. The election officers accepted her vote, showing that the responsibility did not lie with her
    E. The previous law was wrong

158. Why did Judge Ward move the case to Ontario County?
    A. The judge was against Anthony's case
    B. The judge believed that the jury could have been prejudiced in Monroe County
    C. The judge believed that the jury was biased
    D. The judge disagreed with the content of Anthony's lectures
    E. The judge was based in the Ontario County

Passage 31-Birds

The following Passage is found in a book on nature published in 1899:

Five women out of every ten who walk the streets of Chicago and other Illinois cities, says a prominent journal, by wearing dead birds upon their hats proclaim themselves as lawbreakers. For the first time in the history of Illinois laws, it has been made an offence punishable by fine and imprisonment, or both, to have in possession any dead, harmless bird except game birds, which may be possessed in their proper season. The wearing of a tern, or a gull, a woodpecker, or a jay is an offence against the law's majesty, and any policeman with a mind rigidly bent upon enforcing the law could round up, without a written warrant, a wagon load of the offenders any hour in the day and carry them off to the lockup. What moral suasion cannot do, a crusade of this sort undoubtedly would.

Thanks to the personal influence of the Princess of Wales, the osprey plume, so long a feature of the uniforms of a number of the cavalry regiments of the British army, has been abolished. After Dec. 31, 1899, the osprey plume, by order of Field Marshal Lord Wolseley, is to be replaced by one of ostrich feathers. It was the wearing of these plumes by the officers of all the hussar and rifle regiments, as well as of the Royal Horse Artillery, which so sadly interfered with the crusade inaugurated by the Princess against the use of osprey plumes. The fact that these plumes, to be of any marketable value, have to be torn from the living bird during the nesting season induced the Queen, the Princess of Wales, and other ladies of the royal family to set their faces against the use of both the osprey plume and the aigrette as articles of fashionable wear.

159. In 1899:
   A. Women across the USA could be prosecuted for owning ornamental dead birds
   B. There was a significant rise in female arrests in America
   C. Possession of a dead gull could lead to trouble
   D. Americans responded to law by citing the use of jays as ornamentation unfashionable
   E. Delinquency across America increased

160. Ostrich feathers were seen as preferable to osprey plumes because:
   A. Ostriches are less intelligent birds
   B. Ostriches are killed for their meat, so one might as well use their feathers
   C. Queen Elizabeth has an especial love of ospreys
   D. Harvesting osprey plumes was seen as an inhumane process
   E. Ostrich feathers were of superior quality

161. Which of the following is false, based on the passage?
   A. Games birds could be possessed by citizens of Illinois all year round
   B. Possessing a bird was not illegal in every circumstance
   C. Wearing a woodpecker could lead to police action
   D. Possessing certain birds could lead to a fine or imprisonment
   E. All policemen would take action against any person wearing a prohibited bird

162. Banning Osprey plumes in the UK's army was difficult because:
   A. Many uniforms required them
   B. The Princess did not have the authority to implement the ban
   C. Her ultimate support was predominately female, and thus, their concerns seemed to have no relevance from the male domain of the army
   D. It would be hard to differentiate between other regiments within the army, who were already wearing ostrich feathers
   E. The production process of osprey plumes was easier

163. Which of the following could NOT be legally owned in Illinois, according to the passage?
   A. A live bird intended for personal ornamentation
   B. A live bird intended for fighting
   C. A dead bird of prey that had violently attacked you
   D. Feathered garments
   E. None of the above

Passage 32-Books

Gutenberg's father was a man of good family. Very likely the boy was taught to read. But the books from which he learned were not like ours; they were written by hand. A better name for them than books is 'manuscripts,' which means handwritings.

While Gutenberg was growing up, a new way of making books came into use, which was a great deal better than copying by hand. It was what is called block printing. The printer first cut a block of hardwood the size of the page that he was going to print. Then he cut out every word of the written page upon the smooth face of his block. This had to be very carefully done. When it was finished, the printer had to cut away the wood from the sides of every letter. This left the letters raised, as the letters are in books now printed for the blind. The block was now ready to be used. The letters were inked, the paper was laid upon them and pressed down. With blocks, the printer could make copies of a book a great deal faster than a man could write them by hand. But the making of the blocks took a long time, and each block would print only one page.

Gutenberg enjoyed reading the manuscripts and block books that his parents and their wealthy friends had, and he often said it was a pity that only rich people could own books. Finally, he determined to contrive an easy and quick way of printing.

Gutenberg, indeed, found this way and made the first movable-type printing press in Europe, with pieces from lead, tin, and antimony. Crucially. Gutenberg's innovation was confirmed by the production of the 'Gutenberg Bible'; the first major book printed with the movable-type printing press. With the invention came cheaper and higher quality books, thereby encouraging the development of printing presses across Europe, which in turn increased the number of books in supply. A new feature of the movable-type printing press was the advent of oil-based ink, which was more durable. This has been termed the 'Gutenberg Revolution'.

164. Which of the following reasons can be inferred from the above passage to explain Gutenberg's desire to create a new way of printing?
    A. It was a lucrative business to go into
    B. He wanted to make text more accessible
    C. He was tired of waiting for each book to be hand-written or block pressed and wanted quicker access to literature
    D. He found the current books too costly for him to continue his reading habit
    E. He wanted to spread his ideas across Europe

165. Which of the following is **NOT** mentioned as a downside to block printing?
    A. It exhausts the carver
    B. It is intricate and demands attention to detail
    C. It is a lengthy process
    D. An individual block has limited utility
    E. It requires the attention of an individual

166. Which of the following statements is definitely true according to the above passage?
    A. Gutenberg was taught to read as a boy
    B. Gutenberg's father belonged to the aristocracy
    C. Block printing was the predominant book manufacturing process whilst Gutenberg was growing up
    D. Gutenberg's family was somewhat social
    E. Gutenberg was deeply religious

167. Which of the following is false, based on the passage?
    A. Printing with the block process was a simple task of inking up the prepared block and pressing it down on a piece of paper, to make one page of the text
    B. The movable-type printing press was the first of its kind in Europe
    C. Oil based ink was not used in block printing
    D. The ink lasted for longer in books made in the movable type printing press
    E. More books were produced

168. Which of the following statements are **NOT** supported by the above passage?
    A. Manuscripts were beautifully crafted
    B. 'Manuscripts' is an appropriate name for what it describes
    C. Block printing is an appropriate name for what it describes
    D. Having well off friends was a good way to expand your reading
    E. It is probable that Gutenberg was educated

Passage 33- Norway

The following is taken from a book about Norway published in 1909:

'In a country like Norway, with its vast forests and waste moorlands, it is only natural to find a considerable variety of animals and birds. Some of these are peculiar to Scandinavia. Some, though only occasionally found in the British Isles, are not rare in Norway; whilst others (more especially among the birds) are equally common in both countries.

There was a time when the people of England lived in a state of fear and dread of the ravages of wolves and bears, and the Norwegians of the country districts even now have to guard their flocks and herds from these destroyers. Except in the forest tracts of the Far North, however, bears are not numerous, but in some parts, even in the South, they are sufficiently so to be a nuisance and are ruthlessly hunted down by the farmers. As far as wolves are concerned, civilisation is, fortunately, driving them farther afield each year and only in the most out-of-the-way parts are they ever encountered nowadays. Stories of packs of hungry wolves following in the wake of a sleigh are still told to the children in Norway, but they relate to bygone times—half a century or more ago, and such wild excitements no longer enter into the Norsemen's lives.

Yet, less ferocious animals give the people trouble enough, and amongst these may be mentioned the lynx and the wolverine, or glutton, each of which will make his supper off a sheep or a goat if he gets the chance. Of the two, the lynx is perhaps the worse poacher and his proverbial sharpness renders him difficult to catch. Not so the glutton, who, if he succeeds in crawling through a hole in the fence of a sheepfold, stuffs himself so full that he cannot get out again. I think that most of us would rather be called lynx-eyed than gluttonous, and certainly a lynx is a much handsomer beast than a glutton.

With the exception of the rabbit, all our English animals are found in Norway—the badger, fox, hare, otter, squirrel, hedgehog, polecat, stoat, and the rest of them. But besides these, there are little Arctic foxes and Arctic hares with bluish-grey coats in the summer and snowy-white ones in the winter. This change of colour is a provision of nature, rendering these particular animals, and some birds also, almost invisible among the snows. The ermine is another instance of this. In summer, he is just an ugly little brown stoat; but in winter, he comes out in pure white, with a jet-black tip to his tail, a skin worth a lot of money.'

169. Which of the following is best supported by the above passage?
    A. The variety of birds and animals to be found in Norway is unique to that country
    B. The variety of birds and animals to be found in Norway is common to all European countries
    C. By having forests, a country is more likely to have a variety of birds and animals
    D. England and Norway have similar geographical features
    E. All Norwegian animals can be found in England

170. English people are described as:
    A. Having been anxious about certain animals
    B. Sceptical of bears
    C. Living in fear of wolves
    D. Developmentally behind the Norwegians
    E. Worth a lot of money

171. Bears are described as:
    A. Hunting
    B. Scavenging
    C. Damaging
    D. Man-eating
    E. Almost invisible

172. Bears are also:
    A. Numerous in all forest tracts
    B. Numerous throughout the North
    C. Numerous throughout the South
    D. At risk in parts of Norway
    E. In parts of England

173. The Passage suggests:
    A. The movement of wolves to the out-of-reach parts of Norway is beneficial
    B. Wildlife currently threats Norwegian children
    C. Regret at the loss of adventures
    D. Norsemen particularly respect their natural surroundings
    E. It would be preferable to be gluttonous

174. Which of the following is an opinion?
    A. It is better to be called lynx-eyed than gluttonous
    B. The ermine changes colour
    C. Norway has vast forests and waste moorlands
    D. There are a variety of birds
    E. English animals can be found in Norway

Passage 34- Colonialism

Most of the colonists who lived along the American seaboard in 1750 were the descendants of immigrants who had come in fully a century before; after the first settlements, there had been much less fresh immigration than many latter-day writers have assumed. According to Prescott F. Hall, "the population of New England ... at the date of the Revolutionary War ... was produced out of an immigration of about 20,000 persons who arrived before 1640," and we have Franklin's authority for the statement that the total population of the colonies in 1751, then about 1,000,000, had been produced from an original immigration of less than 80,000.

Even at that early day, indeed, the colonists had begun to feel that they were distinctly separated, in culture and customs, from the mother-country and there were signs of the rise of a new native aristocracy, entirely distinct from the older aristocracy of the royal governors' courts. The enormous difficulties of communication with England helped to foster this sense of separation.

The round trip across the ocean occupied the better part of a year, and was hazardous and expensive; a colonist who had made it was a marked man—as Hawthorne said, "the petit maître of the colonies." Nor was there any very extensive exchange of ideas, for though most of the books read in the colonies came from England. The great majority of the colonists, down to the middle of the century, seem to have read little save the Bible and biblical commentaries, and in the native literature of the time, one seldom comes upon any reference to the English authors who were glorifying the period of the Restoration and the reign of Anne. "No allusion to Shakespeare," says Bliss Perry, "has been discovered in the colonial literature of the seventeenth century, and scarcely an allusion to the Puritan poet Milton." Benjamin Franklin's brother, James, had a copy of Shakespeare at the *New England Courant* office in Boston, but Benjamin himself seems to have made little use of it, for there is not a single quotation from or mention of the bard in all his voluminous works. "The Harvard College Library in 1723," says Perry, "had nothing of Addison, Steele, Bolingbroke, Dryden, Pope, and Swift, and had only recently obtained copies of Milton and Shakespeare....Franklin reprinted 'Pamela' and his Library Company of Philadelphia had two copies of 'Paradise Lost' for circulation in 1741, but there had been no copy of that work in the great library of Cotton Mather."

175. Which of the following is true according to the passage?
   A. Over half of the 1750 colonists that lived on the American seaboard had genetic links to immigrants who had arrived a century ago
   B. Most of the books on board ships were Bibles and biblical commentaries
   C. The colonists had poor communication skills
   D. The colonists disliked the English
   E. Many colonists visited England after immigrating to America

176. Which of the following statements is supported by the above passage?
   A. According to Hall, America's population at the date of the Revolutionary War could be entirely traced back to 20,000 immigrants
   B. The population in the 1751 colonies was over ten times the original immigration that moved there
   C. According to Hall, in 1751, the population in the American colonies was one million
   D. According to Hall, 80,000 people led to a population of 1,000,000
   E. Most of the population on the American seaboard immigrated there

177. According to the passage, the new aristocracy that existed in the colonies was:
   A. Similar to the England's
   B. Similar to European aristocratic systems in general
   C. Not based in royal governors' courts
   D. Not based on genetic lines
   E. Based on custom

178. Which of these is **NOT** given as a reason for the sense of separation from England?
   A. Travel between America and England was costly
   B. The English saw the early colonists as backwards
   C. Travel between America and England was slow
   D. Travel between America and England was dangerous
   E. Lack of an exchange of ideas

179. What did the author mean by the word 'allusion' in the passage?
   A. An implication
   B. An illusion
   C. Deference
   D. Thoughts
   E. Reference

Passage 35 - Rorschach

When discussing his famous character Rorschach, the antihero of 'Watchmen', Moore explains, 'I originally intended Rorschach to be a warning about the possible outcome of vigilante thinking. But an awful lot of comic readers felt his remorseless, frightening, psychotic toughness was his most appealing characteristic – not quite what I was going for.' Moore misunderstands his own hero's appeal within this quotation: it is not that Rorschach is willing to break little fingers to extract information, or that he is happy to use violence, that makes him laudable. The Comedian, another 'superhero' within the alternative world of Watchmen, is a thug who has won no great fan base; his remorselessness (killing a pregnant Vietnamese woman), frightening (attempt at rape), psychotic toughness (one only has to look at the panels of him shooting out into a crowd to witness this) is repulsive, not winning. This is because The Comedian has no purpose: he is a nihilist, and as a nihilist, denies any potential meaning to his fellow man, and so to the comic's reader. Everything to him is a 'joke', including his self, and consequently, his own death could be seen as just another gag.

Rorschach, on the other hand, does believe in something: he questions if his fight for justice 'is futile?' then instantly corrects himself, stating 'there is good and evil, and evil must be punished. Even in the face of Armageddon I shall not compromise in this.' Jacob Held, in his essay comparing Rorschach's motivation with Kantian ethics, put forward the postulation that 'perhaps our dignity is found in acting as if the world were just, even when it is clearly not.' Rorschach then causes pain in others not because he is a sadist, but because he feels the need to punish wrong and to uphold the good, and though he cannot make the world just, he can act according to his sense of justice - through the use of violence.

180. Which of the following best describes 'Watchmen'?
    A.   A book that contains only vicious characters
    B.   An expression of despair when contemplating an imperfect world
    C.   An example of how an author's intentions are not always realised
    D.   A book that accidentally glamorises violence
    E.   A book which highlights the ideal hero

181. Which of the following best accords with the view of the author?
    A.   All heroes use a minimum of violence
    B.   No hero uses violence
    C.   Heroes aim to uphold good
    D.   A hero should have a sense of purpose
    E.   It is impossible to be both a comedian and a hero

182. Which of the following best articulates the view put forward by Jacob Held?
    A.   We find dignity through just actions
    B.   If one decides to behave as though the world is fair, this may lead to a discovery of self-worth
    C.   It is shameful to view the world as corrupt
    D.   Self-value can only be found in madness
    E.   All means should be used in the pursuit of justice, including violence

183. What does the passage above argue?
    A.   Rorschach breaking little fingers is preferable to the comedian attempting to rape somebody
    B.   The comedian's depressing sense of humour has made him unpopular
    C.   Rorschach is not actually violent
    D.   Rorschach is popular because his aggressive behaviour has a moral intent, and is not just violence
    E.   Violence is not a pre-requisite for heroism

184. What does the word 'nihilist' mean in the context of the passage?
    A.   Someone who believes there is no meaning to life
    B.   Someone who is full of anger at the corruption of society
    C.   Someone who is narcissistic
    D.   Someone who hates other people
    E.   Someone who treats everything as a joke

Passage 36 – Gambler's Fallacy

The gambler's fallacy is a logical fallacy, where an independent event becomes more predictable the more it is repeated and, of course, takes place in the context of gambling. It can be demonstrated with the example of a dice being thrown. On the first throw yielding a score of 5, the second yielding a 5, what is the probability of another 5 coming up? Some may think it nearly impossible. Indeed, the odds are low when considering all 3 throws together but when considering just that third throw, the odds are still 1 out of 6, just as it was for the first and second throws. Crucially, each throw is independent of each other, so a previous throw has no bearing on the next one. Therefore, contrary to instinct, the independent event becomes no less independent on a second try. However, this is a frequently made mistake by gamblers.

Another name for the gambler's fallacy is the Monte Carlo fallacy, which is named after a famous example of it occurring at the Monte Carlo Casino in 1913. The night proceeded as a normal one until it was noticed that the roulette ball had fallen on black for a number of rounds. As it kept on coming on black, more and more gamblers were putting their money in - surely the ball would fall on red soon? It didn't. 26 spins in a row, in fact, fell on black, leading to losses in the millions. Gamblers believed that the odds were stacked in their favour, even after a few black runs. However, as a mathematical probability, it makes no difference how many blacks or reds there are: at each round, there's always a 50% chance of it being either black or white.

The basic laws of probability are taught at GCSE Mathematics level, so it should not be a problem for those who paid attention at school, which raises a question as to why it remains such a prevalent issue. Indeed, research has shown that sub-conscious processes may have a role to play. In the Journal of Experimental Psychology, researchers conducted a study of participants on the gambler's fallacy. The participants were split into two groups: the experimental group and the control group. The gambler's fallacy was explicitly explained to the experimental group and they were told not to rely on previous runs when making guesses. The control group was told nothing about this, though. The researchers then showed the participants a re-shuffled deck of cards and asked them to guess which shape would come next in the sequence. The results showed that the responses did not differ between the different groups, thereby indicating that the experimental group was not swayed by knowledge of the gambler's fallacy.

In particular, though, Roney and Trick showed in 2003 that grouping events, such that the next event appears as if it comes at the beginning of the next sequence, may overcome the gambler's fallacy. In the study, participants were shown a sequence of 6 coin tosses, with the final 3 flips being heads. One group were asked to predict the seventh flip. The other group was asked to predict the first event for the next sequence. Those in the former group tended to predict tails more. Accordingly, viewing independent events as a 'beginning' rather than as part of a sequence can help abate the gambler's fallacy.

185. Which of the following best explains the word 'fallacy'?
    A. A gambling addiction
    B. A belief
    C. A mistaken belief
    D. When a dice falls independently of the previous dice
    E. A lie

186. Which of the following is the best definition for the Monte Carlo fallacy?
    A. A gambling addiction
    B. The events in Monte Carlo in 1913
    C. The view that a previous event influences a future event
    D. The view that a future event is thrown into doubt by a previous event
    E. The circumstances around an independent event

187. Why did gamblers place money on the red in the Monte Carlo Casino incident in 1913?
    A. They knew they were going to get a windfall
    B. The red was more popular than the black in those times
    C. 26 spins in a row were on black
    D. They thought that the fact of the ball falling on black so many times increased their odds of winning on red
    E. They disagreed with the gambler's fallacy

188. Which of the following can be assumed about the research in the Journal of Experimental Psychology?
    A. The experimental group did not listen to what the researchers said to them
    B. The participants represented all areas of society
    C. The gambler's fallacy was tested
    D. Participants were shown at least one card
    E. The researchers were biased

189. Which of the following can be inferred from the third paragraph?
    A. The gambler's fallacy cannot be solved
    B. The gambler's fallacy was proved to exist by that research
    C. The research took place after 1913
    D. Teaching probability does not necessarily alleviate the gambler's fallacy
    E. Researchers conducted a study of participants

190. In the final paragraph, why did those in the first group tend to predict tails more?
    A. There was a head in the time before
    B. They did not believe in the gambler's fallacy
    C. They believed in the gambler's fallacy
    D. The gambler's fallacy
    E. A matter of probability

Passage 37 – Human Organ Sales

In the year 2013/14, over 4,000 organ transplants took place, 1,146 of which consisted of donations from living individuals. Despite recent year-on-year increases, the current rate of organ transplants is not enough to satisfy demand for them, and this has severe consequences: approximately three people die in the UK every day from not having an organ transplant.

NHS Blood and Transplant suggest that there needs to be a revolution in social attitudes towards donation but such a revolution needs a radical change in tactics. A public campaign to increase awareness of joining the organ donation register has taken heed. Indeed, any member of the public can register to donate their organs in the event of death. However, take up has been relatively low. In order to help counter the shortage, the Welsh government have adopted a system of presumed consent such that a person is deemed to consent to their organs being donated unless they have indicated otherwise (in which case, their organs will not be donated). Again, though, whether such a scheme will alleviate the shortage is unclear.

Accordingly, it has been proposed that a marketplace for organs be developed, where people are allowed to provide organs and be paid for them. This would acknowledge that the shortage is a major public health problem and encourage donors to come forward. In response to such suggestions in the past, opponents have maintained that it is immoral, but they tend only to involve either no or superficial consideration of the countervailing policy considerations and critically do not address the shortage of organs. Indeed, one can actually query whether it is moral to allow people to die in circumstances in which death can be avoided.

The positive case for allowing the sale of organs is generally based on the existence of a shortage, but there are other useful considerations too. The libertarian argument is based on the view that adults exercising their free will should be able to decide what to do with their own bodies. Accordingly, should they wish to give their organs in return for a price, there should not be an objection to it unless there are strong reasons against it. Freedom and respect for personal autonomy demand no less.

A regular assertion is that allowing the sale of organs will exploit the poor; they would be in the most need to cash in on their spare organs which could be at the expense of their health. However, they can still give their organs at present via a donation – why should there be a difference in the health risks between getting money for giving your organ and getting nothing for it? If the current risk framework for living transplants was used to determine whether one can sell their organs, giving an organ for money will not be more dangerous and riskier. In fact, providing money for organs is fairer than the current system – the providers of organs are compensated for the organ, the health risks, and their time. The involvement of money should not make it immoral – the surgeons carrying out the transplant, the transportation officers, and nurses are all paid for carrying out a transplant but no one would argue they're immoral.

Far from being immoral, the individual providing the organ would be saving someone else's life. Providing economic compensation should not make a difference. While opponents suggest that the poor would be exploited, it is incumbent on them to elaborate on how this is a different form of exploitation to working in high-risk occupations such as deep-sea diving, in the military or in a coal mine. The health risks of these occupations far outweigh the risk undertaking an operation to remove an organ, such as a kidney. Far from being exploited, people, however rich they are, should be able to give, should they want to.

Of course, this is not to say that any market in organs should be unregulated. In fact, quite the contrary is suggested by proponents of this. A heavily regulated market and safeguards to ensure fully informed consent and protection for the vulnerable are vital. Moreover, the system as proposed by Erin and Harris (2003) would fit well with the current system. There would only be one buyer of organs, the National Health Service, and organs could be distributed in accordance with existing priority rules, based on need and fairness.

This does not make the medical profession less caring or infringe on morality, it is, in fact, our moral imperative to ensure that we protect as many lives as possible. On a consequentialist perspective, the benefits of a market for human organs are clear but the drawbacks are not.

191. In which of the following circumstances would the libertarian argument not support the sale of organs?
    A. Where it is not provable that organ transplants would save lives
    B. Where it is against nature
    C. For those who do not give informed consent
    D. Where there is a risk to the donor
    E. When there is not a shortage

192. Which of the following is true?
    A. The Welsh government impose organ donation on individuals after their death
    B. Organs that are sold will go to the highest bidder
    C. Organ donation is moral
    D. A person can refuse to have their organs donated after death in Wales
    E. A marketplace for organs would be unethical

193. Which of the following is an argument in the passage?
    A. People should be able to sell their organs
    B. Opponents have maintained that it is immoral
    C. A market for organs should not exist
    D. The libertarian argument should prevail over all others
    E. The health risk of organ donation is outweighed by the benefit to the recipient

194. Which of the following is an opinion?
    A. The poor would be exploited
    B. In 2013/14, over 4,000 organ transplants took place
    C. There is a shortage of organs
    D. There are not enough donors to satisfy demand
    E. A public campaign has taken place to encourage organ donations

195. Which of the following is not advanced as an argument, implicit or explicit, by the author?
    A. Poorer individuals should be allowed to donate their organs should they wish
    B. There is little difference between selling an organ and working in a high-risk job
    C. A human organ market would be fair
    D. It would be better for the poor as opposed to the rich, to donate their organs
    E. The health risks are not increased for an organ transplant which is sold

Passage 38 – Free Speech

If someone says something you don't like, what do you do? Stop them from speaking or let them speak, but state your disagreement with them? In day-to-day conversations, either option does not appear to make much difference but on a larger scale, the repercussions for freedom of expression are enormous.

Either course of action has different consequences and implications for society at large. Take the former option and the other person won't be heard – in the event of the latter option, that person may be heard and the public may disagree with it but at least the public would be able to make up their own minds on the matter. Were the public to agree with the position, surely that is a right people have, regardless of your opinion.

Student unions in British Universities have been particularly vociferous in clamping down on speakers who might offend students. Indeed, many question the extent to which a person has a right to offend another. However, as Louise Richardson has pointed out, education is not comfortable. On the contrary, it is about tackling ideas and arguments you don't like, "confronting the person you disagree with and trying to change their mind. This isn't a comfortable experience but it is a very educational one."

Student unions are in a position of significant power in university circles, in being able to organise protests and decide on who speaks in their debates. A number of student unions operate a 'no-platform' policy, where an individual is prohibited from speaking at events. This may be due to the individual holding extremist views, but someone has to make a judgement as to what counts as extremist. This can potentially be a serious infringement of a person's freedom of speech if misused.

Indeed, student union leaders have taken contentious actions recently. For example, controversial writer, Germaine Greer, gave a speech at Cardiff University. While this was accompanied with a protest, the university's student union also started a petition to ban the writer from speaking in the first place. Most individuals may well disagree with Greer's arguments, which indeed are extreme.

What was more surprising, however, was that Peter Tatchell, a gay rights campaigner, was declared transphobic and racist by the National Union of Student's LGBT (Lesbian, Gay, Bisexual and Transgender) officer, who refused to share a platform with him. The officer justified her statement because Tatchell signed an open letter denouncing the NUS' no platform policy and supporting the free speech of individuals such as Germaine Greer. This shows a surprising mind-set in the student's union. If mere opposition alone to the no platform policy makes one subject to the no platform policy, surely that is an extremist policy in itself. Indeed, I, the author of this piece, would not be given a platform to speak on this basis by the student unions.

The increasing tendency for student unions to censor speakers who may offend them is concerning. Certainly, many may not agree with controversial speakers. However, blocking their free expression not only infringes on free speech but reduces the flow of ideas. It epitomises the undemocratic position that the views of those in powerful positions is more important than the views of others. It is only through allowing controversial speakers, and for people to hear a variety of views that one can be sure they have reached an informed view. Restricting freedom of speech, thus, reveals a lack of respect for the personal autonomy of individuals and their freedom to make up their own minds.

If those who seek to ban controversial speakers instead work to formulate counter-arguments, they will find a richer and more fulfilling public discourse.

'Offensiveness' for the purpose of the so-called 'no-platform' policy is vague and can lead to speakers being arbitrarily cut off from discourse at the behest of those leading student unions. Who are they to say whether something, in particular, is offensive? And even if a view offends a few individuals, it should be through debate that it is countered and not by shutting the debate off.

196. What unstated assumption was made by the author in stating that he would not be given a platform to speak?
    A. He would not be allowed to speak by student unions
    B. He agrees with Germaine Greer's views
    C. He would have the same fate as Peter Tatchell
    D. He is a campaigner for gay rights
    E. The officer's views on the no-platform policy are reflective of the student unions

197. Regarding freedom of speech?
    A. It is a human right
    B. Freedom of speech may be countered in circumstances where there are offensive views
    C. Peter Tatchell should not have been blocked from speaking
    D. Respect for personal autonomy requires freedom of speech
    E. There should not have been protests against Germaine Greer's speech

198. Which of the following is an opinion?
    A. Tatchell signed an open letter denouncing the no-platform policy
    B. 'Offensiveness' has a number of different meanings
    C. A number of student unions operate a 'no-platform' policy
    D. Students protested against Germaine Greer's talk
    E. The increasing tendency for student unions to censor speakers who may offend them is concerning

199. Which of the following follows from the passage?
    A. Students are extremists
    B. Students are against free speech
    C. Many students are against free speech
    D. Student unions express the views of the student body
    E. Some student unions have been trying to prevent certain individuals from speaking

Passage 39 - Bering Fur Seal Arbitration

The Bering Fur Seal Arbitration was one of the earliest international environmental law cases between the United States and Great Britain. The conflict arose once the US bought Alaska from Russia and thereafter granted a monopoly to the Alaskan Commercial Company through granting the company the exclusive right of sealing on the Pribilof Islands, at the edge of US territorial waters in the Bering Sea. In the 1880s, however, Canadian ships engaged in killing seals in the water, an activity known as pelagic sealing, just outside US waters. This depleted seal stocks inside US waters and on US land, which led to an international conflict between the United States and Great Britain, acting on behalf of Canada.

The United States wanted the upkeep of fur seals and, after the failure of negotiations with Great Britain, an arbitration tribunal was set up to resolve the issue. In order to be successful, the US needed to assert some type of right to the seals outside their territory. Among the arguments advanced by them at the arbitration, it was claimed that the US had the exclusive jurisdiction to the Bering Sea but this argument was flatly rejected. Alternatively, the US put forward the proposition that they were acting for the benefit of mankind generally to keep up the seal stocks. Additionally, they suggested that sealing on the land, which the US engaged in, was right but that pelagic sealing was illegitimate. This argument, convenient though it was, had never been recognised in any system of law, international or domestic and would have amounted creating new law. Hence, while their counsel displayed ingenuity, they were effectively countered by the counsel for Great Britain, who ended up persuading the tribunal.

In accordance with the terms of the arbitration treaty, the tribunal laid down a number of regulations for the purpose of preserving the fur seal stocks, which were binding on both Great Britain and the United States. The main regulation was the creation of a 60-mile exclusion zone, where pelagic sealing was not allowed except at certain times and in certain ways. One such regulation prohibited the use of firearms and explosives.

200. Why did the US bring the case against Great Britain?
    A.   Great Britain had responsibility for Canada at the time
    B.   Canadian ships were killing the US' seals
    C.   Canadian ships had breached US waters
    D.   The seal stocks were being depleted
    E.   As they had a monopoly over their seals

201. Who did the arbitration tribunal find in favour of?
    A.   The United States
    B.   Canada
    C.   Great Britain
    D.   Alaska
    E.   No one

202. Why was the US' argument convenient?
    A.   It fitted in well with their case
    B.   The US was sealing on land and Canada's was in the water
    C.   It was not held to be convenient
    D.   It created a new law
    E.   It gave them exclusive jurisdiction

203. Which of the following is true?
    A.   Sealing was prohibited
    B.   The US could not seal on the Pribilof islands
    C.   The US were acting for the benefit of mankind
    D.   Great Britain was more restricted in sealing in the exclusion zone
    E.   Great Britain lost the case

204. Which of the following would the regulation allow?
    A.   British ships to seal in the exclusion zone at restricted times
    B.   US ships to seal in the exclusion zone at restricted times
    C.   US ships to seal on the land on the Pribilof islands
    D.   Sealing at all hours, so long as it was without a firearm or explosive
    E.   No sealing at all

Passage 40 – Footballers' Pay

In 2012-13, the average yearly salary of a Premier League footballer was £1.6m, while at the same time, the average salary in the UK stood at £26,500. The footballers at the top of the tree earn significantly above the average footballer even, with Wayne Rooney on a reported £13 million contract with Manchester United, thereby earning him over 9 times the average every week. This may seem a lot for kicking a ball and running around a field. Footballers tend to enjoy extravagant lifestyles, which have been brought into public focus in recent times given the tough economic conditions that have affected the rest of society.

In 2008, the Labour Government's Sports minister labelled footballers' salaries as obscene, pointing out the difference when compared to the ordinary worker. The Archbishop of York believes that there must be higher taxes for them. What makes a footballer's occupation more worthwhile to society than that of a doctor or nurse, who goes about saving lives? It is indeed arguable that their success is in a big part due to the talent they have from birth, and that is entirely down to luck. It may, thus, seem unfair that footballers earn over a hundred times more than nurses or doctors, many of whom have to work late shifts and make life-and-death decisions and seem to contribute more value to society.

On the contrary, though, the workings of the free market can be used to explain the seemingly excessive salaries of Premier League footballers. Put simply, there aren't many people with the skills of such footballers – they are in short supply – and at the same time, many people – in fact, millions – want to watch them, either at a football match or on television. Consequently, television broadcasters are willing to pay mammoth sums to have the TV rights to football matches and companies are willing to pay millions to sponsor mainstream football teams. These give the teams the enormous purchasing power with which to pay their players large sums. Given that each team wants to win, they will want the best players and thus, will bid against each other for them, thereby inflating their wages even further.

However, if it's just a matter of luck, is it fair that footballers are paid significantly more than doctors? Indeed, supporters of the status quo point to the fact that their respective wages are determined by the market makes it fair – their wage is reflective of their value to society. Millions of people excitedly tune in to watch football and it is, therefore, argued by supporters that society, by this choice, determines the footballer's wage.

On the contrary, though, opponents point out that the market-determined wage (or the market value attributed to one's skills) is mainly determined by luck and not entirely by oneself. Accordingly, the wage differential is not particularly fair given that someone earning a low wage, such as a junior doctor, may well work much harder than a person, such as a footballer, earning a significantly higher wage.

However, it is wrong to focus this debate solely on footballers. It requires significant political discourse as it goes to the heart of a fundamental part of society: the free market system. How, whether and to what extent this should be reformed are matters for careful political debate. The market system does not just lead to footballers enjoying large salaries, but other sports stars, celebrities in general, bankers, lawyers, chief executives etc. earn wages that are significantly above that of the average too. The position of footballers should be seen as not much different from these other professions. It is, thus, incoherent to consider the position of footballers' salaries and not consider society as a whole

Crucially, 45% income tax is already paid on earnings over £150,000 and it is up to democratically elected governments to consider whether taxation suffices or if a wholesale change of the pay system is warranted, such as imposing 'wage-caps'. When considering the latter, however, it is worth bearing in mind that it is not inexorable that the reduction would go to lower earners and not boost the company's own profits.

205. What do the opponents and supporters agree on?
    A. What the wage rate for footballers and nurses should be
    B. That fairness has different meanings depending on the context
    C. That the existing settlement is fair
    D. That fairness has a role to play in the wage debate
    E. That fairness leads to the current outcome

206. Which of the following is not inconsistent with the arguments of <u>both</u> the opponents and supporters of footballers' wages?
    A. Footballers benefit from luck to have their talent
    B. Footballers have a fair wage
    C. Footballers' wage should be determined by the free market
    D. It is unfair that nurses are paid less
    E. Footballers should be paid less

207. Which of the following is not inconsistent with the author's argument in the passage?
    A. One's hard work is a factor for success
    B. It is obvious that the market-based system needs to be overhauled
    C. Fairness is a definitive concept
    D. Wage differentials reflect hard work
    E. Footballers don't pay tax

208. What does the word 'inexorable' mean in this context?
    A. Unnecessary
    B. Mindless
    C. Sporadic
    D. Random
    E. Unavoidable

209. Which of the following is argued by the author?
    A. The supporters of footballers' wages are incorrect
    B. The opponents of footballers' wages are incorrect
    C. Top footballers are in short supply
    D. It is incoherent to only consider footballers' salaries and not society as a whole
    E. 45% income tax is already paid on earnings over £150,000

Passage 41- The Underground

We are once again facing the possibility of major tube strikes across the capital. It seems that the strikes which occurred last summer were not enough to convince Boris Johnson, Mayor of London, to work with the unions rather than against them in his mission to get the night tube rolled out across the city. The national media and the public at large have again declared themselves outraged at the possibility of further strikes. People fear the misery and inconvenience caused by the last 24-hour walk-out. The Prime Minister has also condemned the strikes as 'unacceptable and unjustifiable'. In fact, a Tory London Assembly member has called for a 'Dad's Army' of retired former Tube workers to 'prop up' the Underground during the industrial action. But is all this outrage really justified or should we think again?

I would ask you to consider how you might feel if your employment contract were suddenly changed without your consent. I would further ask how you would feel if that contract were changed such that your working hours became significantly less convenient without additional remuneration. That is the very real situation in which many tube staff now find themselves. These people do some of the most vital work in London – and it's often a thankless task. We ought to be supportive of them, rather than resentful of trying to ensure they work in suitable conditions and are adequately remunerated for their indispensable services to this country. It seems ridiculous that when junior doctors were striking, there was so little public anger but when tube staff strike, people are so incensed. The double standards here are quite incredible.

Much of the national media have already responded to the story that there may be further strikes – largely with real anger. This is especially the case with more conservative newspapers, tending to fight any action by unions that they possibly can. For example, a scathing article was yesterday published by the Daily Mail. Interestingly, social media, as unrepresentative as it may be, has offered a more sympathetic view. More than 10,000 people have retweeted "The night tube can wait, people should be paid to work late." Others have posted messages in support of tube workers on Facebook, as well as posts highlighting the difficulties faced by those who are forced to work late. Of course, we should hesitate before using social media as a useful poll of public opinion, but it is encouraging that a significant proportion of people are being exposed to less conservative views about the issue.

The right-wing press will always be keen to encourage us to disparage unions, and many of us may have unwittingly taken on the views promulgated in such media. I would suggest, however, that we might benefit from rethinking such views. A more sympathetic viewpoint would encourage us all to be more grateful for those who operate Transport for London every day – and such a viewpoint might even help Boris Johnson to finally get the night tube plan moving.

210. What is the main argument of the Passage?
    A. That the Mayor is being unfair
    B. That we should be more sympathetic to tube drivers and workers
    C. That tube staff should work harder
    D. That strikes are pointless
    E. The night tube is unnecessary

211. What is the inconsistency being highlighted in the Passage?
    A. That the public are not angry about junior doctors' strikes but are angry about tube strikes
    B. That most Londoners don't want the night tube
    C. That the night tube will cost a lot of money but not be worth it
    D. That the public support teachers' strikes but not tube strikes
    E. The right-wing press only disparages unions

212. What view of social media is taken in the Passage?
    A. That most people on it are right-wing
    B. That people on it are cruel
    C. That it may offer a skewed view of public opinion
    D. That it offers a very accurate view of public opinion
    E. People are very active on social media

213. What does the word 'conservative' mean in the Passage?
    A. People who were privately educated
    B. People who are uncaring humans
    C. People who are members of a specific political party
    D. People who are on the right-wing of a spectrum of thought
    E. Unwillingness to accept industrial change

214. What is suggested by the final paragraph?
    A. That we should not listen to newspapers
    B. That we should not listen to Boris Johnson
    C. That we should re-evaluate some of our beliefs
    D. That we should think before we tweet
    E. A different approach will help Boris Johnson to retain his position

Passage 42 - Politics

Not enough people join political parties anymore and it's damaging our political process. Although membership of two major parties surged after the 2015 election, still less than 10% of the UK population are members. There are three ways in which this damages our political process: many people are completely uninformed about politics and have no idea, for instance, as to how to distinguish between left-wing and right-wing parties; unpopular leaders run parties and political parties only speak for a small proportion of the electorate. Because the majority of people refuse to make a choice, the minority makes that choice for them.

A large proportion of the UK population are completely disengaged from politics. A recent survey showed that over 70% of people have never read a political manifesto in their entire life. When the most recent election rolled around, many of my friends had no idea to vote for because they had no idea about what their political convictions were. This is troubling because it means that none of them have ever bothered to vote at a general or local election. Where's the democracy in that? It's not necessarily the case that these friends of mine don't have views on how the country should work – they do – but they have no idea which parties represent those ideas. If they investigated the policies and ideals of the different political parties, read their election manifestos, and then joined political parties, they would not be in such a predicament. As far as politics is concerned, in this day and age, ignorance is certainly not bliss.

Jeremy Corbyn took control of the Labour Party in autumn 2015 despite being very unpopular with the majority of the electorate. Although he obtained a very significant majority within the Labour Party, this is not representative of the rest of the UK's views. Most consider his pacifism to be untenable and found Hilary Benn's approach much more persuasive. However, a silent majority refused to participate in the labour elections, some due to laziness, and others due to the mentality (prevalent amongst many) that their favourite candidate would be selected regardless of whether they voted or not. If more of those who identify with the left had joined Labour and voted in their leadership election, perhaps a more appropriate leader would have been selected.

It is very easy for political parties to speak only for a small proportion of the electorate when so many people do not get involved with them. This fact is glaringly obvious, but, surprisingly, has been missed by many people. Young people often complain that politicians do not care about them or their interests – but, of course, it is very easy for that to be the case when so few of them join parties and force politicians to consider that their interests really do matter. It's vitally important for the young to engage if, for example, they ever want a chance of reversing the Conservative party's tuition fee increases, which so many of them seem to resent.

215. What are the political views of the author?
    A. Conservative
    B. Liberal Democrat
    C. Labour
    D. There is not enough information to tell
    E. Communist

216. Which of the following is an argument made in favour of joining political parties?
    A. Corbyn is a pacifist
    B. If more people joined, inappropriate leaders would be less likely to get selected
    C. Tuition fees are expensive
    D. The author's friends have political views
    E. Many people are uninformed about politics

217. What unstated assumption is being made?
    A. Engagement with politics is a good thing
    B. Corbyn is bad at his job
    C. A lot of the UK population are not engaged with politics
    D. Young people complain a lot
    E. Corbyn's views are not representative of many people's views

218. What is meant by untenable?
    A. Pointless
    B. Unable to stand up against attack
    C. Unfair on some people
    D. Annoying
    E. Understandable

219. What does the writer seek to demonstrate in using the following statistic: "less than 10% of the UK population are members (of political parties)"?
    A. That their friends do not care about politics
    B. That Corbyn is a pacifist
    C. That Corbyn got a large majority of votes in the Labour leadership election
    D. That a lot of people are disengaged from politics
    E. That political activism is more popular elsewhere in Europe

Passage 43 – Gender Equality

Gender equality has, for as long as we can remember, been a serious and troubling issue. Recent studies have shown that a pay gap still exists and many women report incidents of discrimination experienced at work. However, in the UK, there is a growing willingness to tackle the issue of gender equality and this is reflected both through the passing and enforcement of anti-discrimination legislation and through the implementation of various policies by employers in the public and private sectors. In the ongoing struggle for gender equality, one author suggests that the approach to tackling differences between men and women has failed...

"It is sometimes possible to identify existing differences, such as responsibility for childcare and legislate to accommodate for those. It would seem that this might be a major step forward in the struggle for gender equality. Many employers in the private sector now offer schemes which are designed to assist women; for instance, partnership schemes (where an employer buys many places at local nurseries near the workplace for their employees to use) are becoming more and more common. A large proportion of employers also offer women with children flexible working hours and childcare vouchers. In the public sector, assistance to women with children also abounds. Government bodies have been known to buy places with childcare providers before and after school so that employees' children of a relevant age can attend breakfast and after-school clubs. Some may even provide care during the school holidays, which can prove a particularly hard time for employees to get childcare cover. By offering assistance to women who want childcare so they can return to work, governments would seem to be advancing the struggle for gender equality.

However, if such privileges are given only to women, the idea that only women should be responsible for childcare may be perpetuated. In one case, a father was denied access to his employer's subsidised childcare scheme because it was only for female staff. The employer initially maintained that the general policy was justified, notwithstanding the fact that, in that particular case, the father was the sole carer for the child. Fortunately, the employer reached an agreement with that particular employee. However, doubtless many other fathers will find themselves excluded from childcare vouchers and nursery partnership programmes in a similar manner. The aim of the subsidised childcare policy was to allow women to continue working, rather than leaving their jobs after having a child. This was an exception to a general policy of gender neutrality, specifically to tackle the under-representation of women in the banking sector. As the policy was one of "affirmative action", it was not illegal. In theory, this might seem very beneficial for women. However, from the introduction of the policy to the time of the case, there had only been a 7% increase in the number of female employees, and it was not clear whether this was caused by the policy.

Emphasising women's differences and adjusting for them may actually have done more harm than good in this case, therefore, especially if the idea that childcare is a women's issue was perpetuated."

220. What does the word 'perpetuated' mean here?
    A. Caused to continue
    B. Defeated altogether
    C. Partially dealt with
    D. Made stronger
    E. Made more widespread

221. What does the writer think about the 7% increase?
    A. That it means the policy was a success
    B. That it justified the cost of subsidised childcare
    C. That it does not necessarily show that the policy made a difference
    D. That most people didn't want it
    E. That it is a significant figure

222. What is the negative outcome of the policy of subsidising childcare for women only?
    A. Men may be resentful and angry
    B. It may be expensive
    C. Women may go back into the workplace
    D. It may reinforce the idea that women are responsible for childcare
    E. The policy is random in its operation

223. What was the aim of the policy of subsidising childcare for women only?
    A. To support women because they are poorer than men
    B. To help them continue working after childbirth
    C. To force them to continue working after childbirth
    D. To stop men taking advantage of free childcare
    E. To give women as many life advantages as men

224. Which of these ideas can be inferred from the Passage?
    A. Highlighting gender differences may not help gender equality
    B. Men will always try to take advantage of schemes aimed at women
    C. Highlighting gender differences always help gender equality
    D. Gender neutrality is a bad thing
    E. It is impossible to find a fair way of dealing with gender differences

Passage 44 -An Extract from a Piece on Policing:

In 1829, when Sir Robert Peel was Home Secretary, the first Metropolitan Police Act was passed and the Metropolitan Police Force was established. Previously, three different groups had strived to protect London; the Bow Street Patrols, the police force constables who were under the control of the magistrates, and the River Police. However, overall, before 1829, the law enforcement had been severely lacking in organisation. As London expanded during the 19th century, the issue of maintaining law and order became more and more pertinent.

The original mandate of Peel's Metropolitan Police placed the greatest emphasis on the prevention of crime. The police sought to prevent the violent crimes which were common in London at that time – including robberies and murders. More petty thefts and public disorder were amongst the less serious crimes which the force also aimed to pre-empt. This was done mainly through visible patrol. During the 19th century, a detective function was added to the job. The two strategies of preventative patrol and criminal investigation continued to form the core of policing for over a century. Recent decades have seen important developments in policing, many of which were prompted by the pressures under which all criminal justice agencies were placed by governments keen to secure value for money. Changes aimed at combating the way the police are viewed by the public have also been important, especially in the wake of recent criticisms of the conduct of officers and accusations of police brutality: new developments in policing are now almost always as much about police legitimation as they are about police effectiveness.

The most significant change occurring in the function and nature of policing is the pluralisation of the system. There has been a dramatic increase in the number of different groups which form part of our law enforcement. The extent of this change has been so great that leading commentators have argued that 'future generations will look back on our era as a time when one system of policing ended and another took its place'. The non-Home Office forces include, on the public side: specialist nationalist forces; police sections of broader public organisations and municipal policing; and on the private side, they include voluntary self-policing or civil policing.

One major change has been the commercial security sector growing and taking over some role traditionally filled by the Home Office police. For those who can afford it, provision of security will be increasingly privatised, either in residential areas or in the 'mass private property' sector where more and more middle-class leisure and work takes place. In this sense, the police are changing in response to a changing world. The growth of the non-Home Office sector, in particular, commercial and municipal policing is a compensatory response to the financial crisis of the state. It also has to do with the growth of mass private property – shopping centres, office plazas and so forth. These give rise to considerable security issues because they are open to the public. These developments don't quite fit with the traditional view of police on public streets – and so private policing takes over.

225. Which of the following is merely an assertion?
    A. New developments in policing are almost always as much about police legitimation as they are about police effectiveness
    B. The most significant change occurring in the function and nature of policing is the pluralisation of the system
    C. These developments don't quite fit with the traditional view of police on public streets – and so private policing takes over
    D. The original mandate of Peel's Metropolitan Police placed the greatest emphasis on the prevention of crime
    E. The police are changing in response to a changing world

226. What reason is given for the major changes in policing in the first paragraph?
    A. The growth of mass private property
    B. Peel's Metropolitan Police had become old-fashioned
    C. Visible patrol wasn't working
    D. Governments being keen to secure value for money
    E. Public pressure to change the structure of the police force

227. What evidence is cited for the claim that "the most significant change occurring in the function and nature of policing is the pluralisation of the system"?
    A. A list of different private and public police forces
    B. The statement of leading commentators
    C. An explanation of what pluralisation means
    D. The fact that voluntary self-policing is happening
    E. The growth of the commercial security sector

228. Which of these is not included in the Passage in support of the statement that commercial security is becoming increasingly important?
    A. The fact that there has been a fiscal crisis in the state
    B. The increase in mass private property
    C. An increased number of private security firms
    D. The security issues generated by large shopping centres and office complexes
    E. New security issues

229. What is the main point of the Passage?
    A. To persuade the reader that we have a problem with policing
    B. To inform the reader of changes taking place within the police
    C. To inform the reader of the dangers of spending time in areas of mass private property
    D. To persuade the reader to join the police
    E. To inform the reader of how changes in the public sector will affect the private sector

Passage 45 – International Law

The international legal system lacks one body which has authority to legislate for all states. Some states are not members of the United Nations, and even those which are, are only bound by that which they have agreed to be bound to. Essentially, for most states, co-operation with international law is a matter of political goodwill, rather than one of legal or physical compulsion. This would seem to suggest that there cannot be any kind of successful international legislative process: but the commentators seem all to agree that violations of international law are rare.

Because there is no independent body to legislate for all states, it can be difficult to establish what exactly constitutes international law. However, there is a consensus that various sources, however—principally treaties between states—are authoritative statements of international law. International law involves four sources of law: international conventions; international custom; general principles of law and judicial teachings. These are set out in Article 38(1) of the Statute of the International Court of Justice 1945. An example of custom is the practice relating to ambassadors. For thousands of years, countries have given protection to ambassadors. As far back as the ancient Greek and Roman world, ambassadors from another country were not harmed whilst on their diplomatic missions, even if the countries they represented were at war with the country they were located in or travelling through. Throughout history, many countries have also publicly stated that they believe that ambassadors should be given this sort of protection. Therefore, today, if a country harmed an ambassador it would constitute a violation of customary international law. Treaties are the most common source today; they are similar to contracts between countries; promises between States are exchanged, put into formal writing, and signed. States may query the interpretation or implementation of a treaty, but the written provisions of a treaty are considered to be binding. Treaties can address any number of topics, such as trade relations, like the North American Free Trade Agreement, or control of nuclear weapons, such as the Nuclear Non-Proliferation Treaty. Treaties may be multilateral or bilateral - although more treaties now are bilateral and, thus, bind only the two states which sign them.

The fact that the international legislative process works reasonably well is often met with surprise. This is perhaps because breaches of international law are so well-publicised, or because many people imagine that international law only governs war: the area where most breaches of the law occur. Steve Smith argues that the fact that "the law is normally observed... receives limited attention because most people are more interested in the occasions on which it is broken. And yet these occasions are very rare." It is important to recognise that in the day-to-day business of states, the great majority of treaties are observed. Such treaties include rules on international postage and telephone calls: areas which function without issue all the time. This shows that the legislative process does work reasonably well: the majority of states do obey the majority of the law.

230. What is the paradox highlighted in the Passage?

A. States are only bound by that which they have agreed to but some states are not members of the UN

B. There is no one body which can authoritatively legislate for all states but violations of international law are rare

C. International law involves four sources but one is more common than others

D. People are surprised that the international legislative process works well but they are not interested in the law

E. Most violations of international law concern war

231. What can we infer about how the nature of treaties has changed?

A. More of them are now bilateral; in the past more were multilateral

B. They have become a lot more popular

C. They used to be contained in a completely different statute

D. They have become far less useful since World War Two

E. More treaties are observed nowadays than in the past

232. What is the central argument of the piece?

A. International law is not very useful

B. Most people are only interested in breaches of international law

C. International telephone calls function without issue

D. The international legislative process actually works well

E. People do not believe that international law works

233. Which of these is not presented as a reason for why people are surprised to hear that the international legislative process functions well?

A. Breaches of international law are heavily publicised

B. Some people think international law is all about war

C. International law does not have a good PR system

D. People are most interested in the breaches of international law

E. People do not focus on the areas of life in which international law functions well

234. What point are the rules on international postage and telephone calls used to illustrate?

A. That the great majority of treaties are observed

B. That the great majority of treaties are not observed

C. That people are most interested in the breaches of international law

D. That treaties are the most common source of international law

E. That treaties relating to postage and telephone calls are amongst the most successful

Passage 46 – The Nature of law

For many years, academics have debated and written on the topic of whether law is, by nature, coercive. It is important to find an answer to the question of whether or not law is coercive because, as the Utilitarians argued, we need to stop those who make law from abusing their power. If, in making law, the legislature in-fact coerces people into certain behaviours, this is much more important, because their power is much greater. This situation can be contrasted with one in which every man is like Austin's 'puzzled man' – who wishes to follow the law voluntarily, if he can only find out what it is. If, as it will be argued here, most people choose to follow the law, at least some of the time, because they will experience negative consequences otherwise, then keeping law makers in check is much more important and needs to be taken very seriously and should be at the centre of our focus. Thus, finding an answer to this question is important firstly because it helps us respond appropriately to law makers.

A second reason it is important to find an answer to the question of whether or not law is essentially coercive is that it helps to understand how people respond to the law, and why it is that, most of the time, people are in the habit of obeying it (as Austin states). This is crucial because law should be made for the good of society, and, therefore, for the people who make up that society. If law is coercive, and that is why people obey it, law makers can better choose appropriate sanctions, at least in the criminal law, so that people continue to do so. For instance, if a particular area of law is frequently broken – such as traffic laws or laws relating to fly-tipping – an apt solution might be to increase the penalty for breach of such laws in order to bring about the fear necessary for the coercion effect. If, however, law was not coercive and most people were like the aforementioned 'puzzled man', then appropriate sanctions for breaches might well be different. For example, we might implement lighter sentences – such as community service orders, or even dispense with sanctions altogether. However, it is here argued that a substantial portion of society obeys the law because of fear of sanction and therefore, serious sanctions are needed for a breach of the criminal law. To put it frankly, we do not live in a 'society of angels', and many individuals would readily admit that they would freely break a number of provisions were it not for the hefty penalties attached to breach.

The criminal law is the clearest example of law being like commands (as argued by Austin) issued by the legislature, and despite Hart's criticism of Austin, he too concedes this. The criminal law tells people how they are expected to behave and that sanctions will be imposed if they breach these standards of behaviour. This is totally coercive in nature.

235. What does the author think is the most important reason to find an answer to the question of whether or not law is coercive?

A. It helps to understand how people respond to the law
B. We need to stop those who make law from abusing their power
C. The criminal law tells people how they are expected to behave
D. There is not enough information in the Passage to know
E. To verify Austin's theory

236. What point do Hart and Austin agree on?
   A. The criminal law is the clearest example of law being like commands
   B. Most people choose to follow the law because they will experience negative consequences otherwise
   C. The criminal law has a lot of sanctions
   D. Most people do obey the law
   E. Serious sanctions are needed for breach of the criminal law

237. Which of these statements is closest to what the Utilitarians argued?
   A. People who make law enjoy using sanctions
   B. It's important to ensure those who make law don't abuse the power they have
   C. Lawmakers are all very puzzled
   D. Law is only coercive when it makes people feel puzzled
   E. All of the above

238. What is 'coercive' used to mean in the Passage?
   A. The law is used to make people puzzled
   B. The law is used to make people do things
   C. The law is used to make people happy
   D. The law is used to make people commit crime
   E. The law is used to show people what to do

239. Why does the author think that major sanctions are needed for breaches of the criminal law?
   A. People obey the law a lot of the time
   B. People only obey the law because they fear sanctions
   C. People do not fear sanctions enough and criminal law should scare them
   D. The author does not think this at all
   E. None of the above

Passage 47 – Multinational Companies

New management structures in multinational companies are big news for those wanting to make a career move. Increasingly, consultancies are advising such companies to hire from within their own firms rather than bring in new people from outside. The logic behind this appears to be that long-term employees will have had time to come up with reliable new ideas rather than high-risk ones which are more likely if one hires from outside the company. For instance, the Fast Company Group have written an article warning against hiring 'tons of people quickly'. The idea seems to be that focusing on long-standing employees - employees with 'shared interest, preferences and priorities' - will be a better strategy for building up a successful company. The dominant criteria for hiring, according to this consultancy, should be whether the new individual fits with the character, ethos, and ideas of the firm.

It's unfortunate, however, that this means there will be so much less innovation in such companies. Sahara's London office, for example, has only hired three people from outside the company at management level in the last two years. In the two years prior to that, it hired fifteen. The big changes in approach – like introducing drones for delivery – that they have been working on recently almost all came from those who did not start their working lives at Sahara. It's concerning, therefore, to those of us who enjoy the big changes and crazy new business ideas, that there will be so much less cross-stimulation of ideas in big businesses.

The statistics for other large firms are similarly bleak. A recent survey revealed that many had either slowed down their hiring or were planning on doing so in the coming months.

If multinational companies are going to be making safer choices as a result of their new structures and hiring policies, there is one way in which we might all benefit significantly: we're much less likely to go crashing into another financial crisis. The crash was, of course, caused by lots of risky decisions being made by lots of banks. Examples include the granting of mortgages to sub-prime candidates, both in the field of residential estates and in that of commercial estates. Other risky decisions included the loans made to financial markets (up to 32% of the money created by the big banks at that time). Avoiding a repeat of that could mean that we all avoid a great deal of misery.

240. What is the viewpoint of the author about the new structures overall?
    A. Very against them – only presents negative arguments

B. Very in favour of them – only presents positive arguments

C. Neutral – presents both positive and negative arguments

D. She does not understand them

E. She believes that they are risky

241. Which of the following is an argument in favour of the new management structures?

A. We're less likely to experience a financial crisis

B. We might get more drones for delivery services

C. There will be less cross-stimulation in business

D. Fewer people will be hired by new companies

E. Unemployment will reduce

242. What can be gathered from the statistics about Sahara's recent hiring policy?

A. They are worried about the financial crisis

B. They have been listening to the advice of consultancies

C. They don't want to use drones after all

D. They have run out of money for new staff

E. They do not want new ideas

243. Which idea does not appear in the text?

A. The recession

B. New delivery methods

C. Entry level staff

D. Consultancy firms

E. Reliable ideas

244. What can be inferred from what the author says about the financial crisis?

A. It caused a lot of misery

B. It was caused by drones

C. It was because of consultancies

D. It wasn't actually all that bad

E. None of the above

Passage 48 – Women and the law

There is an enduring belief that female offenders are protected from the full rigours of the law.

However, there is a significant amount of evidence that women actually suffer more in the criminal justice system than men do, however. 40% of the women sentenced in the Crown Court are now being given custodial sentences. This is in contrast to the figure from eight years ago when it was less than a quarter. This is not because, according to the Home Office, female offending has become more serious or more violent. Indeed, there is little reason to suggest that this is the case. It may be, instead, a result of the 'enduring belief' that female offenders are treated too leniently by the criminal justice system. The justice system often attempts to respond to allegations of discrimination by changing its practices – it seems that, in this case, false reports or beliefs have influenced the treatment of women by the judiciary over the years.

Moreover, female offenders, especially those sent to prison, are best viewed as an especially disadvantaged group: 70% have an addiction, double the proportion for men. A higher proportion of women in prison have suffered abuse or neglect as children and a higher proportion self-harm or commit suicide in prison. Indeed, the Prisoners' Reform Trust recently reported that 'Women in prison are highly likely to be victims as well as offenders. More than half (53%) report having experienced emotional, physical or sexual abuse as a child.' Indeed, the Corston report of 2007 identified a myriad of problems with the treatment of more vulnerable female offenders by the prison system – one being the frequency of strip searches and the absence of adequate sanitation. Furthermore, prison itself is harder for women because they tend to be held further from home (because there are fewer women's prisons). This means that women in prisons often feel more isolated from family and friends and, as a result, less able to cope with the harshness of prison life than their male counterparts. One should not underestimate the impact of the geographical location where imprisonment is concerned. It should also be noted that women also more likely to lose their children to care – perhaps because a third of women in prison are lone parents. This can be extremely detrimental to their mental health – especially where the child is the only family they have and they are serving a long prison sentence and potentially face a battle for custody of the child when they eventually leave prison.

There have also been significant failures in programmes designed for women. When women are the subject of "special penal treatment", the result is often the development of "benevolently repressive regimes" which emphasise dependency and traditional femininity, and most crucially, fail to aid rehabilitation. Such programmes tend to be structured around assumed characteristics and needs of women, rather than evidence. Thus, it is argued, women may actually suffer more at the hands of the criminal justice system than men do. There is no place to argue they are treated too leniently.

245. What is the main argument of the Passage?
   A.   Female offenders are protected from the full rigours of the law
   B.   Arguments that women are treated too leniently are wrong

    C.   Female offending has become more serious

    D.   Prison is easier for women

    E.   Men are treated harshly by the criminal justice system

246. Which of these is not given as evidence about how women are treated in the criminal justice system?

    A.   Statistics on losing children to care

    B.   Statistics on addiction

    C.   Statistics on the types of crimes committed by women

    D.   Statistics on how many women are sent to prison

    E.   Information about the location of prisons for women

247. What is the reason for the 'enduring belief' that women are treated more leniently in the criminal justice system?

    A.   A lot of women have addiction problems

    B.   Women tend to be held further from home

    C.   Women do not commit a lot of crime

    D.   We do not have enough evidence to say

    E.   None of the above

248. What can we infer from the information about programmes designed for women?

    A.   They do not help women very much

    B.   They make women less likely to lose their children to care

    C.   They mean women are treated more leniently

    D.   They help women to change

    E.   They aid rehabilitation

249. What proportion of men in prison have an addiction problem?

    A.   35%

    B.   20%

    C.   70%

    D.   We do not have enough information to say

    E.   30%

Passage 49 - Prisons

Prisons in the UK are failing. The reality is that the use of imprisonment is less a reflection of crime rates than an outcome of a country's political, economic, and institutional conditions. Neo-liberal nations, including the USA and the UK, imprison at much higher rates than countries where there is a more egalitarian culture, with more of a welfare state. We are over-burdening the taxpayer by throwing people into prison and paying to keep them there, rather than allowing them to undertake rehabilitation schemes within the community.

There is little evidence that prisons are effective in terms of deterrence or reform at a general level, although certain programs and certain individuals sometimes show positive effects. Despite the positive effects on the minority of prisoners, the high rates of re-offending overall are, according to Clark, stark evidence of the failure of the system. Clarke told BBC Radio 4's Today programme that the prison system was failing to deter a 'criminal underclass' who revert back to committing "more crime" as soon as they come out of prison because underlying problems such as drug abuse and mental illness issues were not being tackled by the prison system. Another issue was the lack of support for those able to change their ways who were striving (unsuccessfully) to obtain some form of legal employment. The prison has, in fact, been found to have an overall negative effect – destabilising family ties, disrupting employment opportunities on a long-term basis, stigmatising, and otherwise deskilling its population.

29% of the prison population are serving sentences of over 4 years but often such prisoners are not dangerous at all. Rather, they have committed lots of property-related offences. The public are not benefitted from keeping these people in prison. In relation to these types of prisoners, Clarke said 'What has been happening is there has been a huge increase in the number of people in prison, which is not only bad value for money, but even more importantly, I don't think that is the right way to keep on protecting people ... some of my critics think you should put more and more people in prison for longer and longer and longer. I personally don't think that is the best way of protecting society. It seems that a thoughtless Victorian-style 'bang em' up' mentality has prevailed in the UK.'

Moreover, it goes without saying that prison is a truly bleak experience for many. This means that some prisoners who might be reformed on the outside become too miserable or depressed to engage with the programmes being offered to them in prison. In England and Wales in 1995, the government introduced an incentives and earned privileges (IEP) scheme. Prisoners on 'basic' – the lowest tier of the IEP scheme – are subject to quite an austere regime. Their in-cell TV is removed, they have reduced opportunities for mixing with other prisoners and they are forced to wear prison uniform. Making people miserable is not the way to bring about reform. We can never hope to cultivate something positive from such negative conditions.

250. What is the main argument of the passage?
    A. Some individuals benefit from prison
    B. Prisons in the UK are better than the US
    C. Prisons in the UK are not working well
    D. Prisons make people sad
    E. We should not punish people

251. What can we infer from the fact that prison may destabilise family ties and disrupt employment opportunities?
    A. Prison may actually make rehabilitation harder
    B. Some people do well in prison
    C. Prison makes rehabilitation easy
    D. The public are benefitted by keeping lots of people in prison
    E. Prisons always reduce social mobility

252. What does the author think about the IEP scheme?
    A. That it makes people happy and thereby hinders rehabilitation
    B. That it makes people sad and thereby hinders rehabilitation
    C. That it makes people sad and thereby helps rehabilitation
    D. That it makes people happy and thereby helps rehabilitation
    E. It makes people sad and thereby aptly punishes them

253. What is an unstated assumption of the passage?
    A. Making people miserable is not the way to bring about reform
    B. Prisons have a positive effect
    C. Prisons have a negative effect
    D. Rehabilitation should be a goal of the prison system
    E. Everyone deserves to be able to maintain their family ties

254. Which of these is not offered as a criticism of keeping too many people in prison?
    A. It may cause fights within prisons
    B. It costs the taxpayer a lot of money
    C. Prison is not helping to rehabilitate many people
    D. Keeping non-dangerous people in prison does not benefit the public
    E. All of the above

Passage 50 – Sleep

Sleeping badly? We all are these days. The regularity with which I hear complaints from friends and colleagues that try as they might, they just cannot get to sleep, leads me to think we have something of a non-sleeping epidemic at the moment. Recently, the Guardian reported that up to 51% of British people struggle to get to sleep – and women are three times more likely than men to suffer from this problem. I'm not saying we're *all* insomniacs for the purposes of medical science – I'm no doctor – but certainly I've noticed something of a pattern, and I'm not alone in that. Professor Colin Espie of Glasgow University said that "Insomnia affects people's quality of life during the day, not just their sleep at night," he said. "Indeed, the survey data show significant effects across different aspects of personal functioning. Living with poor sleep and its consequences is not only very common, but it is in all likelihood degrading Britain's health. This is not a trivial matter. It's time for the NHS to pay attention to the scientific evidence that persistent poor sleep elevates the risk of developing new illnesses. This has been shown in disorders such as diabetes, but also very convincingly in depression." In light of the severity of this problem, and as someone who does rather well with sleep, I've devised some top tips for getting more kip...

Firstly, don't eat immediately before going to bed. You will feel too full and heavy to drift off into the land of nod. Eating too close to bedtime can also cause reflux problems which will either wake you up at night or leave you with a very sore chest the morning after.

Secondly, always have water with you in your bedroom – it's refreshing. It'll stop you from getting up within 15 minutes of trying to sleep just because you've become parched. Another benefit is that when you inevitably feel dehydrated in the morning, water is there with you already! Be careful, though, you might knock it over.

Thirdly – and this is important - keep your room cool. It's amazing to me that so many people have their bedrooms heated up to the very last moment before they attempt to sleep. This will only serve to keep you awake! A mild drop in temperature has been found to induce sleep. Heller says, "If you are in a cooler [rather than too-warm] room, it is easier for that to happen." But if the room becomes uncomfortably hot, you are more likely to wake up.'

Fourthly, make sure your room is very dark. If that's not possible, consider investing in a sleeping mask. It'll help you to sleep if you have fewer visual distractions.

Fifth and finally, consider doing some light exercise before bed. Outdoing some stretches can help you get into a nice relaxed state before you get into bed.

I hope this helps you all sleep much better. Leave your comments below with tips and tricks which work for you!

255. Which of these is not offered as a benefit of having water in your bedroom?
    A. It will refresh you
    B. It's useful when you wake up
    C. You might knock it over
    D. You won't become thirsty
    E. It will stop you from having to get up

256. Which of these is a direct contradiction?
    A. We are all sleeping badly these days but the author is not a doctor
    B. We are all sleeping badly these days but the author does well with sleep
    C. We're not all insomniacs but some people are
    D. The author is not a doctor but has tips for sleeping better
    E. None of the above

257. Which of these is an example of hyperbole?
    A. "Sleeping badly? We all are these days"
    B. "I'm no doctor – but certainly I've noticed something of a pattern"
    C. "I'm not saying we're all insomniacs for the purposes of medical science"
    D. "Always have water with you in your bedroom – it's refreshing"
    E. "Stretching out can help you get into a nice relaxed state"

258. Which does the author think is the best piece of advice?
    A. The first piece
    B. The third piece
    C. The fourth piece
    D. We don't have enough information to say
    E. The fifth piece

259. What might we infer about the author's occupation?
    A. She is a doctor
    B. She is a nurse
    C. She is a blog writer
    D. She sells water
    E. She is a sleep therapist

Passage 51 - Statehood

An extract from a passage on statehood:

Recently, debate has swirled around whether Palestine ought to be recognised as a state. This issue ought to be settled in a fair and neutral manner - by reference to the traditional criteria for statehood. Clearly, if the Palestinian entity fails to satisfy the traditional legal criteria for statehood, it cannot be recognised as a sovereign state. Eligibility for recognition does not depend on whether an entity wishes to satisfy the criteria for statehood, but on whether it meets those standards as a matter of fact. It is a well-established principle that unless an entity can show that, in practice, it meets the four criteria of statehood, recognition of statement must be withheld. As Kelsen has stated, 'a state violates international law and, thus, infringes upon the rights of other states if it recognises as a state a community which does not fulfil the requirements of international law.' Similarly, Lauterpacht has argued that the recognition of an entity which is not legally a state: '...is a wrong...because it constitutes an abuse of the power of recognition. It acknowledges as an independent state a community which is not, in law, independent and which does not, therefore, fulfil the essential conditions of statehood. It is, accordingly, a recognition which an international tribunal would declare not only to constitute a wrong but probably also to be in itself invalid.'

The four criteria for statehood contained in Article 1 of the 1933 Montevideo Convention are: a territory; a permanent population; a government which exercises authority and has the right to do so; and the capacity to enter into legal relations. It is submitted that all four are useful and are relatively complete as a set, meaning that the criteria do not need to be re-written.

There is no limit on the minimum territorial area necessary and a state may exist even when there are other claims to its territory – as Israel did in 1948. It is self-evident that we expect states to be territorial entities and it is clear that this is an appropriate criterion for statehood. States are required to have a population with a reasonable prospect of sustaining itself: because Western Sahara does not have this, it has not become a state. States are necessarily legal entities made up of people: it is evident that this is an appropriate criterion for a state.

The 'government' requirement has two elements: the actual exercise of authority, and the right to do so. This may not necessarily be control without outside help: when Republic of the Congo (now Democratic Republic of Congo) became a state, it needed UN assistance for control. The government need not be 'legitimate' or 'democratic' – it just needs to be in effective control. This is a necessary criterion because, without it, other states would not be able to interact with the state in question. Capacity to enter into relations with other states is essentially 'independence' from any other state. Taiwan is not a state because it lacks independence from China. The two main elements here are the separate existence of an entity with reasonably coherent frontiers and the fact that the entity is not subject to the authority of any other state or group of states. This is a necessary criterion for statehood because it permits the state to interact with other states and be an actor on the world stage: ensuring the principle of equality of states.

260. What is the passage arguing?
  A. That the current criteria for determining statehood are inappropriate
  B. That Taiwan deserves independence
  C. That the current criteria for determining statehood are appropriate
  D. That government is not a good criterion for determining statehood
  E. That Western Sahara should be a state

261. Why is Taiwan not a state?
  A. It lacks a government
  B. It lacks a permanent population
  C. It isn't very important
  D. It lacks the capacity to enter into legal relations
  E. It lacks boundaries

262. What does the author think about the fact that government need not be 'legitimate' or 'democratic'?
  A. She thinks it's immoral
  B. She thinks it's a really good thing because it's neutral
  C. She thinks it's unfair on Taiwan
  D. She does not offer a viewpoint
  E. None of the above

263. Which of the following is an argument given in favour of the permanent population requirement?
  A. It is necessary that states are entities made up of people
  B. There wouldn't be a government without a population
  C. There wouldn't be an army without a population
  D. Western Sahara is not a state
  E. There needs to be a population over which to exercise control

264. Which criterion lacks any real argument in favour of its existence?
  A. Permanent population
  B. Territory
  C. Government
  D. Capacity to enter into legal relations
  E. All of the above

Passage 52- Sex Education

Recently, the Educational Select Committee has been calling for sex education to be made compulsory in schools. This may well be necessary, however, more pressing, in my view, is the argument that sex education in schools has got to change. The current system leaves children totally confused and unequipped for dealing with relationships. The whole system ought to be built around teaching healthy and happy relationships, with the biology section firmly as a secondary point to that. A report has warned that some children are able to name body parts but yet do not properly understand what was meant by giving consent. Admittedly, the proportion of young people citing school lessons as their main source of information about sexual matters has increased; however, according to the National Survey of Sexual Attitudes and Lifestyles from 2010-12, 61 percent of children still get their information elsewhere.

This means that as it is, children are going to the internet or their peers to glean more information about sexual relationships and this is leading to unstable adolescents. MPs were told it is now "normal" for 14-year-olds to pose in bras for social media photos and that around one in three 15-year-olds had sent someone a naked photo of themselves. Change must come in order to address problems such as teen pregnancy, STIs, sexual exploitation, and cyberbullying. Concurring with this view, Graham Stewart said, 'It's important that school leaders and governors take PSHE seriously and improve their provision by investing in training for teachers and putting PSHE [personal, social, health and economic education] lessons on the school timetable. Statutory status will help ensure all of this happens.' The Tory MP added that 'young people have a right to information that will keep them healthy and safe' and said that changes to the system could help children 'live happy and healthy lives'.

I propose that we change the curriculum so that the first few lessons are based on different kinds of relationships and how to have them in a healthy way. If it's really necessary, we could even start with learning about happy friendships. After that, we could proceed on to romantic relationships and marriage and co-habitation. Teachers should explain conflict resolution, compromise, and communication to children, as well as parenthood. All these things are more useful to young children than talking about sperm and eggs. Moreover, children would not be traumatised because they are forced to learn, instead of things entirely inappropriate for them to know.

I admit that children do need to learn about the biology of sex – but this should not be pushed as the most important element of their lessons. The classes themselves would be better labelled 'relationship education' than 'sex education' anyway because that is what the lessons should focus on. The truth, of course, is that even adults struggle with issues like compromise and communication. Children would benefit enormously from learning about these things: and it would create adults who were much better prepared for the world of love and relationships.

265. What is the main argument of the passage?
    A. We should teach children communication skills
    B. We should change how sex education is taught
    C. We should rename sex education classes
    D. We should stop children going on the internet
    E. We should ban the teaching of sex education

266. What concession does the author make in the third paragraph?
    A. She admits children do need to learn about biology
    B. She agrees we could start classes by learning about friendship
    C. She says we should teach children about marriage
    D. She says we should change the curriculum
    E. She admits that it would be impossible to avoid teaching sex education

267. Which of these arguments is not made in favour of a new teaching approach?
    A. Children would enjoy such classes more
    B. They would mean children were more able to have healthy relationships
    C. The classes would be more useful to children
    D. Children would not be traumatised
    E. Children would be more prepared for love

268. What does the author regard as the relationship between teaching about relationships and about the biology of sex?
    A. We shouldn't teach biology at all; only relationships
    B. Teaching biology is more important
    C. Teaching relationships is more important
    D. We shouldn't teach relationships at all; only biology
    E. None of the above

269. What does the author think is leading to unstable adolescents?
    A. Too many sex education classes
    B. The fact that children go to the wrong sources for sex education
    C. Children do not have happy friendships
    D. Children do not know how to communicate
    E. Children being traumatised and confused by hearing about sex

Passage 53 – Life at work

The UK workforce has undoubtedly become more diverse as society itself has changed and become more diverse. For instance, recent years have seen an increase in the numbers of women at work, the numbers of single parents; reports of UK demographics have also revealed that we have a much larger ageing population than before. Domestic and family responsibilities have also shifted over the years and in response, legislation has evolved and changed to offer both men and women the opportunity to request flexible working which better suits modern conditions and families.

The need to provide parents with more flexible working patterns was first identified following the issue of the Government's Green Paper – "Work and Parents: Competitiveness and Choice". The Government then set up the Work and Parents Taskforce to consider the issue of working parents and flexible working. They submitted their final report – "About Time: Flexible Working" at the end of 2001. Since then, the report has been updated to extend the recommendation for a right to request flexible working to all employees.

It is also promising that many businesses have started to recognise that there are myriad benefits in allowing their staff to have flexible working hours. Such arrangements may include working flexi-time, job shares, and part-time work. This is especially useful for new parents, who want to return to work but not do as much as they had previously. It may also be used by those who want to phase into retirement but not stop working too abruptly, or by those struggling with low-level health problems.

One of the major benefits from the perspectives of businesses is that their employees become much more productive. This may be for several reasons. Firstly, their employees may simply be grateful to the business for allowing them to plan their lives as they wish. It could also be that staff working fewer hours have more energy to give to the business, more progressive companies believe that by allowing workers time off to deal with personal issues and childcare, there is less likelihood of employees having to resort to calling in sick. Alternatively, they may have greater job satisfaction because they are happier in their jobs, meaning they give more to their jobs. Some have suggested that employees become more productive simply because they feel if they do not, flexible working hours will be taken away from them. This is unlikely, however, because businesses which offer flexi-time have largely invested in the scheme wholesale.

Flexible working hours are especially useful for those who have just had children. Given that the government has recently made movements towards allowing shared parental leave, businesses which are allowing a flexible return for their employees are essentially ahead of the curve. It may soon be that all businesses are required to have this policy – no bad thing, if you consider how many fathers wish to be more involved with the early weeks of their children's lives. Businesses might also consider introducing nurseries if they wish to accommodate new parents further.

270. What does the author think about the possibility of all businesses having to offer flexible returns to work for new parents?
    A. She thinks it's positive
    B. She thinks it's negative
    C. She is neutral
    D. There is not enough information to say
    E. She thinks it has pros and cons

271. Which group does the author think benefit most from flexible working?
    A. Ill people
    B. Old people
    C. People who have just had children
    D. There is not enough information to say
    E. Disabled people

272. Which reason does the author think is unlikely to explain why employees become much more productive when working flexi-time?
    A. Employees may be grateful
    B. They may have greater job satisfaction
    C. They may have more energy
    D. They may feel flexi-time will be taken away if they don't work hard
    E. They may feel more determined

273. What benefit can we be certain that businesses gain from offering flexi-time?
    A. They are ahead of the curve
    B. Their staff is more productive
    C. Their employees have more energy
    D. They make more money
    E. Their employees feel indebted to them

274. What is the purpose of the piece?
    A. To inform the reader of the new trend of flexi-time
    B. To persuade the government to make flexi-time compulsory
    C. To persuade new parents to take flexi-time
    D. To persuade people not to retire abruptly
    E. To persuade trade unions to campaign for flexi-time

Passage 54 – The Economy

The global downturn has resulted in soaring rates of unemployment and dramatic cuts to wages. This has brought about a stark decline in living standards across the country and has even, according to one commentator this week, brought about a decline in cardiovascular health. Given the severity of the crisis, it is no wonder that commentators are now putting pressure on the government in the hope of spurring them on to more effective action.

Some commentators have been arguing this week that due to continuing global economic uncertainty, it would be appropriate for governments to invest much more seriously in large-scale infrastructure projects. These might include the construction of new homes, development of the railway and road systems, or work on major bridges. It is well-known that such projects do a great deal to stimulate the economy because the projects employ a lot of people in the building. More significant, however, is that they employ even more people in the planning and sales of such projects. Moreover, a lot of people will spend money in response to these new projects – for example, buying flats, further helping to stimulate the economy. One wonders, then, why the government is choosing not to invest much further. Economic recovery will not happen by itself.

Environmental groups have expressed concern at the pressure being put on the government to invest in infrastructure, saying that the carbon footprint of such projects is huge. It is true that the building of new roads alone has a significant impact on noise pollution, water pollution, habitat destruction, and local air pollution. An increased number of cars on the roads also has the potential to affect climate change. However, it seems that the time to worry about climate change is not right now. Of course, environmental groups are concerned for people's welfare, but they must realise that people are becoming poorer and suffering due to the economic downturn *right now* and that ought to be our top priority at the moment. Climate change is more of a long-term issue and after we have averted the current economic crisis, we will have the opportunity and the resources to turn our attention towards it.

The proposals have also faced criticism from people living in greenbelt areas. They are worried that land near their homes will inevitably be built on and ruin their views. This is a possibility and it would be foolish to deny it. However, theirs is a selfish attitude to take. The best way to stimulate the economy has to be big infrastructure projects, and people living in the greenbelt need to make sacrifices like the rest of us have had to.

House building companies have, to everyone's surprise, been sitting on the fence about this issue. They are of course keen to take on new contracts but have expressed concern about how fast they might be expected to roll out new projects. This is unacceptable – such companies ought to be leading the way in attempting to save the UK from a double dip recession.

275. What does the author think about the relationship between the number of people employed to build large-scale infrastructure projects and the number of people involved in planning and selling those projects?
    A. She thinks they are equally important
    B. She thinks the number employed to build is more important
    C. She thinks the number employed in planning and selling is more important
    D. There is not enough information to say
    E. She doesn't outline any relationship between the two

276. What is the unstated assumption in the second paragraph?
    A. A lot of people will be employed thanks to big infrastructure projects
    B. The government ought to help with stimulating the economy
    C. The economic recovery will not happen without help
    D. It's good for people to live in flats
    E. There is currently global uncertainty

277. Which of the groups in the passage does the author agree with?
    A. The commentators arguing for infrastructure investment
    B. Environmental groups arguing against it
    C. House building companies which are unsure
    D. People living in green areas arguing against it
    E. People living in green areas arguing for it

278. What effect does the author think infrastructure projects would have on the economy?
    A. It will annoy people in greenbelt areas
    B. It will mean more money is going into the economy and lead to growth
    C. It will not impact the economy
    D. It will cause pressure on house building companies
    E. It will reduce the price of houses

279. Which group does the author think are not considering other people?
    A. The commentators arguing for infrastructure investment
    B. Environmental groups arguing against it
    C. House building companies which are unsure
    D. People living in green areas arguing against it
    E. The government

Passage 55 - Tourism

Tourism is seriously big business in this country. The website, visitengland.com, has presented research this week on where and how the UK makes an income from tourism. Cardiff had more visitors than the rest of Wales combined, with Bristol and Manchester also proving popular with international visitors. The World Tourism rankings can also offer insight as to the impact of tourism on economies in the world; the rankings are compiled by the United Nations World Tourism Organisation (UNWTO) as part of their *World Tourism Barometer* publication. This publication is released three times throughout the year. In the publication, UN regions are ranked by three factors: the number of international visitor arrivals, by the revenue generated by inbound tourism, and by the expenditures of outbound travellers. In 2014, there were over 51 million international tourist arrivals to the Middle East, an increase of 5.0% over 2013. In 2014, there were over 582 million international tourist arrivals to Europe, an increase of 3.0% over 2013. We also discovered that London is the world's most popular tourist destination!

The most profitable tourist attraction in London is Madame Tussauds, bizarrely enough. People queued on average two hours last summer to get into a museum which is essentially just wax models of celebrities. And yet, it made £5 million of profit in 2015. It is astonishing that Madame Tussauds received three times as many visitors as the Science Museum last year – but apparently that is what the people want. There really is no accounting for taste.

It is a concern, however, that some councils have responded to this news by announcing plans to divert more money into advertising their cities to tourists. Given that vital services are being cut due to lack of local authority funding, councils should not be investing in tourism. This is because it is such a high-risk industry. Holidays are luxuries, rather than necessities. During recessions or times of economic downturn, families cut back their spending on holidays. This can mean that the tourism industry suffers the most when a crash or any sort of crisis occurs. Political relations between different nations can also have an impact on the level of tourism that one nation receives. The potential income to be obtained through tourism is also very seasonal. Much of the UK, London notwithstanding, receives virtually no tourism in the winter months. Families wishing to take their children abroad also tend to only travel during the long school holidays.

280. Which city in the UK receives the most visitors?

A. Bristol
B. Manchester
C. London
D. Cardiff
E. The passage does not say

281. What point is the author making in the third paragraph?
    A. London gets tourists in winter
    B. Councils are making a risky move
    C. Councils are making a good move
    D. Everywhere gets tourists in winter
    E. Even London receives no tourism in winter

282. Which idea is not discussed in the passage?
    A. Tourists going to the beach
    B. The recession
    C. Tourists going to museums
    D. A report on tourism
    E. Investment in tourism

283. Which of the following ideas is supported by the passage?
    A. Other countries would benefit from copying the UK's approach
    B. London is a very nice place to live
    C. London is the most fun place to visit in the UK
    D. Madame Tussauds is not as interesting or enriching as people think
    E. Madame Tussauds is a good place to visit

284. What does the author mean by saying that the income from tourism is 'very seasonal'?
    A. That there is income all year round
    B. That no money can be made outside summer
    C. That the income depends on the seasons
    D. That people spend more money in winter
    E. No income is made in winter

Passage 56 – The drinks industry

The most profitable drinks industry in the USA is the soft drinks industry, with brands like Coca-Cola raking in most of their sales revenue there; whereas in Europe, it's the alcohol industry. This has caused some worries for Europe; Alcohol-related harm is a major public health concern in the EU as it is accountable for over 7% of all ill health and early deaths. Even moderate alcohol consumption has been found to increase the long-term risk of certain heart conditions, liver diseases, and cancers, and frequent consumption of large amounts can, of course, lead to dependence.

Young people are particularly at risk of short-term effects of drunkenness, which include accidents and violence, with alcohol-related deaths accounting for around 25% of all deaths in young men aged between 15 and 29. This revelation has also received a great deal of attention on social media, with American teens tweeting to call Europeans 'alcoholics'. However, although Europe must, undoubtedly, decrease its consumption of alcohol, the real reason for the difference between the European and American statistics is simply the high sugar taxes imposed on much of Europe and the high alcohol taxes imposed in the US. Tax, it seems, makes all the difference.

There is more sugar in American fizzy drinks than their European equivalents. In fact, there is even less in Asian markets. American senators have had very little joy in trying to tackle this problem because drinks companies provide financial support to so many of those in politics in the US. In the UK, at least, we cannot imagine what that would be like. But the truth is, the American politicians have their hands tied on this issue. In the UK and EU more broadly, sugar is also becoming more unpopular with consumers. Brands are seeking to offer more appealing sweeteners in soft drinks, with a growing media focus on the health hazards of "hidden" sugar in soft drinks. Firms are, therefore, increasingly investing in 'reduced sugar' options. A new set of pocket squashes was, for example, created by brands such as Robinsons Squash'd and Oasis Mighty Drops in 2014, with these products all containing no added sugar. However, consumers remain concerned about artificial sweeteners too.

American teenagers are not allowed to consume alcohol until they are 21 years old – an idea which many Europeans find truly bizarre. The French and Italians, in particular, allow young people to start drinking at a very young age as long as they are accompanied by their parents. The French – very meritoriously – take children with them to dinner parties and slowly allow them to start drinking alcohol in a sensible way. The British are yet to catch on to this. As a result, the Americans have a significant problem with the drinking habits of young people. They've attempted to tackle this by slapping a heavy tax on alcohol – but this appears to have made little difference to those in their twenties. The Europeans have taken the same approach with fizzy drinks, which has been somewhat more effective.

285. To what does the author attribute the difference between the USA and Europe in terms of drinks industries?
    A. Sugar levels
    B. Tax
    C. The fact Europeans are alcoholics
    D. She does not attribute it to anything
    E. Branding

286. Which market has the most sugar in their fizzy drinks?
    A. We do not know
    B. European market
    C. American market
    D. Asian market
    E. None of the above

287. What does the author mean when she says that "American politicians have their hands tied"?
    A. They find it difficult to tackle the sugar levels in fizzy drinks
    B. They cannot make alcohol more popular
    C. They cannot communicate with UK politicians on this issue
    D. They cannot get Asian markets to put more sugar in their drinks
    E. Legislation prevents them from tackling the issue of sugar consumption

288. With regards to alcohol, which country's approach does the author commend?
    A. The USA
    B. The UK
    C. The Italians
    D. The French
    E. The Chinese

289. What impact has the tax on alcohol in the USA had on the drinking habits of young people?
    A. They have stopped drinking entirely
    B. They drink a lot more as a rebellion
    C. It has reduced their drinking significantly
    D. It has not had much effect at all
    E. It has increased their consumption of cheaper brands of alcohol

Passage 57 – Age Imbalance

There are statistics which suggest that in many of the world's least developed countries, close to half the population are under the age of 20. Such countries have not experienced the same decrease in birth rates that we have seen in the more developed West. One reason may be that less developed countries depend on farming for food and wealth. More children mean more potential farm workers, and more food for everyone (and more money). In less developed countries, generally, children are less likely to survive into adulthood (disease, etc.). So having more children is a way of ensuring that a parent has, at least, some children to take care of them in their old age. Lastly, people may not always have access to birth control in less developed countries. So although here in the UK we can drive down to the pharmacy for condoms, people in rural India don't have that option. Consequently, the demographics here in the West are quite different. In fact, in the UK, more than 20% of the population are over the age of 75 – a proportion which in itself is not sustainable. 10 million people in the UK are over 65 years old. The latest projections are for 5½ million more elderly people in 20 years' time and the number will have nearly doubled to around 19 million by 2050. Within this total, the number of very old people grows even faster. There are currently three million people aged more than 80 years and this is projected to almost double by 2030 and reach eight million by 2050. France finds itself in a similar position.

In countries such as Japan, the percentage of the population aged over 75 is set to hit almost 30%. This is concerning because governments are likely to struggle to pay for their ageing populations. Much of today's public spending on benefits is focused on elderly people. 65% of Department for Work and Pensions benefit expenditure goes to those over working age, equivalent to £100 billion in 2010/11 or one-seventh of public expenditure. According to a parliament publication, 'continuing to provide state benefits and pensions at today's average would mean additional spending of £10 billion a year for every additional one million people over working age.' As a direct result of this issue, those in the workforce will have to retire much later. It may also be that pensions are cut, along with social care. There is serious concern that the ageing population could lead to a reduction in government support for care homes.

Most alarmingly, in countries such as Niger, there will simply not be enough money to fund education for all these young people. It is widely accepted that education is the most efficient route for development and for Niger, which routinely finds itself in the bottom 10% of the world's countries on development indexes, education of its youth could make a huge difference. But the surge in births means this simply will not be possible. Other issues preventing development may include the relative lack of healthcare and climate change.

290. Which consequence can we be sure will result from Japan's ageing population?
  A. A cut in government support for care homes
  B. A cut to pensions
  C. A cut to social care
  D. People working for more of their lives
  E. None of the above

291. What reason does the author give for the difference between the birth rates between the developed West and developing countries?
  A. Differing employment opportunities
  B. She does not give a reason
  C. In developing countries, children are a financial asset
  D. In developed countries, people are too career-focused to have children
  E. Lack of money for contraception

292. Which country is the author most concerned for?
  A. Niger
  B. The UK
  C. Japan
  D. France
  E. Nigeria

293. What is the unstated assumption in the last paragraph?
  A. Education leads to development
  B. Development is a good thing in itself
  C. Niger is very undeveloped
  D. The West should help Niger
  E. Climate change is more important than healthcare

294. What can we infer from the second paragraph about an ageing population?
  A. They work hard
  B. Families of elderly people are not looking after them properly
  C. Governments only care about elderly people
  D. They are very expensive
  E. They are able to care for grandchildren

Passage 58 – Jobs for graduates

Joining a tech start-up is the number one trendy move for top graduates across the UK and the US today. Graduates nowadays are gravitating towards app development and other technology markets. Indeed, these industries are welcoming graduates with open arms; opportunities in search engine optimisation and in pay per click advertising. However, many graduates do not properly think through the options and opt for these roles all too quickly. It's depressing to see young people just go and do something because it seems 'cool' rather than because it's actually of any use. They swan around thinking they are so on trend and yet most of them produce nothing of value whatsoever. Further – a lot of their jobs are pretty dull in practice.

More interesting, in my view, is a career in the arts – for instance, a career as an art gallery curator. This job involves the managing of collections of paintings and objects. Curators also undertake art research, identifying and cataloguing paintings and other items. Moreover, they interact with the public, answering visitors' queries and giving talks about the art collections they manage. The job also involves organising displays and negotiation funding for the sale or loan of paintings to the art gallery. An alternative job in the arts is that of a museum curator. These curators manage collections of objects of historical or scientific interest. They ensure that these objects are stored in the right conditions, give talks to groups, and liaise with staff in other museums. They also have to have good finance management skills as they are given the responsibility of managing museum budgets. One unfortunate consequence of the recession has been the cutting of government provision to the arts. Regardless, it is an exciting career path for those who have the strength of character to overcome the difficulties the area is currently facing. The opportunity to create and to help others to do the same has to be one the most exciting things anyone could do in their twenties.

Jobs in finance may seem like good ideas for graduates but jobs in publishing actually offer more opportunities for the truly talented. This is because publishers are much more rigorous in their hiring procedures – and rightly so – and as such, get a much better crop of graduates. One type of job in publishing is that of the commissioning editor. These individuals commission or buy new authors, book titles, or ideas for publication. You'll choose books or media products that you think will sell well. You'll also monitor the performance of published titles. If you love reading and you're able to combine that with market research and spot market trends, then this could be the role for you. Content creation is another option for those who enjoy writing; firms are crying out for graduates to produce content which will enable them to promote their businesses. Graduates hoping to produce or edit content will do well in this fast-growing industry. The point here is that although I'm sure you've been told that jobs in the industries that graduates typically go into have a lot to offer, you should not close your mind to other options. Think a little bigger!

295. What is the attitude of the author towards those who join start-ups?

    A.   She looks down on them

    B.   She admires them

    C.   She thinks they are selfish

    D.   She doesn't understand their motivations

    E.   She is accepting of them

296. What is the unstated assumption in the passage?

    A.   The best jobs are well paid

    B.   The best jobs are in tech

    C.   The best jobs are interesting

    D.   The best jobs are charitable

    E.   All of the above

297. Which pair of jobs does the author draw a direct contrast between?

    A.   Jobs in finance and jobs in publishing

    B.   Jobs in finance and jobs in the arts

    C.   Jobs in the arts and jobs in publishing

    D.   Jobs in the publishing and jobs in technology

    E.   Jobs in technology and jobs in government

298. What view does the author take on hiring procedures in publishing firms?

    A.   She does not give a view

    B.   She thinks they're not as good as hiring procedures in finance

    C.   She thinks they're too hard

    D.   She thinks they're the right approach

    E.   She thinks that they are unfair

299. What is the aim of the article?

    A.   To recruit people into publishing

    B.   To recruit people into the arts

    C.   To get students to consider less traditional career paths

    D.   To get students to work hard at university

    E.   To argue that careers in arts are the best

Passage 59- The Theory of Social Revolutions – Brooks Adams

Were all other evidence lacking, the inference that radical changes are at hand might be deduced from the past. In the experience of the English-speaking race, about once in every three generations a social convulsion has occurred; and probably such catastrophes must continue to occur in order that laws and institutions may be adapted to physical growth. Human society is a living organism, working mechanically, like any other organism. It has members, a circulation, a nervous system, and a sort of skin or envelope, consisting of its laws and institutions. This skin, or envelope, however, does not expand automatically, as it would had Providence intended humanity to be peaceful, but is only fitted to new conditions by those painful and conscious efforts which we call revolutions. Usually, these revolutions are warlike, but sometimes they are benign, as was the revolution over which General Washington, our first great "Progressive," presided, when the rotting Confederation, under his guidance, was converted into a relatively excellent administrative system by the adoption of the Constitution.

Taken for all in all, I conceive General Washington to have been the greatest man of the eighteenth century, but to me, his greatness chiefly consists in that balance of mind which enabled him to recognise when an old order had passed away, and to perceive how a new order could be best introduced. Washington was, in his way, a large capitalist, but he was much more. He was not only a wealthy planter, but he was an engineer, a traveller, to an extent a manufacturer, a politician, and a soldier, and he saw that, as a conservative, he must be "Progressive" and raise the law to a power high enough to constrain all these thirteen refractory units. For Washington understood that peace does not consist in talking platitudes at conferences, but in organising a sovereignty strong enough to coerce its subjects.

The problem of constructing such a sovereignty was the problem which Washington solved, temporarily at least, without violence. He prevailed not only because of an intelligence and elevation of character which enabled him to comprehend, and to persuade others, that, to attain a common end, all must make sacrifices, but also because he was supported by a body of the most remarkable men whom America has ever produced. Men who, though doubtless in a numerical minority, taking the country as a whole, by sheer weight of ability and energy, achieved their purpose.

Yet, even Washington and his adherents could not alter the limitations of the human mind. He could postpone, but he could not avert the impact of conflicting social forces. In 1789, he compromised, but he did not determine the question of sovereignty. He eluded an impending conflict by introducing courts as political arbitrators, and the expedient worked more or less well until the tension reached a certain point. Then it broke down, and the question of sovereignty had to be settled in America, as elsewhere, on the field of battle.

300. What is meant by the phrase 'social convulsion'?
    A. A catastrophe
    B. An attack
    C. A period of change
    D. Development of human society
    E. A revolution

301. Which of the following ideas is the skin metaphor used to illustrate?
    A. Human beings are comparable to skin
    B. Human society grows over time
    C. Progression is inevitable
    D. There are now too many laws
    E. Change does not come about peacefully and organically

302. Which of the following would most weaken the author's argument in relation to General Washington?
    A. Washington's supporters went on to achieve great things independently
    B. Many intelligent men opposed Washington
    C. Washington was known as a genius
    D. Most of Washington's supporters had little experience or charisma
    E. Washington's supporters produced ingenious solutions

303. Which of the following is not used to characterise Washington?
    A. Successful
    B. Sacrificial
    C. Persuasive
    D. Traveller
    E. Energetic

304. Why, according to the author, was Washington unable to prevent a revolution?
    A. He was too disorganised
    B. He was able only to postpone it
    C. Because it is impossible for the human mind to devise a means of doing so
    D. He didn't face the problem head-on
    E. He commanded the support of only a minority

Passage 60 - Culture and anarchy - Matthew Arnold

The whole scope of the essay is to recommend culture as the great help out of our present difficulties; culture being a pursuit of our total perfection by means of getting to know, on all the matters which most concern us, the best which has been thought and said in the world, and, through this knowledge, turning a stream of fresh and free thought upon our stock notions and habits, which we now follow staunchly but mechanically, vainly imagining that there is a virtue in following them staunchly which makes up for the mischief of following them mechanically. This, and this alone, is the scope of the following essay.

I say again here, what I have said in the pages which follow, that from the faults and weaknesses of bookmen a notion of something bookish, pedantic, and futile, has got itself more or less connected with the word culture, and that it is a pity we cannot use a word more perfectly free from all shadow of reproach. And yet, futile as are many bookmen, and helpless as books and reading often prove for bringing nearer to perfection those who use them, one must, I think, be struck more and more the longer one lives to find how much, in our present society, a man's life of each day depends for its solidity and value on whether he reads during that day. More and more, he who examines himself will find the difference it makes to him, at the end of any given day, whether or not he has pursued his avocations throughout it without reading at all.

This, however, is a matter for each man's private conscience and experience. If a man without books or reading, or reading nothing but his letters and the newspapers, gets nevertheless a fresh and free play of the best thoughts upon his stock notions and habits, he has got culture. He has got that for which we prize and recommend culture; he has got that which at the present moment we seek culture that it may give us. This inward operation is the very life and essence of culture, as we conceive it.

Nevertheless, it is not easy so to frame one's discourse concerning the operation of culture, as to avoid giving frequent occasion to a misunderstanding whereby the essential inwardness of the operation is lost sight of. We are supposed, when we criticise by the help of culture some imperfect doing or other, to have in our eye some well-known rival plan of doing, which we want to serve and recommend. Thus, for instance, because I have freely pointed out the dangers and inconveniences to which our literature is exposed in the absence of any centre of taste and authority like the French Academy, it is constantly said that I want to introduce here in England an institution like the French Academy.

I have indeed expressly declared that I wanted no such thing; but let us notice how it is just our worship of machinery, and of external doing, which leads to this charge being brought; and how the inwardness of culture makes us seize, for watching and cure, the faults to which our want of an academy inclines us, and yet prevents us from trusting to an arm of flesh, as the Puritans say – from blindly flying to this outward machinery of an academy, in order to help ourselves. For the very same culture and free inward play of thought which shows us how the Corinthian style, or the whimsies about the One Primeval Language, are generated and strengthened in the absence of an academy, shows us, too, how little any academy, such as we should be likely to get, would cure them.

305. What is the main argument of the passage?
    A. Culture should be pursued because it will give us fresh perspective
    B. Reading makes a big difference to the value of one's day
    C. We do not need an academy
    D. We worship machinery
    E. It doesn't matter how you obtain culture, so long as you obtain it

306. The author mentions the 'charge' brought against him to illustrate what idea?
    A. That culture is inward
    B. That an academy is desirable
    C. That an academy is futile
    D. That we only care about external actions
    E. That his critics do not understand him

307. What is meant by the phrase 'inward operation'?
    A. One's inner thoughts
    B. Emotions
    C. That culture manifests its effects only in our thoughts
    D. That culture is powerful
    E. That culture leads to secret change

308. Why, according to the author, would an academy be unhelpful?
    A. It is outward in its operation
    B. It is inward in its operation
    C. It is puritan
    D. It would be too large
    E. We are all already cultured enough without one

309. What is meant by the phrase 'a matter for each man's private conscience and experience'?
    A. It should be up to the individual to choose how to get culture
    B. Culture is subjective
    C. Culture is objective
    D. People feel guilty without culture
    E. All of the above

Passage 61 - Laughter: An Essay on the Meaning of Comic

Henri Bergson, member of the Institute Professor at the College de France

What does laughter mean? What is the basal element in the laughable? What common ground can we find between the grimace of a merry-andrew, a play upon words, an equivocal situation in a burlesque, and a scene of high comedy? The greatest of thinkers, from Aristotle downwards, have tackled this little problem, which has a knack of baffling every effort, of slipping away and escaping only to bob up again, a pert challenge flung at philosophic speculation. Our excuse for attacking the problem in our turn must lie in the fact that we shall not aim at imprisoning the comic spirit within a definition. We regard it, above all, as a living thing. We shall confine ourselves to watching it grow and expand. Passing by imperceptible gradations from one form to another, it will be seen to achieve the strangest metamorphoses. We shall disdain nothing we have seen.

And maybe we may also find that we have made an acquaintance that is useful. For the comic spirit has a logic of its own, even in its wildest eccentricities. It has a method in its madness. It dreams, I admit, but it conjures up, in its dreams, visions that are at once accepted and understood by the whole of a social group. Can it then fail to throw light for us on the way that human imagination works, and more particularly social, collective, and popular imagination? Begotten of real life and akin to art, should it not also have something of its own to tell us about art and life?

At the outset, we shall put forward three observations which we look upon as fundamental. The first point to which attention should be called is that the comic does not exist outside the pale of what is strictly <u>human</u>. A landscape may be beautiful, charming and sublime, or insignificant and ugly; it will never be laughable.

Here I would point out, as a symptom equally worthy of notice, the <u>absence of feeling</u> which usually accompanies laughter. Indifference is its natural environment, for laughter has no greater foe than emotion. I do not mean that we could not laugh at a person who inspires us with pity, for instance, or even with affection, but in such a case we must, for the moment, put our affection out of court and impose silence upon our pity. It is enough for us to stop our ears to the sound of music, in a room where dancing is going on, for the dancers at once to appear ridiculous. How many human actions would stand a similar test? Should we not see many of them suddenly pass from grave to gay, on isolating them from the accompanying music of sentiment? To produce the whole of its effect, then, the comic demands something like a momentary anaesthesia of the heart. Its appeal is to intelligence, pure and simple.

This intelligence, however, must always remain in touch with other intelligences. And here is the third fact to which attention should be drawn. Our laughter is always the laughter of a group. It may, perchance, have happened to you, when seated in a railway carriage or at table d'hôte, to hear travellers relating to one another stories which must have been comic to them, for they laughed heartily. Had you been one of their company, you would have laughed like them; but, as you were not, you had no desire whatever to do so.

To understand laughter, we must put it back into its natural environment, which is society, and above all must we determine the utility of its function, which is a social one. Laughter must answer to certain requirements of life in common. It must have a <u>social</u> signification.

310. The writer uses the word 'basal' to mean:
    A. Essential
    B. Notable
    C. Undeveloped
    D. Elusive
    E. Significant

311. What is the central motivation for the writer's exploration of the topic of laughter?
    A. To avoid imprisoning the concept within a definition
    B. To watch it achieve 'strange metamorphoses'
    C. To throw light on human functioning and behaviour
    D. To demonstrate that laughter has 'a method in its madness'
    E. The fact that other great thinkers have failed to reach a conclusion

312. What is the conclusion of the second paragraph?
    A. We can only laugh at those we pity by momentarily ignoring that sentiment
    B. Indifference is laughter's natural environment
    C. It is noteworthy that absence of feeling accompanies laughter
    D. If we stop our ears to the sound of the accompanying music, the dancers immediately appear ridiculous
    E. Music determines the sentiments of dancers

313. Which of the following is a metaphor used by the author?
    A. The accompanying music of sentiment
    B. Its appeal is to intelligence
    C. Akin to art
    D. Pass from grave to gay
    E. The comic does not exist outside the pale of what is strictly human

314. Which one of the following is not used to characterise laughter?
    A. It is the laughter of a group
    B. Social
    C. Sentimental
    D. Illuminating
    E. A living thing

Passage 62 - The God-Idea of the Ancients

Eliza Burt Gamble

Nowhere is the influence of sex more plainly manifested than in the formulation of religious conceptions and creeds. With the rise of male power and dominion, and the corresponding repression of the natural female instincts, the principles that originally constituted the God-idea gradually gave place to a Deity better suited to the peculiar bias that had been given to the male organism. An anthropomorphic God, like that of the Jews, whose chief attributes are power and virile could have had its origin only under a system of masculine rule.

Religion is especially liable to reflect the vagaries and weaknesses of human nature; and, as the forms and habits of thought connected with worship take a firmer hold on the mental constitution than do those belonging to any other department of human experience, religious conceptions should be subjected to frequent and careful examination in order to perceive, if possible, the extent to which we are holding on to ideas which are unsuited to existing conditions.

In an age when every branch of inquiry is being subjected to reasonable criticism, it would seem that the origin and growth of religion should be investigated from beneath the surface and that all the facts bearing upon it should be brought forward as a contribution to our fund of general information. As well might we hope to gain a complete knowledge of human history by studying only the present aspect of society, as to expect to reach reasonable conclusions respecting the prevailing God-idea by investigating the various creeds and dogmas of existing faiths.

Doubtless, the worship of a female energy prevailed under the matriarchal system and was practised at a time when women were the recognised heads of families and when they were regarded as the more important factors in human society. After women began to leave their homes at marriage, and after property, especially land, had fallen under the supervision and control of men, the latter, as they manipulated all the necessaries of life and the means of supplying them, began to regard themselves as superior beings, and later, to claim that as a factor in reproduction, or creation, the male was the more important. With this change, the ideas of a Deity also began to undergo a modification. The dual principle necessary to creation, and which had hitherto been worshipped as an indivisible unity, began gradually to separate into its individual elements, the male representing spirit, the dominant moving or forming force in the creation processes. A little observation and reflection will show us that during this change in the ideas relative to a creative principle, or God, descent and the rights of succession which had hitherto been reckoned through the mother were changed from the female to the male line, the father having in the meantime become the only recognised parent.

315. What assumption does the author make in the first paragraph?
    A. There is no God
    B. The Jewish God does not exist
    C. The Jewish ideas of God did not come about through divine revelation
    D. The Jewish God is a product of patriarchy
    E. In a more equal society, we would have a gender neutral conception of God

316. Why does the author believe that religious conceptions are in special need of being subjected to careful examination?
    A. Religious views and rites have a very strong hold on the mental constitution
    B. In order to determine whether we are holding on to ideas that are unsuited to existing conditions
    C. They have a huge impact on the structure of society
    D. They afford insight into our fund of information about society
    E. They can influence the way in which women are perceived

317. What comes closest to the author's view of the relationship between religion and society?
    A. Religion determines the structure of society
    B. Religion and society evolve together
    C. There is no relationship between the two
    D. The structure of society determines our religious views
    E. Different societies have different conceptions of religion

318. Which of the following is merely an assertion?
    A. Religious conceptions should be subjected to frequent and careful examination
    B. Religion is liable to reflect the vagaries and weaknesses of human nature
    C. The origin and growth of religion should be investigated
    D. Nowhere is the influence of sex more plainly manifested than in the formulation of religious conceptions and creeds
    E. An anthropomorphic god like that of the Jews could have had its origin only under a system of masculine rule

319. What is stated to be the origin of contemporary attributions of might to a Deity?
    A. The valuing by men of male qualities over female qualities
    B. The weaknesses of human nature
    C. The repression of the female instincts
    D. The rise of male dominance
    E. (a) The value by men of male qualities over female qualities AND (d) The rise of male dominance

Passage 63 - The Nature of Goodness

George Herbert Palmer

In undertaking the following discussion, I foresee two grave difficulties. My reader may well feel that goodness is already the most familiar of all the thoughts we employ, and yet he may at the same time suspect that there is something about it perplexingly abstruse and remote. Familiar it certainly is. It attends all our wishes, acts, and projects as nothing else does so that no estimate of its influence can be excessive. When we take a walk, read a book, make a dress, hire a servant, visit a friend, attend a concert, choose a wife, cast a vote, enter into business, we always do it in the hope of attaining something good. The clue of goodness is accordingly a veritable guide of life. On it depend actions far more minute than those just mentioned. We never raise a hand, for example, unless with a view to improve in some respect our condition. Motionless we should remain forever, did we not believe that by placing the hand elsewhere we might obtain something that we do not now possess. Consequently, we employ the word or some synonym of it during pretty much every waking hour of our lives.

But while thus familiar and influential when mixed with action, and just because of that very fact, the notion of goodness is bewilderingly abstruse and remote. People, in general, do not observe this curious circumstance. Since they are so frequently encountering goodness, both laymen and scholars are apt to assume that it is altogether clear and requires no explanation. But the very reverse is the truth. Familiarity obscures. It breeds instincts and not understanding. So in woven has goodness become with the very web of life that it is hard to disentangle. We cannot easily detach it from encompassing circumstance, look at it nakedly, and say what in itself it really is. Never appearing in practical affairs except as an element, and always intimately associated with something else, we are puzzled how to break up that intimacy and give to goodness independent meaning. It is as if oxygen were never found alone, but only in connection with hydrogen, carbon, or some other of the eighty elements which compose our globe. We might feel its wide influence, but we should have difficulty in describing what the thing itself was. Just so if any chance a dozen persons should be called on to say what they mean by goodness, probably not one could offer a definition that he would be willing to hold to for fifteen minutes.

It is true; this strange state of things is not peculiar to goodness. Other familiar conceptions show a similar tendency, and just about in proportion, too, to their importance. Those that count for most in our lives are least easy to understand. What, for example, do we mean by love?

For ordinary purposes probably it is well not to seek to understand it. Acquaintance with the structure of the eye does not help seeing. To determine beforehand just how polite we should be would not facilitate human intercourse. And possibly a completed scheme of goodness would rather confuse than ease our daily actions. Science does not readily connect with life. For most of us all the time, and for all of us most of the time, instinct is the better prompter. But if we mean to be ethical students and to examine conduct scientifically, we must evidently at the outset come face to face with the meaning of goodness. This word being the ethical writer's chief tool, both he and his readers must learn its construction before they proceed to use it.

320. Which of the following qualities is not attributed to goodness?
    A. A motivator of action
    B. All-pervasive
    C. In woven
    D. Curious
    E. Independent

321. Why, according to the author, is the nature of goodness so difficult to grasp?
    A. Because it is a very familiar concept
    B. Because humans have difficulty in describing concepts
    C. Because we cannot employ scientific means to work out real-life ideas
    D. Because things which mean the most to our lives are the most difficult to understand
    E. Because we have difficulty in describing what it is, though we feel its wide influence

322. Which comes closest to what is meant by the phrase '*familiarity obscures*'?
    A. Familiarity with a concept prevents us from exploring its true meaning
    B. The familiarity of a concept prevents us from detaching it from the circumstances in which we find it and giving it independent meaning.
    C. Widespread concepts do not have independent meanings
    D. It is difficult to understand familiar concepts because they are of great significance to our lives
    E. We relate to familiar concepts by instinct and not through rational thought

323. Why does the author use the statement '*acquaintance with the structure of the eye does not help seeing*'?
    A. To show that science has little application in everyday life
    B. To show that certain forms of knowledge can hinder us in day to day life
    C. To show that there is no point in studying the concept of goodness in depth
    D. To demonstrate that knowledge of a topic does not assist in the practical application of that topic to everyday life
    E. None of the above

324. What is the main conclusion of the final paragraph?
    A. An understanding of the concept of goodness is crucial for ethical students
    B. We need to understand goodness in order to become better people
    C. A completed scheme of goodness would confuse everyday life
    D. Scientists need not preoccupy themselves with matters of everyday life
    E. The word '*goodness*' is the ethical writer's chief tool

Passage 64 - Marriage and Love

Eliza Burt Gamble

The popular notion about marriage and love is that they are synonymous, that they spring from the same motives, and cover the same human needs. Like most popular notions, this also rests not on actual facts, but on superstition.

Marriage and love have nothing in common; they are as far apart as the poles; they are, in fact, antagonistic to each other. No doubt some marriages have been the result of love. Not, however, because love could assert itself only in marriage; much rather is it because few people can completely outgrow a convention. There are today large numbers of men and women to whom marriage is naught but a farce, but who submit to it for the sake of public opinion. At any rate, while it is true that some marriages are based on love, and while it is equally true that in some cases love continues in married life, I maintain that it does so regardless of marriage, and not because of it.

On the other hand, it is utterly false that love results from marriage. On rare occasions, one does hear of a miraculous case of a married couple falling in love after marriage, but on close examination, it will be found that it is a mere adjustment to the inevitable. Certainly growing used to each other is far away from the spontaneity, the intensity, and beauty of love, without which the intimacy of marriage must prove degrading to both the woman and the man.

That marriage is a failure none but the very stupid will deny. One has but to glance over the statistics of divorce to realise how bitter a failure marriage really is. Nor will the stereotyped Philistine argument that the laxity of divorce laws and the growing looseness of woman account for the fact that: first, every twelfth marriage ends in divorce; second, that since 1870, divorces have increased from 28 to 73 for every hundred thousand population; third, that adultery, since 1867, as ground for divorce, has increased 270.8 percent; fourth, that desertion increased 369.8 percent.

Henrik Ibsen, the hater of all social shams, was probably the first to realise this great truth. Nora leaves her husband, not—as the stupid critic would have it—because she is tired of her responsibilities or feels the need of woman's rights, but because she has come to know that for eight years she had lived with a stranger and borne him children. The moral lesson instilled in the girl is not whether the man has aroused her love, but rather is it, "How much?" The important and only God of practical American life: Can the man make a living? Can he support a wife? That is the only thing that justifies marriage. Gradually this saturates every thought of the girl; her dreams are not of moonlight and kisses, of laughter and tears; she dreams of shopping tours and bargain counters. Can there be anything more humiliating, more degrading than a life-long proximity between two strangers? No need for the woman to know anything of the man, save his income. As to the knowledge of the woman—what is there to know except that she has a pleasing appearance?

But the child, how is it to be protected, if not for marriage? After all, is not that the most important consideration? The sham, the hypocrisy of it! Marriage, protecting the child, yet thousands of children destitute and homeless. So long as love begets life no child is deserted, or hungry, or famished for the want of affection. I know this to be true. I know women who became mothers in freedom by the men they loved. Few children in wedlock enjoy the care, the protection, the devotion free motherhood is capable of bestowing.

325. In the final paragraph, what is meant by the phrase '*in freedom*'?
    A. Being unmarried
    B. Freedom from the dictates of society
    C. Freedom from social stigma
    D. Freedom from male presence
    E. Economic freedom

326. What does the author imply in the fifth paragraph is ultimately responsible for unsuccessful marriages?
    A. Use of a flawed criteria for determining who to marry
    B. Marrying a person unsuited to you
    C. A feeling of estrangement from one's spouse
    D. The women's rights movement
    E. Women's need for greater freedom

327. What, according to the author, is the relationship between marriage and love?
    A. Marriage always springs from love
    B. Marriage stifles love
    C. There is no relationship between the two
    D. After a long time, marriage can bring about love
    E. Marriages break down when the love is lost

328. What assumption is made by the author in the final paragraph?
    A. The parents of the destitute children are married
    B. That children of married couples are happier than those of unmarried couples
    C. That unmarried couples are able to love children more than married couples
    D. That society is doing nothing to help children
    E. That the underlying rationale for marriage is the protection of children

329. What is the main conclusion in the passage?
    A. Marriage cannot protect children
    B. Marriage has been unable to bring about love
    C. Most marriages will end in divorce
    D. Abolishing the institution of marriage is the best way to protect children
    E. The institution of marriage is a failure

Passage 65 - Minorities versus Majorities

Emma Goldman

If I were to give a summary of the tendency of our times, I would say, Quantity. The multitude, the mass spirit, dominates everywhere, destroying quality. Our entire life—production, politics, and education—rests on quantity, on numbers. The worker who once took pride in the thoroughness and quality of his work has been replaced by brainless, incompetent automatons who turn out enormous quantities of things, valueless to themselves, and generally injurious to the rest of mankind. Thus, quantity, instead of adding to life's comforts and peace, has merely increased man's burden.

In politics, naught but quantity counts. In proportion to its increase, however, principles, ideals, justice, and uprightness are completely swamped by the array of numbers. In the struggle for supremacy, the various political parties outdo each other in trickery, deceit, cunning, and shady machinations, confident that the one who succeeds is sure to be hailed by the majority as the victor. That is the only God—success. As to what expense, what terrible cost to character, is of no moment. We have not far to go in search of proof to verify this sad fact. Never before did the corruption, the complete rottenness of our government stand so thoroughly exposed. Yet when the crimes of that party became so brazen that even the blind could see them, it needed but to muster up its minions, and its supremacy was assured.

The oft-repeated slogan of our time is, among all politicians, the Socialists included, that ours is an era of individualism, of the minority. Only those who do not probe beneath the surface might be led to entertain this view. Have not the few accumulated the wealth of the world? Are they not the masters, the absolute kings of the situation? Their success, however, is due not to individualism, but to the inertia, the cravenness, the utter submission of the mass. The latter wants but to be dominated, to be led, to be coerced. As to individualism, at no time in human history did it have less chance of expression, less opportunity to assert itself in a normal, healthy manner.

Need I say that in art we are confronted with the same sad facts? One has but to inspect our parks and thoroughfares to realize the hideousness and vulgarity of the art manufacture. Certainly, none but a majority taste would tolerate such an outrage on art. False in conception and barbarous in execution, the statuary that infests American cities has as much relation to true art, as a totem to a Michael Angelo. Yet that is the only art that succeeds. The true artistic genius, who will not cater to accepted notions, who exercises originality, and strives to be true to life, leads an obscure and wretched existence. His work may someday become the fad of the mob, but not until his heart's blood had been exhausted; not until the pathfinder has ceased to be, and a throng of an idealess and visionless mob has done to death the heritage of the master.

I know so well that as a compact mass, the majority has never stood for justice or equality. It has suppressed the human voice, subdued the human spirit, and chained the human body. In other words, the living, vital truth of social and economic well-being will become a reality only through the zeal, courage, the non-compromising determination of intelligent minorities, and not through the mass.

330. What is the conclusion of the first paragraph?
    A. Quantity is prioritised over and has destroyed quality
    B. Workers who took pride in their work have been replaced by brainless automatons
    C. Goods produced nowadays are generally injurious to us
    D. Politics now focuses on quantity
    E. We must focus on quality in order to improve wellbeing

331. Which of the following is a metaphor employed by the author?
    A. False in conception and barbarous in execution
    B. Success is the only god
    C. The mob
    D. Its supremacy was assured
    E. Suppressed the human voice

332. Which of the following would most weaken the argument in the fourth paragraph?
    A. The head of the city council has sole responsibility for commissioning statues and works of arts to be erected in public places
    B. Artistic geniuses are all acknowledged upon their deaths
    C. Many artistic geniuses have been acknowledged just immediately prior to their deaths
    D. Most people enjoy abstract art
    E. The statues and works of art in public places are reviewed every five years

333. Which of the following lines most strongly supports/advances the author's rejection of the 'oft-repeated slogan' in the third paragraph?
    A. Have not the few accumulated the wealth of the world?
    B. Only those who do not probe beneath the surface might be led to entertain this view
    C. Their success, however, is due not to individualism, but to the inertia, the cravenness, the utter submission of the mass
    D. As to individualism, at no time in human history did it have less chance of expression
    E. None of the above

334. In the second paragraph, what, according to the writer, is the relationship between morals/character in politics and democracy?
    A. Democracy allows those with poor morals to thrive
    B. Democracy encourages trickery
    C. Democracy helps to root out politicians of poor moral character
    D. There is no relationship
    E. Democracy destroys the morals of political parties

Passage 66 - What is a novel?

F. Marion Crawford

"What is a novel?" A novel is a marketable commodity, of the class collectively termed "luxuries," as not contributing directly to the support of life or the maintenance of health. The novel, therefore, is an intellectual artistic luxury in that it can be of no use to a man when he is at work, but may conduce to peace of mind and delectation during his hours of idleness.

Probably, no one denies that the first object of the novel is to amuse and interest the reader. But it is often said that the novel should instruct as well as afford amusement, and the "novel-with-a-purpose" is the realisation of this idea. The purpose-novel, then, proposes to serve two masters, besides procuring a reasonable amount of bread and butter for its writer and publisher. It proposes to escape from my definition of the novel in general and make itself an "intellectual moral lesson" instead of an "intellectual artistic luxury." It constitutes a violation of the unwritten contract tacitly existing between writer and reader. A man buys what purports to be a work of fiction, a romance, a novel, a story of adventure, pays his money, takes his book home, prepares to enjoy it at his ease and discovers that he has paid a dollar for somebody's views on socialism, religion, or the divorce laws.

Such books are generally carefully suited with an attractive title. The binding is as frivolous as can be desired. The bookseller says it is "a work of great power," and there is probably a sentimental dedication on the flyleaf to a number of initials to which a romantic appearance is given by the introduction of a stray "St." and a few hyphens. The buyer is possibly a conservative person, of lukewarm religious convictions, whose life is made "barren by marriage, or death, or division"—and who takes no sort of interest in the laws relating to divorce, in the invention of a new religion, or the position of the labour question. He has simply paid money, on the ordinary tacit contract between furnisher and purchaser, and he has been swindled, to use a very plain term for which a substitute does not occur to me. Or say that a man buys a seat in one of the regular theatres. He enters, takes his place, preparing to be amused, and the curtain goes up. The stage is set as a church, there is a pulpit before the prompter's box, and the Right Reverend the Bishop of the Diocese is on the point of delivering a sermon. The man would be legally justified in demanding his money at the door, I fancy, and would probably do so, though he might admit that the Bishop was the most learned and edifying of preachers. In ordinary cases, the purpose-novel is a simple fraud, besides being a failure in nine hundred and ninety-nine cases out of a thousand.

What we call a novel may educate the taste and cultivate the intelligence; under the hand of genius it may purify the heart and fortify the mind; but it has no right to tell us what its writer thinks about the relations of labour and capital, nor to set up what the author conceives to be a nice, original, easy scheme of salvation, any more than it has a right to take for its theme the relative merits of the "broomstick-car" and the "storage system," temperance, vivisection, or the "Ideal Man" of Confucius. Lessons, lectures, discussions, sermons, and didactics generally belong to institutions set apart for especial purposes and carefully avoided, after a certain age, by the majority of those who wish to be amused. The purpose-novel is an odious attempt to lecture people who hate lectures, to preach at people who prefer their own church, and to teach people who think they know enough already. It is an ambush, a lying-in-wait for the unsuspecting public, a violation of the social contract—and as such, it ought to be either mercilessly crushed or forced by law to bind itself in black and label itself "Purpose" in very big letters.

335. Which of the following is merely an assertion?
    A. The purpose-novel, then, proposes to serve two masters
    B. The purpose-novel is an odious attempt to lecture people
    C. The man would be legally justified in demanding his money at the door
    D. Probably, no one denies that the first object of the novel is to amuse and interest the reader
    E. The purpose-novel constitutes a violation of the unwritten contract tacitly existing between writer and reader

336. Which comes closest to the point being made by the bishop example in paragraph three?
    A. People shouldn't be tricked into reading a book, regardless of how otherwise beneficial that book is
    B. Purpose-novels aim to preach at people
    C. Purpose-novels often concern subjects in which the reader has no interest
    D. The purpose-novel is a failure in nine hundred and ninety-nine cases out of a thousand.
    E. A contract to purchase a novel is the same as any other contract

337. What is meant by the phrase '*the purpose-novel purports to serve two masters*'?
    A. The purpose-novel aims to educate and provide an income for the author
    B. The purpose-novel aims to entertain and to provide an income for the author
    C. The purpose-novel aims to entertain and to educate the reader
    D. The purpose-novel aims to provide income for the writer and for the publisher
    E. The purpose-novel aims to lecture on religion and society

338. Which of the following topics is the author likely to view as inappropriate for a novel?
    A. Romance
    B. Comedy
    C. Politics
    D. Sibling bickering
    E. Love triangles

339. Which of the following would most strengthen the author's argument?
    A. A recent survey found that most people pick a novel on the basis of its title alone
    B. A few purpose-novels are known to have been very entertaining
    C. Research has shown that fewer and fewer people are reading nowadays
    D. Works of non-fiction are highly in demand nowadays
    E. Religious and political books are often bestsellers

Passage 67 - We the Media

Dan Gillmor

We freeze some moments in time. Every culture has its frozen moments, events so important and personal that they transcend the normal flow of news. Americans of a certain age, for example, know precisely where they were and what they were doing when they learned that President Franklin D. Roosevelt died. Another generation has absolute clarity of John F. Kennedy's assassination. And no one who was older than a baby on September 11, 2001, will ever forget hearing about, or seeing, airplanes exploding into skyscrapers. In 1945, people gathered around radios for the immediate news and stayed with the radio to hear more about their fallen leader and about the man who took his place. Newspapers printed extra editions and filled their columns with detail for days and weeks afterward. Magazines stepped back from the breaking news and offered perspective. September 11, 2001, followed a similarly grim pattern. We watched—again and again—the awful events. Consumers of news learned about the attacks, thanks to the television networks that showed the horror so graphically. Then we learned some of the how's and why's as print publications and thoughtful broadcasters worked to bring depth to events that defied mere words. Journalists did some of their finest work and made me proud to be one of them.

But something else, something profound, was happening this time around: news was being produced by regular people who had something to say and show, and not solely by the "official" news organizations that had traditionally decided how the first draft of history would look. This time, the first draft of history was being written in part, by the former audience. It was possible—it was inevitable—because of new publishing tools available on the Internet. Another kind of reporting emerged during those appalling hours and days. Via emails, mailing lists, chat groups, personal web journals—all nonstandard news sources—we received valuable context that the major American media couldn't, or wouldn't, provide.

We were witnessing—and in many cases were part of—the future of news.

In the 20th century, making the news was almost entirely the province of journalists; the people we covered, or "news-makers"; and the legions of public relations and marketing people who manipulated everyone. The economics of publishing and broadcasting created large, arrogant institutions—call it Big Media, though even small-town newspapers and broadcasters exhibit some of the phenomenon's worst symptoms.

Big Media, in any event, treated the news as a lecture. We told you what the news was. You bought it, or you didn't. You might write us a letter; we might print it. (If we were television and you complained, we ignored you entirely unless the complaint arrived on a libel lawyer's letterhead.) Or you cancelled your subscription or stopped watching our shows. It was a world that bred complacency and arrogance on our part. It was a gravy train while it lasted, but it was unsustainable.

Tomorrow's news reporting and production will be more of a conversation or a seminar. The lines will blur between producers and consumers, changing the role of both in ways we're only beginning to grasp now. The communication network itself will be a medium for everyone's voice, not just the few who can afford to buy multimillion-dollar printing presses, launch satellites, or win the government's permission to squat on the public's airwaves. This evolution—from journalism as a lecture to journalism as a conversation or seminar—will force the various communities of interest to adapt. Everyone, from journalists to the people we cover to our sources and the former audience, must change their ways. The alternative is just more of the same.

We can't afford more of the same. We can't afford to treat the news solely as a commodity, largely controlled by big institutions. We can't afford, as a society, to limit our choices. We can't even afford it financially because Wall Street's demands on Big Media are dumbing down the product itself. There are three major constituencies in a world where anyone can make the news. Once largely distinct, they're now blurring into each other.

340. What is implied by the line: *'no one who was older than a baby on September 11, 2001, will ever forget hearing about, or seeing, airplanes exploding into skyscrapers'*?
    A. No one will ever forget the events of 9/11
    B. 9/11 was a highly televised and widely discussed tragedy
    C. 9/11 was an extremely destructive event
    D. 9/11 is of little significance to American history
    E. Children were badly affected by the events of 9/11

341. Which of the following is an argument against *'Big Media'*?
    A. Under 'Big Media', people were told what the news was
    B. The economics of publishing and broadcasting created 'Big Media'
    C. Big Media provides a well-rounded view of news
    D. The dominance of Big Media allows for media complacency and ignorance in manner in which news is presented
    E. There are other potential sources of news

342. What, according to the author, first triggered the change in the identity of the publishers of news?
    A. 9/11
    B. The financial unsustainability of Big Media
    C. Society's refusal to continue to treat media as a lecture
    D. The blurring of the lines between producers and consumers
    E. The availability of new publishing tools on the internet

343. What is stated to be the future effect of the evolution to journalism as a conversation?
    A. Big Media will be entirely eradicated
    B. Different interest groups will be forced to adapt
    C. News reporting will become more of a conversation
    D. The way in which our community functions will improve
    E. There will be no effect/change

344. What purpose does the repetition of *'we can't afford'* in the final paragraph serve?
    A. It makes clearer the author's argument
    B. It is employed as a device to convince the reader of the truth of the author's propositions
    C. It reinforces the idea that change is necessary
    D. It reinforces the idea that we can no longer support Big Media financially
    E. It suggests that the progression of broadcasting is no longer under our control

Passage 68 - Foreword for Authorama

Jonathan Dunn

Throughout history, all literature was in the public domain, but, in the United States, "intellectual property" is traded as if it were some sort of tangible commodity. This is especially shameful when one considers that the public domain is precisely what drives the advancement of society. As the technology to promulgate and store information increases, so too does the ability to use that information as a framework for future advances. It is unfortunate that as the physical obstacles are overcome, legal ones are created to replace them.

Intellectual property rights simply do not exist outside of man's legislation, and this type of law is, in my opinion, akin to protectionism. Let me explain: on one hand we constantly endeavour to improve transportation and the moving of goods. But as the obstacles to trade are eliminated, we find that trade is increased. In order to "protect" our labour, taxes or tariffs are placed on the products that are thus exchanged. It is essentially the same as building a highway between two cities in order to make travel less expensive and then charging a toll that entirely replaces the expense saved by the highway. In the end, it brings no increase in efficiency. The parallel is that while we have advanced methods of storing and promulgating information, we replace any advantage gained in that respect with legislation that restricts the flow of information (such as our oppressive copyright laws). On the one hand, the laws of nature no longer inhibit us from accumulating knowledge, but on the other hand, the laws of men make it more difficult than ever.

Until the digital revolution, intellectual property was rarely separated from its physical manifestation. By that, I mean that if you wanted to read a book, you bought the physical product and the price of the intellectual property was hidden within the price of its materials. But when the medium and the actual content became separated, suddenly the issue of intellectual property came into being. But how can a product be sold without a transfer of something? If I purchase an e-book, it costs the publisher nothing to sell one to someone else because, from the one original copy, an infinite amount of copies can be grafted. It costs something to produce the original e-book, perhaps, but I am merely buying a copy of it. First, we had a fiat currency, which the government can conjure up at whim and promulgate for profit, and now we also have a fiat product with which publishers can do the same.

The only argument against the public domain is the protection of the writers and artists and programmers who create the work in the first place. It should be noted, however, that in many cases they are not even the ones who own the copyrights. A poverty-stricken musician could, perhaps, argue that he needed copyrights to survive, but how can a corporation of people who did not produce the work in the beginning argue the same? "But", one says, "the artist sold the copyright to the corporation by his own will, and that is how he supports himself." Perhaps, but that is assuming that a piece of intellectual property – in essence, a thought – can be traded as if it were a physical entity as if there were no difference between it and a piece of land or a car. Yet, by its very definition, it is something that is not tangible, something that has no value outside of its communication with a human brain.

The basic question is this: is a thought a piece of property, is an entity that exists beyond its physical manifestation the same as one which only is its physical manifestation? If I buy a chair, I am paying for the materials and the labour that went into the individual chair. But if I buy an e-book or mp3, what am I paying for? There is no cost for materials and the same labour that went into making what I bought also went into what everyone else bought. If that were the measure of its price, e-books would cost an insignificant amount.

I will conclude by saying that it is shameful that there are so many children (and adults) around the world who have received an insufficient education for no other reason than that they couldn't afford it when all the knowledge of the world could be given away freely in a digital format. Knowledge is power, as they say, and right now knowledge is kept from the people through the oppression of copyright laws and the forces that maintain them.

345. Which of the following is not stated as an argument in favour of the public domain?
    A. The public domain drives the advancement of society
    B. Man-made laws make it more difficult than ever to accumulate information
    C. The abolition of copyright laws would enable the education of children
    D. Intellectual property is, by definition, something that is not tangible
    E. The abolition of copyright laws would empower the people

346. What is the main premise of the author's argument against the trading (for value) of intangible property?
    A. Thought cannot be viewed as something physical
    B. No costs or extra labour are incurred in the marginal production of such property
    C. Making intangible property widely available would advance the free flow of knowledge
    D. E-books are currently too expensive
    E. The wide availability of intangible property would remedy the problem of insufficient education

347. What point is made by the 'protectionism' example in the second paragraph?
    A. Restrictive laws tend to be inefficient
    B. Copyright laws offset the advantages we have gained in the storing and accumulating information
    C. Eliminating obstacles to trade is a useful means of increasing trade
    D. Society cannot advance if protectionist attitudes prevail
    E. The effects of protectionist behaviour are oppressive

348. Which of the following is merely an assertion?
    A. Intellectual property laws are akin to protectionism
    B. It is shameful that there are so many children (and adults) around the world who have received an insufficient education for no other reason than that they couldn't afford it
    C. The public domain is precisely what drives the advancement of society
    D. Intellectual property [should not] be traded as if it were a physical entity
    E. Knowledge is power

349. Which of the following is a logical conclusion of the argument in paragraph five?
    A. People should not be able to rent property for money
    B. Aeroplane tickets should be free
    C. Music CDs should be free
    D. Recipe books should be free
    E. None of the above

Passage 69 - History

Ralph Waldo Emerson

There is one mind common to all individual men. Every man is an inlet to the same and to all of the same. He that is once admitted to the right of reason is made a freeman of the whole estate. What Plato has thought, he may think; what a saint has felt, he may feel; what at any time has befallen any man, he can understand. Who hath access to this universal mind is a party to all that is or can be done, for this is the only and sovereign agent.

Of the works of this mind, history is the record. Its genius is illustrated by the entire series of days. Man is explicable by nothing less than all his history. Without hurry, without rest, the human spirit goes forth from the beginning to embody every faculty, every thought, and every emotion, which belongs to it, in appropriate events. But the thought is always prior to the fact; all the facts of history pre-exist in the mind as laws. Each law, in turn, is made by circumstances predominant and the limits of nature give power to but one at a time. A man is the whole encyclopaedia of facts. The creation of a thousand forests is in one acorn, and Egypt, Greece, Rome, Gaul, Britain, America, lie folded already in the first man. Epoch after epoch, camp, kingdom, empire, republic, democracy, is merely the application of his manifold spirit to the manifold world.

This human mind wrote history, and thus must read it. The Sphinx must solve her own riddle. If the whole of history is in one man, it is all to be explained from individual experience. There is a relation between the hours of our life and the centuries of time. Of the universal mind, each individual man is one more incarnation. All its properties consist in him. Each new fact in his private experience flashes a light on what great bodies of men have done, and the crises of his life refer to national crises. Every revolution was first a thought in one man's mind, and when the same thought occurs to another man, it is the key to that era. Every reform was once a private opinion, and when it shall be a private opinion again it will solve the problem of the age.

The fact narrated must correspond to something in me to be credible or intelligible. We, as we read, must become Greeks, Romans, Turks, priest and king, martyr and executioner; must fasten these images to some reality in our secret experience, or we shall learn nothing rightly. What befell Asdrubal or Caesar Borgia is as much an illustration of the mind's powers and depravations as what has befallen us. Each new law and political movement have meaning for you. This throws our actions into perspective; and as crabs, goats, scorpions, the balance and the waterpot lose their meanness when hung as signs in the zodiac, so I can see my own vices without heat in the distant persons of Solomon, Alcibiades, and Catiline.

All that Shakespeare says of the king, yonder slip of a boy that reads in the corner feels to be true of himself. We sympathise in the great moments of history, in the great discoveries, the great resistances, the great prosperities of men;—because there, law was enacted, the sea was searched, the land was found, or the blow was struck, for us, as we ourselves in that place would have done or applauded.

We have the same interest in condition and character. We honour the rich because they have externally the freedom, power, and grace that we feel to be proper to man, proper to us. So all that is said of the wise man by Stoic or Oriental or modern essayist, describes to each reader his own idea, describes his unattained but attainable self.

350. What, according to the author, is the main benefit of studying history?
- A. It allows us to learn about the lives of prominent historical figures
- B. It enables reform
- C. It has a meaning for us because it enables us to learn more about ourselves
- D. It is a puzzle which is interesting to solve
- E. It allows us to see what great men have done

351. What is meant by the statement: '*there is one mind common to all of mankind*'?
- A. People tend to 'follow the crowd'
- B. All humans have shared characteristics and values
- C. We all remember historical facts and are affected by them
- D. We can see our own flaws in the actions of others
- E. Our biological makeup is largely identical

352. Which of the following is a metaphor employed by the author?
- A. The blow was struck
- B. Crabs, goats, scorpions, the balance and the waterpot lose their meanness when hung as signs in the zodiac
- C. The Sphinx must solve her own riddle
- D. The whole of history is in one man
- E. A man is the whole encyclopaedia of facts

353. Which of the following best sums up the author's main argument?
- A. We sympathise in the great moments of history
- B. All humans have shared characteristics
- C. We should not judge wrongdoers as we ourselves have the same propensity for wrongdoing
- D. We can learn about and improve ourselves through a study of history
- E. Everyone aspires to have certain characteristics

354. What can be inferred from the author's repeated use of '*we*' in the final paragraph?
- A. Generalised statements can be made about the attitudes and values of all human beings
- B. Everyone deserves human rights
- C. Humans are identical
- D. All of the above
- E. None of the above

Passage 70 - Self-Reliance

Ralph Waldo Emerson

What I must do is all that concerns me, not what the people think. This rule, equally arduous in actual and in intellectual life, may serve for the whole distinction between greatness and meanness. It is harder because you will always find those who think they know what your duty is better than you know it. It is easy in the world to live after the world's opinion; it is easy in solitude to live after our own but the great man is he who in the midst of the crowd keeps with perfect sweetness the independence of solitude.

The objection to conforming to usages that have become dead to you is that it scatters your force. It loses your time and blurs the impression of your character. If you maintain a dead church, contribute to a dead Bible-society, vote with a great party either for the government or against it, spread your table like base housekeepers, —under all these screens I have difficulty to detect the precise man you are: and of course so much force is withdrawn from your proper life. Most men have bound their eyes with one or another handkerchief and attached themselves to some one of these communities of opinion. This conformity makes them not false in a few particulars, authors of a few lies, but false in all particulars. Their every truth is not quite true. Their two is not the real two, their four not the real four; so that every word they say chagrins us and we know not where to begin to set them right. Meantime nature is not slow to equip us in the prison-uniform of the party to which we adhere. We come to wear one cut of face and figure and acquire by degrees the gentlest asinine expression. There is a mortifying experience in particular, which does not fail to wreak itself also in the general history; I mean "the foolish face of praise," the forced smile which we put on in company where we do not feel at ease in answer to conversation which does not interest us. The muscles, not spontaneously moved but moved by a low usurping wilfulness, grow tight about the outline of the face with the most disagreeable sensation.

For nonconformity, the world whips you with its displeasure. And therefore, a man must know how to estimate a sour face. The by-standers look askance on him in the public street or in the friend's parlour. If this aversion had its origin in contempt and resistance like his own, he might well go home with a sad countenance; but the sour faces of the multitude, like their sweet faces, have no deep cause, but are put on and off as the wind blows and a newspaper directs. Yet is the discontent of the multitude more formidable than that of the senate and the college? It is easy enough for a firm man who knows the world to brook the rage of the cultivated classes. Their rage is decorous and prudent, for they are timid, as being very vulnerable themselves. But when to their feminine rage the indignation of the people is added, when the ignorant and the poor are aroused, when the unintelligent brute force that lies at the bottom of society is made to growl and mow, it needs the habit of magnanimity and religion to treat it godlike as a trifle of no concernment.

The other terror that scares us from self-trust is our consistency; a reverence for our past act or word because the eyes of others have no other data for computing our orbit than our past acts, and we loath to disappoint them. But why should you keep your head over your shoulder? Why drag about this corpse of your memory, lest you contradict somewhat you have stated in this or that public place? Suppose you should contradict yourself; what then? It seems to be a rule of wisdom never to rely on your memory alone, scarcely even in acts of pure memory, but to bring the past for judgment into the thousand-eyed present, and live ever in a new day.

355. Which of the following is, according to the author, the most difficult?
    A. To live according to one's opinion in solitude
    B. To follow one's own opinion whilst living amongst others
    C. To formulate one's own opinion
    D. To be given sour looks
    E. To conform to the world's expectations

356. What is most challenging about non-conformity according to the author?
    A. Disapproval of masses at bottom of society
    B. Being shunned by the privileged classes
    C. It saps one's life force
    D. It becomes hard for others to tell what sort of person you really are
    E. We end up hurting the feelings of others

357. Which of the following is an argument (by the author) against '*reverence for our past acts*'?
    A. It is difficult, in any case, to maintain consistency with one's past actions
    B. Whether or not we disappoint others is irrelevant
    C. Our priority shouldn't be to please others
    D. Being predictable is foolish
    E. Reflecting on the past is always unhelpful

358. Which of the following conclusions is best supported by the passage above?
    A. We should refuse to follow social conventions
    B. We should always reflect on our behaviour
    C. We should decide for ourselves if a convention is worth following
    D. We should not join religious groups or political parties
    E. It is best to avoid incurring the wrath of the masses

359. To what effect does the author employ the phrase '*corpse of your memory*'?
    A. To show that remembrance of past events is largely unhelpful for life in the present
    B. To encourage the reader to consider whether memories are in fact burdensome
    C. To show that recalling the past is tedious
    D. To show that memories of past events are often unpleasant
    E. To portray memory as the ghost of past events

Passage 71 - Seven Discourses on Art

Joshua Reynolds

The principal advantage of an academy is, that, besides furnishing able men to direct the student, it will be a repository for the great examples of the art. These are the materials on which genius is to work, and without which the strongest intellect may be fruitlessly or deviously employed. By studying these authentic models, that idea of excellence which is the result of the accumulated experience of past ages may be at once acquired, and the tardy and obstructed progress of our predecessors may teach us a shorter and easier way. The student receives at one glance the principles which many artists have spent their whole lives in ascertaining; and, satisfied with their effect, is spared the painful investigation by which they come to be known and fixed. How many men of great natural abilities have been lost to this nation for want of these advantages? They never had an opportunity of seeing those masterly efforts of genius that at once kindle the whole soul, and force it into sudden and irresistible approbation.

Raffaelle, it is true, had not the advantage of studying in an academy; but all Rome, and the works of Michael Angelo in particular, were to him an academy. On the site of the Capella Sistina, he immediately, from a dry, Gothic, and even insipid manner, which attends to the minute accidental discriminations of particular and individual objects, assumed that grand style of painting, which improves partial representation by the general and invariable ideas of nature.

Every seminary of learning may be said to be surrounded with an atmosphere of floating knowledge, where every mind may imbibe somewhat congenial to its own original conceptions. Knowledge, thus obtained, has always something more popular and useful than that which is forced upon the mind by private precepts or solitary meditation. Besides, it is generally found that a youth more easily receives instruction from the companions of his studies, whose minds are nearly on a level with his own, than from those who are much his superiors; and it is from his equals only that he catches the fire of emulation.
Impressed as I am, therefore, with such a favourable opinion of my associates in this undertaking, it would ill become me to dictate to any of them. But as these institutions have so often failed in other nations, and as it is natural to think with regret how much might have been done and how little has been done, I must take leave to offer a few hints, by which those errors may be rectified, and those defects supplied. These, the professors and visitors may reject or adopt as they shall think proper.

I would chiefly recommend that an implicit obedience to the rules of art, as established by the great masters, should be exacted from the <u>young</u> students. That those models, which have passed through the approbation of ages, should be considered by them as perfect and infallible guides as subjects for their imitation, not their criticism.

I am confident that this is the only efficacious method of making a progress in the arts; and that he who sets out with doubting will find life finished before he becomes a master of the rudiments. For it may be laid down as a maxim, that he who begins by presuming on his own sense has ended his studies as soon as he has commenced them. Every opportunity, therefore, should be taken to discountenance that false and vulgar opinion that rules are the fetters of genius. They are fetters only to men of no genius; as that armour, which upon the strong becomes an ornament and a defence, upon the weak and misshapen turns into a load, and cripples the body which it was made to protect.

360. What, according to the author, is the main benefit of having a 'repository for the great examples of the art'?
     A.   Students gain insight into useful principles without having to experiment to discover them for themselves
     B.   Students need not think for themselves
     C.   Students who don't have natural ability are able to thrive nonetheless
     D.   Future generations are able to appreciate and enjoy the great examples
     E.   To ensure that certain styles of art do not die out

361. What is the main premise of the view that students should not focus on criticising works of great masters?
     A.   The works of the great masters are infallible
     B.   They will produce work which departs from the style of the great masters
     C.   Those who presume to be better than the great masters will not be able to learn anything new
     D.   Only students without talent focus on criticism rather than just accepting rules of art
     E.   There has already been enough criticism of these works

362. Which of the following, if true, would most weaken the author's argument?
     A.   Raffaelle produced his best work before gaining access to the works of Michael Angelo
     B.   The works of Michael Angelo are highly valued today
     C.   Contemporaries of Raffaelle produced better work than him
     D.   Raffaelle was noted for his artistic talent even before gaining access to the works of Michael Angelo
     E.   Raffaelle did not esteem the art or architecture of Rome very highly

363. What purpose is served by the 'armour' illustration in the final paragraph?
     A.   To show that rules enhance the skills of those who possess artistic abilities, whilst further hindering those who lack those abilities
     B.   To show that the abilities of weak students can be improved if they dispense with rules
     C.   To show that art rules can protect the talented artist
     D.   To show that military life has similarities to art
     E.   None of the above

364. What is the underlying assumption of the passage?
     A.   Art students wish to improve
     B.   The value of a piece of art is objective
     C.   Students can learn more from the works of art than they can learn on their own
     D.   Students don't want to learn through their own mistakes
     E.   Studying the works of the great masters can improve a student's own work

Passage 72 - Discourses on Art: Discourse IV

Joshua Reynolds

Gentlemen, the value and rank of every art is in proportion to the mental labour employed in it, or the mental pleasure produced by it. As this principle is observed or neglected, our profession becomes either a liberal art or a mechanical trade. In the hands of one man it makes the highest pretensions, as it is addressed to the noblest faculties. In those of another, it is reduced to a mere matter of ornament, and the painter has but the humble province of furnishing our apartments with elegance.

This exertion of mind, which is the only circumstance that truly ennobles our art, makes the great distinction between the Roman and Venetian schools. I have formerly observed that perfect form is produced by leaving out particularities, and retaining only general ideas. I shall now endeavour to show that this principle, which I have proved to be metaphysically just, extends itself to every part of the art; that it gives what is called the grand style to invention, to composition, to expression, and even to colouring and drapery.

Invention in painting does not imply the invention of the subject, for that is commonly supplied by the poet or historian. With respect to the choice, no subject can be proper that is not generally interesting. It ought to be either some eminent instance of heroic action or heroic suffering. There must be something either in the action or in the object in which men are universally concerned, and which powerfully strikes upon the public sympathy.
Strictly speaking, indeed, no subject can be of universal, hardly can it be of general concern: but there are events and characters so popularly known in those countries where our art is in request, that they may be considered as sufficiently general for all our purposes. Such are the great events of Greek and Roman fable and history, which early education and the usual course of reading have made familiar and interesting to all Europe, without being degraded by the vulgarism of ordinary life in any country. Such, too, are the capital subjects of Scripture history, which, besides their general notoriety, become venerable by their connection with our religion.

As it is required that the subject selected should be a general one, it is no less necessary that it should be kept unembarrassed with whatever may anyway serve to divide the attention of the spectator. Whenever a story is related, every man forms a picture in his mind of the action and the expression of the persons employed. The power of representing this mental picture in canvas is what we call invention in a painter. And as in the conception of this ideal picture, the mind does not enter into the minute peculiarities of the dress, furniture, or scene of action, so when the painter comes to represent it, he contrives those little necessary concomitant circumstances in such a manner that they shall strike the spectator no more than they did himself in his first conception of the story.

365. Which of the following is the author likely to view as being of most value?
    A. A rough sketch by a court artist
    B. A sculpture of a hand
    C. A complicated scene planned for 10 years and painted across the ceiling of a huge cathedral
    D. A blank piece of canvas
    E. A detailed picture of a beach, produced by an artist immediately upon seeing noticing the beach

366. What is implied by the third paragraph?
    A. Works of art which depict merely personal subjective experiences are of lesser value
    B. Art should depict instances of heroism
    C. A piece of art is not valuable if a majority of people do not appreciate it
    D. Artists have little scope for choice of the subject of their paintings
    E. The nature of invention differs for the artist and for the poet

367. Which of the following is not stated to be a requirement of good art?
    A. Of universal concern
    B. Embodying a grand style
    C. The product of mental labour or producing mental pleasure
    D. Generally interesting
    E. Expressing general ideas

368. What is the assumption of the argument in the last paragraph?
    A. People's minds do not focus on tiny details
    B. The recording of reality is not the purpose of art
    C. Spectators are easily distracted by details
    D. Painters wish to be inventors
    E. Every painting tells a story

369. The author uses the phrase 'invention in painting' to mean/what meaning does the author give to the phrase?
    A. Inspiring poets and writers through one's paintings
    B. Passing across a message through one's paintings
    C. Developing new innovative ways of painting
    D. Representing a mental image using a painting on canvas
    E. Creating an idea in people's minds

Passage 73 - Patriotism: A Menace to Liberty

Emma Goldman

What is patriotism? Is it love of one's birthplace, the place of childhood's recollections and hopes, dreams, and aspirations? Is it the place where, in childlike naivety, we would watch the fleeting clouds and wonder why we, too, could not run so swiftly? The place where we would count the milliard glittering stars, terror-stricken lest each one "an eye should be," piercing the very depths of our little souls? Is it the place where we would listen to the music of the birds, and long to have wings to fly, even as they, to distant lands? Or the place where we would sit at mother's knee, enraptured by wonderful tales of great deeds and conquests? In short, is it love for the spot, every inch representing dear and precious recollections of a happy, joyous, and playful childhood?

If that were patriotism, few American men of today could be called upon to be patriotic, since the place of play has been turned into a factory, mill, and mine, while deafening sounds of machinery have replaced the music of the birds. Nor can we longer hear the tales of great deeds, for the stories our mothers tell today are but those of sorrow, tears, and grief.

What, then, is patriotism? "Patriotism, sir, is the last resort of scoundrels," said Dr. Johnson. Leo Tolstoy, the greatest anti-patriot of our times, defines patriotism as the principle that will justify the training of wholesale murderers; a trade that requires better equipment for the exercise of man-killing than the making of such necessities of life as shoes, clothing, and houses; a trade that guarantees better returns and greater glory than that of the average workingman.

Gustave Herve, another great anti-patriot, justly calls patriotism a superstition—one far more injurious, brutal, and inhumane than religion. The superstition of religion originated in man's inability to explain natural phenomena. That is, when primitive man heard thunder or saw the lightning, he could not account for either, and therefore concluded that back of them must be a force greater than himself. Similarly, he saw a supernatural force in the rain, and in the various other changes in nature. Patriotism, on the other hand, is a superstition artificially created and maintained through a network of lies and falsehoods; a superstition that robs man of his self-respect and dignity, and increases his arrogance and conceit.

Indeed, conceit, arrogance, and egotism are the essentials of patriotism. Let me illustrate. Patriotism assumes that our globe is divided into little spots, each one surrounded by an iron gate. Those who have had the fortune of being born on some particular spot, consider themselves better, nobler, grander, and more intelligent than the living beings inhabiting any other spot. It is, therefore, the duty of everyone living on that chosen spot to fight, kill, and die in the attempt to impose his superiority upon all the others.

The inhabitants of the other spots reason in like manner, of course, with the result that, from early infancy, the mind of the child is poisoned with blood-curdling stories about the Germans, the French, the Italians, Russians, etc. When the child has reached manhood, he is thoroughly saturated with the belief that he is chosen by the Lord himself to defend HIS country against the attack or invasion of any foreigner. It is for that purpose that we are clamouring for a greater army and navy, more battleships and ammunition. It is for that purpose that America has within a short time spent four hundred million dollars. Just think of it—four hundred million dollars taken from the produce of the people. For surely, it is not the rich who contribute to patriotism. They are cosmopolitans, perfectly at home in every land. We in America know well the truth of this. Are not our rich Americans Frenchmen in France, Germans in Germany, or Englishmen in England? And do they not squander with cosmopolitan grace fortunes coined by American factory children and cotton slaves? Yes, theirs is the patriotism that will make it possible to send messages of condolence to a despot like the Russian Tsar, when any mishap befalls him, as President Roosevelt did in the name of HIS people, when Sergius was punished by the Russian revolutionists.

370. How, according to the author, does patriotism differ from the '*superstition of religion*'?
    A. The latter is maintained through lies
    B. The former has the effect of robbing man of his self-respect
    C. The latter is more widespread
    D. The former is artificially created
    E. The former is more widespread

371. Which one of these is not stated by the author to be a feature of patriotism?
    A. It justifies the training of murderers
    B. It is a superstition
    C. Conceit and arrogance are its essentials
    D. It leads one group to try to impose their superiority on another
    E. It poisons children's minds

372. Which of the following is merely an opinion?
    A. America has within a short time spent four hundred million dollars
    B. The rich are cosmopolitans, perfectly at home in every land
    C. We are clamouring for a greater army and navy, more battleships and ammunition
    D. The stories our mothers tell today are of grief
    E. Sergius was punished by the Russian revolutionists

373. Who, according to the last paragraph, bears the principal detriment of patriotism?
    A. Only factory children and cotton slaves
    B. The rich
    C. Children
    D. The people
    E. The army

374. What word best sums up the way in which the rich are portrayed by the author in the final paragraph?
    A. Non-conformist
    B. Inconsiderate
    C. Hypocritical
    D. Stingy
    E. Selfish

Passage 74 - Friendship

Ralph Emerson Waldo

Friendship, like the immortality of the soul, is too good to be believed. The lover, beholding his maiden, half knows that she is not verily that which he worships; and in the golden hour of friendship, we are surprised with shades of suspicion and unbelief. We doubt that we bestow on our hero the virtues in which he shines. Shall we fear to cool our love by mining for the metaphysical foundation of this Elysian temple? Shall I not be as real as the things I see? If I am, I shall not fear to know them for what really they are. Their true essence is not less beautiful than their appearance, though it needs finer organs for its appreciation.

The law of nature is alternation forevermore. Each electrical state super-induces the opposite. The soul environs itself with friends that it may enter into a grander self-acquaintance or solitude; and it goes alone, for a season, that it may exalt its conversation or society. This method betrays itself along the whole history of our personal relations. The instinct of affection revives the hope of union with our mates, and the returning sense of insulation recalls us from the chase. Thus, every man passes his life in the search after friendship, and if he should record his true sentiment, he might write a letter like this, to each new candidate for his love.

> Dear Friend:—
> If I was sure of thee, sure of thy capacity, sure to match my mood with thine, I should never think again of trifling flaws. Thou art very wise; thy moods are quite attainable; and I respect thy genius; it is to me as yet unfathomed; yet dare I presume in thee a perfect match for me? And so thou art to me a delicious torment. Thine ever, or never.

Yet this form of worship, these uneasy pleasures, and fine pains are for curiosity, and not for life. They are not to be indulged. This is to weave cobweb and not cloth. Our friendships hurry to short and poor conclusions because we have made them a texture of wine and dreams, instead of the tough fibre of the human heart. The laws of friendship are great, austere, and eternal, of one web with the laws of nature and of morals. But we have aimed at a swift and petty benefit, to suck a sudden sweetness. We snatch at the slowest fruit in the whole garden of God, which many summers and many winters must ripen. We seek our friend not sacredly but with an adulterate passion which would appropriate him to ourselves.

375. What is meant by '*is too good to be believed*'?
- A.   Friendship is a beautiful thing
- B.   Friendship is a sham
- C.   We are unrealistic in our expectation of friends
- D.   We view our friends in an extraordinarily positive light
- E.   We are unrealistic in expecting to have perfect friendships

376. Which of the following is a simile employed by the author?
- A.   Their true essence is not less beautiful than their appearance
- B.   Each electrical state super-induces the opposite
- C.   Like the immortality of the soul
- D.   This method betrays itself
- E.   Thou art to me a delicious torment

377. What is meant by the phrase '*we snatch at the slowest fruit in the whole garden of God*'?
- A.   We always want what is forbidden
- B.   We are not patient enough to cultivate a deep friendship and want to obtain it instantly
- C.   We desire things which seem attractive outwardly
- D.   Friendship takes a long time to cultivate
- E.   We make poor friendship choices

378. Which of the following are not presented as contrasting pairs?
- A.   Swift and petty
- B.   Sacredly and passionately
- C.   Eternal laws and sudden sweetness
- D.   Cobweb and cloth
- E.   None of the above

379. What is the main argument of the first paragraph?
- A.   Friendship is too good to be believed
- B.   We attribute qualities to our friends which they do not possess
- C.   Friendship is a sham
- D.   We often falsely worship our friends
- E.   We should not be afraid to confront and explore our friends' true characteristics

Passage 75- Gifts

Ralph Emerson Waldo

The law of benefits is a difficult channel, which requires careful sailing, or rude boats. It is not the office of a man to receive gifts. How dare you give them? We wish to be self-sustained. We do not quite forgive a forgiver. The hand that feeds us is in some danger of being bitten. We can receive anything from love, for that is a way of receiving it from ourselves (hence the fitness of beautiful, not useful things for a gift); but not from anyone who assumes to bestow. We sometimes hate the meat that we eat, because there seems something of degrading dependence in living by it.

He is a good man, who can receive a gift well. We are either glad or sorry at a gift, and both emotions are unbecoming. Some violence, I think, is done, some degradation borne, when I rejoice or grieve at a gift. I am sorry when my independence is invaded, or when a gift comes from such as do not know my spirit, and so the act is not supported; and if the gift pleases me overmuch, then I should be ashamed that the donor should read my heart, and see that I love his commodity, and not him.

This giving is flat usurpation, and therefore when the beneficiary is ungrateful, as all beneficiaries hate all Timons, not at all considering the value of the gift, but looking back to the greater store it was taken from, I rather sympathise with the beneficiary, than with the anger of my lord, Timon. For, the expectation of gratitude is mean and is continually punished by the total insensibility of the obliged person. It is a great happiness to get off without injury and heart-burning, from one who has had the ill luck to be served by you. It is a very onerous business, this of being served, and the debtor naturally wishes to give you a slap. A golden text for these gentlemen is that which I admire in the Buddhist, who never thanks, and who says, "Do not flatter your benefactors."

The reason of these discords I conceive to be, that there is no commensurability between a man and any gift. You cannot give anything to a magnanimous person. After you have served him, he at once puts you in debt by his magnanimity. The service a man renders his friend is trivial and selfish, compared with the service he knows his friend stood in readiness to yield him, alike before he had begun to serve his friend, and now also. Compared with that good-will I bear my friend, the benefit it is in my power to render him seems small. Besides, our action on each other, good as well as evil, is so incidental and at random, that we can seldom hear the acknowledgments of any person who would thank us for a benefit, without some shame and humiliation. We can rarely strike a direct stroke, but must be content with an oblique one; we seldom have the satisfaction of yielding a direct benefit, which is directly received. But rectitude scatters favours on every side without knowing it and receives with wonder the thanks of all people.

380. In the third paragraph, the author's main argument is that:
    A. It is unkind to expect gratitude
    B. Gift givers should not expect gratitude on the part of receivers
    C. Receivers of gifts are angered by givers
    D. All beneficiaries are like Timon
    E. Receivers of gifts can be violent

381. The writer uses the term '*magnanimous*' to mean:
    A. Easy-going
    B. Wealthy
    C. Meek
    D. Generous
    E. Noble

382. Why, according to the author, are we unworthy of being thanked for conferring a benefit on someone else?
    A. Receiving thanks is shameful
    B. We tend to succeed in conferring benefits only by luck
    C. We are obliged, in any case, to be altruistic
    D. The benefits we confer are insignificant
    E. The benefits are unsolicited

383. What is implied by the line '*He is a good man who can receive a gift well*'?
    A. Most people cannot receive a gift well
    B. Those who cannot receive a gift well are bad men
    C. Good men receive more gifts
    D. The author is able to receive gifts graciously
    E. Graciousness is a quality which is not found in abundance amongst people

384. Why might it be acceptable to receive a gift bestowed out of love?
    A. Such givers mean no harm
    B. Gift-giving is one of the main ways of expressing love
    C. The intention of the giver is not to provide for the recipient
    D. Beneficiaries who feel loved are less likely to be ungrateful
    E. Gifts given out of love are more likely to be beautiful

Passage 76- Prudence

Ralph Emerson Waldo

What right have I to write on Prudence, whereof I have little, and that of the negative sort? My prudence consists in avoiding and going without, not in the inventing of means and methods, not in adroit steering, not in gentle repairing. I have no skill to make money spend well, no genius in my economy, and whoever sees my garden discovers that I must have some other garden. Yet, I love facts and hate lubricity and people without perception. Then I have the same title to write on prudence that I have to write on poetry or holiness. We write from aspiration and antagonism, as well as from experience. We paint those qualities which we do not possess. The poet admires the man of energy and tactics; the merchant breeds his son for the church or the bar; and where a man is not vain and egotistic, you shall find what he has not by his praise.

There are all degrees of proficiency in knowledge of the world. It is sufficient to our present purpose to indicate three. One class lives to the utility of the symbol, esteeming health and wealth a final good. Another class live above this mark of the beauty of the symbol, as the poet and artist and the naturalist and man of science. A third class live above the beauty of the symbol to the beauty of the thing signified; these are wise men. The first class have common sense; the second, taste; and the third, spiritual perception. Once in a long time, a man traverses the whole scale, and sees and enjoys the symbol solidly, then also has a clear eye for its beauty, and lastly, whilst he pitches his tent on this sacred volcanic isle of nature, does not offer to build houses and barns thereon reverencing the splendour of the God which he sees bursting through each chink and cranny.

The world is filled with the proverbs and acts and winkings of a base prudence, which is a devotion to matter, as if we possessed no other faculties than the palate, the nose, the touch, the eye, and ear; a prudence which adores the Rule of Three, which never subscribes, which gives never, which seldom lends, and asks but one question of any project—Will it bake bread? This is a disease like a thickening of the skin until the vital organs are destroyed. But culture, revealing the high origin of the apparent world and aiming at the perfection of the man as the end, degrades everything else, as health and bodily life, into means. It sees prudence not to be a several faculty, but a name for wisdom and virtue conversing with the body and its wants. Cultivated men always feel and speak so as if a great fortune, the achievement of a civil or social measure, great personal influence, a graceful and commanding address, had their value as proofs of the energy of the spirit. (But this is not so.)

385. In which of the following does the writer employ irony?
    A. I have little prudence
    B. I have no skill to make money spend well
    C. We paint those qualities which we do not possess
    D. The world is filled with the proverbs
    E. None of the above

386. What purpose is served by the illustration: '*the merchant breeds his son for the church or the bar*'?
    A. It shows that we esteem qualities which we ourselves do not possess
    B. People want their children to have different life experiences to them
    C. Parents often restrict their children's freedom of choice
    D. It shows that there are very few career options for those of certain backgrounds
    E. It suggests that the most prudent career choices lie in the church or the bar

387. What does it mean to: '*live above the beauty of the symbol to the beauty of the thing signified*'?
    A. To seek to find beauty in everything
    B. To be wise
    C. To appreciate nature
    D. To be sensual
    E. To focus on the spiritual significance of physical things

388. Which of the following best summarises the author's main argument?
    A. There are different forms of prudence
    B. Those who have the least prudence aspire to it the most
    C. Prudence involves being spiritual
    D. We don't need to know how to use money economically in order to be prudent
    E. Prudence is difficult to obtain

389. Which of the following words does the author use to signify disapproval?
    A. Disease
    B. The palate
    C. Several
    D. Commanding
    E. Apparent

Passage 77 - Russia's part in the war

M. Shumsky-Solomonov

In discussing Russia's role in the past World War, it is customary to cite the losses sustained by the Russian Army, losses numbering many millions. There is no doubt that Russia's sacrifices were great, and it is just as true that her losses were greater than those sustained by any of the other Allies. Nevertheless, these sacrifices are by far not the only standard of measurement of Russia's participation in this gigantic struggle. Russia's role must be gauged, first of all, by the efforts made by the Russian Army to blast the German war plans during the first years of the War, when neither America, nor Italy, nor Romania were among the belligerents, and the British Army was still in the process of formation.

[Secondly], and this is the main thing, the role played by the Russian Army must be considered also in this respect that the strenuous campaign waged by Russia, with her 180 millions of inhabitants, for three years against Germany, Austro-Hungary and Turkey, sapped the resources of the enemy and thereby made possible the delivery of the final blow. This weakening of the powers of the enemy by Russia was already bound at various stages of the War to facilitate correspondingly the various operations of the Allies. Therefore, at the end of the War three years of effort on the part of Russia, which had devoured the enemy's forces, were destined to enable the Allies to finally to crush the enemy. The final catastrophe of the Central Powers was the direct consequence of the offensive of the Allies in 1918, but Russia made possible this collapse to a considerable degree, having effected, in common with the others, the weakening of Germany, and having consumed during the three years of strenuous fighting countless reserves, forces and resources of the Central Powers.

Could Germany have won the war? A careful analysis of this question brings home the conviction that Germany was very close to victory, and that it required unusual straining of efforts on the part of France and Russia to prevent Germany from "winning out."

The plan of the old Field Marshal, Moltke, was far from worthless. It is a fact that it took from six weeks to two months to mobilise the armed forces of Russia, during which period Russia was unprepared for action. The population of Germany was 70 million and that of Austria-Hungary 52 million, a total of 122 million persons. During these two months of forced inaction, those 122 million of Teutons were faced only by 40 million Frenchmen, for Russia was not yet ready. A threefold superiority in numbers, in addition to an equal degree of military skill, technical equipment, and culture, was bound to crush lone France.

The outcome was different. The concentrated attack upon France failed because of the fact that of the 104 German divisions and the 50 Austrian divisions, only about 92 or 94 divisions were on the scene of action in France. The Russian Army, unprepared for action for another 40 days, nevertheless rushed into East Prussia in an impulse of self-sacrifice and received, in addition, the full strength of the blow from the Austro-Hungarian Army. This generous move on the part of Russia destroyed the Moltke plan and his basic idea "the concentration of *all forces* against France", as a part of the German force had been diverted from that front. The plan collapsed, and the only actual chance that the Germans had of winning a victory was lost with it. Later, when Russia was prepared, when the English Army began to grow and Italy, Romania and America had abandoned their neutrality, Germany's chances for a final victory vanished.

It is the recognition of these facts that should prompt every impartial historian of the War to admit that the self-sacrifice of the unprepared Russian Army during the first days of the War played an enormous role in the only period when Germany had victory almost within her grasp. It is to be regretted that the extraordinary conditions that developed in Russia towards the end of the War are obscuring the true historic role of Russia in the sanguine World struggle. It is just as easy—from an examination of the maps of the first three years of the War, maps which speak only of two principal fronts, the French and the Russian, and no other—to grasp the significance of the gigantic role played in this War by great Russia and the millions of sacrifices she consecrated to the common cause of the Allies. Sadly enough, this only correct criterion of Russia's historic role in the War is becoming more and more obscured from the public opinion of the world.

390. What is the main conclusion in the first paragraph?
   A. Russia's sacrifices were great
   B. The losses sustained by the Russian Army should not be the primary means of gauging their role in the war
   C. Russia made victory possible for the allies
   D. The Russian army's losses numbered millions
   E. Russia's losses were greater than those of the other allies

391. What is the main premise of the argument that Russia had a significant role in the defeat of Germany?
   A. Russia's efforts ensured that the German army was significantly weakened
   A. Russia delivered the final blow to Germany
   B. Russia received the full strength of the blow from the Austro-Hungarian army
   C. Russia rushed into the war although they were unprepared
   D. The Russian army was the only army large enough to confront Germany

392. Which one of the following statements does the least to support the argument that Russia had a significant role in the defeat of Germany?
   A. Maps of the first three years of the War speak only of two principal fronts, the French and the Russian, and no other
   B. Russia destroyed the Moltke plan
   C. Russia weakened the powers of the enemy
   D. The Russian army acted in an impulse of self-sacrifice
   E. The concentrated attack upon France failed because of the fact that of the 104 German divisions and the 50 Austrian divisions, only about 92 or 94 divisions were on the scene of action in France

393. Which of the following, if true, would most significantly weaken the author's argument?
   A. Very few Germans died during conflict with Russia
   B. The Russian army lost more men than the German army
   C. The other allies lost more men than the Russian army
   D. The efforts of the other allied nations were essential to the defeat of Germany
   E. France played a minimal role in the attaining of victory

394. What is implied by the phrase '*It is the recognition of these facts that should prompt every impartial historian of the War to admit…*'
   A. Historians ought to base their views on those facts
   B. Historians are all aware of the facts
   C. The facts alone are very compelling
   D. Only biased historians fail to concede the significance of Russia's role in the war
   E. Many historians have an anti-Russian bias

Passage 78 – Goethe's Theory of Colour: Translator's Preface

Charles Lock Eastlake

English writers who have spoken of Goethe's "Doctrine of Colours," have generally confined their remarks to those parts of the work in which he has undertaken to account for the colours of the prismatic spectrum, and of refraction altogether, on principles different from the received theory of Newton. The less questionable merits of the treatise consisting of a well-arranged mass of observations and experiments, many of which are important and interesting, have thus been in a great measure overlooked. The translator, aware of the opposition which the theoretical views alluded to have met with, intended at first to make a selection of such of the experiments as seem more directly applicable to the theory and practice of painting. Finding, however, that the alterations this would have involved would have been incompatible with a clear and connected view of the author's statements, he preferred giving the theory itself, reflecting, at the same time, that some scientific readers may be curious to hear the author speak for himself even on the points at issue.

In reviewing the history and progress of his opinions and research, Goethe tells us that he first submitted his views to the public in two short essays entitled "Contributions to Optics." Among the circumstances which he supposes were unfavourable to him on that occasion, he mentions the choice of his title, observing that by a reference to optics he must have appeared to make pretensions to a knowledge of mathematics, a science with which he admits he was very imperfectly acquainted. Another cause to which he attributes the severe treatment he experienced was in having ventured so openly to question the truth of the established theory: but this last provocation could not be owing to mere inadvertence on his part; indeed, the larger work, in which he alludes to these circumstances, is still more remarkable for the violence of his objections to the Newtonian doctrine.

There can be no doubt, however, that much of the opposition Goethe met with was to be attributed to the manner as well as to the substance of his statements. Had he contented himself with merely detailing his experiments and showing their application to the laws of chromatic harmony, leaving it to others to reconcile them as they could with the pre-established system, or even to doubt in consequence, the truth of some of the Newtonian conclusions, he would have enjoyed the credit he deserved for the accuracy and the utility of his investigations. As it was, the uncompromising expression of his convictions only exposed him to the resentment or silent neglect of a great portion of the scientific world, so that for a time he could not even obtain a fair hearing for the less objectionable or rather highly valuable communications contained in his book. A specimen of his manner of alluding to the Newtonian theory will be seen in the preface.

It was quite natural that this spirit should call forth a somewhat vindictive feeling and with it not a little uncandid as well as unsparing criticism. "The Doctrine of Colours" met with this reception in Germany long before it was noticed in England, where a milder and fairer treatment could hardly be expected especially at a time when, owing perhaps to the limited intercourse with the continent, German literature was far less popular than it is at present. This last fact, it is true, can be of little importance in the present instance, for although the change of opinion with regard to the genius of an enlightened nation must be acknowledged to be beneficial, it is to be hoped there is no fashion in science, and the translator begs to state once for all, that in advocating the neglected merits of the "Doctrine of Colours," he is far from undertaking to defend its imputed errors. Sufficient time has, however, now elapsed since the publication of this work (in 1810) to allow a calmer and more candid examination of its claims. In this more pleasing task, Germany has again for some time led the way, and many scientific investigators have followed up the hints and observations of Goethe with a due acknowledgment of the acuteness of his views.

395. Why did the writer decide against selecting only certain parts of Goethe's views for showcasing?
    A. In order to maintain honesty
    B. Goethe himself preferred to state his views in their entirety
    C. The views are more convincing in their entirety
    D. It would have obscured the clarity of the author's statements
    E. All of Goethe's views are useful to the practice of painting

396. Which of the following was not stated by Goethe to be a reason for the unfavourable treatment of his work?
    A. His uncompromising expression of his convictions
    B. His choice of title
    C. The fact that he openly ventured to question the truth of the established theory
    D. His lack of mathematical knowledge
    E. None of the above

397. Which of the following, if true, would most strengthen the author's main argument?
    A. Goethe's work met the most success in France
    B. Goethe's contemporaries, who published similar theories in more tentative language, were hailed as geniuses
    C. Most of Goethe's contemporaries were conservative in their approach to science
    D. All of Goethe's work was expressed in uncompromising terms
    E. Most of Goethe's work was unaccepted by the scientific community

398. The author uses the phrase '*fashion in science*' to mean:
    A. Popularity contests in the world of scientists
    B. Changes in scientific opinion that are not based on reason
    C. Periodic changes in scientific opinion
    D. The way in which things are done in the scientific community
    E. Competition amongst scientists

399. Which of the following does the author use to convey his approval?
    A. Popular
    B. Due
    C. Imputed
    D. Milder
    E. Sufficient

Passage 79 - Town Life in the Fifteenth Century

JR Green

There is nothing in England today with which we can compare the life of a fully enfranchised borough of the fifteenth century. The town of those earlier days in fact governed itself after the fashion of a little principality. Within the bounds which the mayor and citizens defined with perpetual insistence in their formal perambulation year after year it carried on its isolated self-dependent life.

The inhabitants defended their own territory, built and maintained their walls and towers, armed their own soldiers, trained them for service, and held reviews of their forces at appointed times. They elected their own rulers and officials in whatever way they themselves chose to adopt, and distributed among officers and councillors just such powers of legislation and administration as seemed good in their eyes. They drew up formal constitutions for the government of the community, and as time brought new problems and responsibilities, more were made, re-made and revised; again their ordinances with restless and fertile ingenuity, till they had made of their constitution a various medley of fundamental doctrines and general precepts and particular rules, somewhat after the fashion of an American state of modern times.

In all concerns of trade, they exercised the widest powers, and bargained and negotiated and made laws as nations do on a grander scale today. They could covenant and confederate, buy and sell, deal and traffic after their own will; they could draw up formal treaties with other boroughs, and could admit them to or shut them out from all the privileges of their commerce; they might pass laws of protection or try experiments in free trade. Often, their authority stretched out over a wide district and surrounding villages gathered to their markets and obeyed their laws; it might even happen in the case of a staple town that their officers controlled the main foreign trade of whole provinces.

Four hundred years later, the very remembrance of this free and vigorous life was utterly blotted out. When Commissioners were sent in 1835 to enquire into the position of the English boroughs, there was not one community where the ancient traditions still lived. There were Mayors, and Town Councils, and Burgesses; but the burgesses were for the most part deprived of any share whatever in the election of their municipal officers while these officers themselves had lost all the nobler characteristics of their former authority. Too often the very limits of the old "liberties" of the town were forgotten; or if the ancient landmarks were remembered at all it was only because they defined bounds within which the inhabitants had the right of voting for a member of Parliament; and in cases where the old boundaries now subsisted for no other reason, it was wholly forgotten that they might ever have had some other origin.

There were, it is true, exceptions to this common apathy and towns like Lynn might still maintain some true municipal life while others like Bristol might yet show a good fighting temper which counted for much in the political struggles of the early nineteenth century. But the ordinary provincial burghers had lost, or forgotten, or been robbed of the heritage bequeathed by their predecessors of the fifteenth century. With the loss of their municipal independence went the loss of their political authority; and the four hundred or so of members whom they sent to Parliament took a very different position there from that once held by their ancestors. In the Middle Ages, the knights of the shire were the mere nominees of the wealthy or noble class, returned to Parliament by the power of the lord's retainers, while the burgesses of the towns preserved a braver and freer tradition. At the time of the Reform Bill, on the other hand, a vast majority of the town members sat among the Commons as dependents and servants of the landed aristocracy, whose mission it was to make the will of their patrons prevail, and who in their corrupt or timid subjection simply handed back to the wealthier class the supreme political power which artisans and shopkeepers and "mean people" of the medieval boroughs had threatened to share with them.

400. Which of the following is stated to be the ultimate reason for the prevailing apathy towards political freedom in towns?
    A. The loss of independent political authority
    B. The manipulation of the voting system
    C. The Reform Bill
    D. Servitude to the landed aristocracy
    E. The loss of municipal independence

401. What is meant by the phrase '*little principality*'?
    a. A small region
    b. A politically independent state
    c. Local government
    d. A tribe
    e. An army

402. In what way was former town life '*free and vigorous*'?
    A. The reform bill had not yet been passed
    B. The citizens could define the bounds of the town
    C. The towns could bargain and negotiate trade laws
    D. They had established peace treaties with other towns for their protection
    E. They could impose their rule on other towns

403. Which of the following, if true, would most weaken the author's argument?
    A. The emergence of political apathy in towns preceded the loss of their municipal independence
    B. Some knights during the Middle Ages were not subject to the control of the lords
    C. Some towns still maintain vigorous municipal life
    D. Some town inhabitants would like to return to the way things were in times past
    E. The reform bill had little impact on the political status of towns

404. Which is the best explanation for why the author uses quotation marks in the final line?
    A. He wishes to challenge those who use the term
    B. In order to use a term commonly used without himself endorsing it
    C. Because he is quoting another person/other persons
    D. To emphasise the difference between the mean people and the landed aristocracy
    E. All of the above

Passage 80 - Counter-Terrorism Laws and Human Rights

Melody Ihuoma

During the 1960s and 70s, terrorism was a contemporary subject due to the conflict between the UK government and the IRA. 'The Troubles' (the name given to the violence) originated in the 1920s and eventually resulted in bombings on the streets of Northern Ireland and occasionally in England. The government felt that in order to prevent mayhem their actions needed to be swift and decisive. Thus, a series of temporary measures were initiated; policemen and soldiers all over Ulster were given the right to stop, question, search and arrest members of the public.

In 2000, the Terrorism Act 2000 was passed as a definitive measure following twenty years of temporary measures. Policemen were given wider stop and search powers and enabled to detain suspects for up to 48 hours without charge. The Act was met with strong criticism as it outlawed certain Islamic fundamentalist groups and this was seen as a portrayal of Islam as a religion that fuels terrorism. This, in turn, made it likely that discrimination would occur in the form of the disproportionate stopping and searching of Asians who were thought to 'look Muslim'.

Although, prior to the September 2001 attacks on Washington D.C, government legislation in the UK had attempted to prevent the occurrence of terrorism; the counter-terrorism strategies had focused a lot of attention on the punishment of terrorists and the criminalisation of new offences following their occurrence. However, the Anti-Terrorism Crime and Security Act 2001 (ATCSA) marked a more firm move towards the 'management of anticipatory risk' (Piazza, Walsh, 2010) which was to characterise the counter-terrorism legislation of the 21st century.

In December 2001, the setting aside of Article 5 of the Human Rights Act (by the Terrorism Act 2000) was found to be incompatible with the Human Rights Act (HRA). Lord Bingham led the House of Lords in the judgement that the law aimed to combat the threat posed to the UK by foreign nationals yet ignored the threat posed by British citizens and was thereby, 'ipso facto discriminatory'. This ruling is a clear indication that counter-terrorism laws have a tendency to cause certain groups of people to suffer the infringement of their human rights more so than others.

Criticism of this piece of legislation also came from the Human Rights Watch. In March 2003, the group stated that the September 2001 attacks resulted in legislation within the UK which undermined fundamental human rights and that the UK was the only member of the Council of Europe to have departed from the ECHR whilst passing counter-terrorism laws.

405. How do the counter-terrorism laws passed after 9/11 differ to their predecessors?
    A.   They punish terrorists more harshly
    B.   They focus more on preventing the occurrence of terrorist attacks
    C.   They were more definitive measures
    D.   They were more discriminatory against certain groups of people
    E.   They were driven by fear

406. What is implied by the fact that the UK was the only council member to depart from the ECHR?
    A.   A departure may not have been necessary for the protection of the UK
    B.   Other nations were under pressure to stay within the terms of the ECHR
    C.   The ECHR is too rigid
    D.   The threat of terrorism to the UK was higher than to the other council members
    E.   The British government cares less about human rights

407. Which of the following are not presented as contrasting pairs in the passage?
    A.   Temporary and definitive measures
    B.   Punishment and management of risk
    C.   The UK and other members of the council of Europe
    D.   UK foreign nationals and British citizens
    E.   The Terrorism Act 2000 and the Human Rights Act

408. Which of the following, if true, would most weaken the author's argument in paragraph four?
    A.   Minority groups such as the Chinese have not been unfairly affected by terrorism legislation
    B.   Many parts of the 2000 Act were compatible with the HRA
    C.   A disproportionately large number of foreign nationals are stopped and searched
    D.   The 2000 Act is the only piece of terrorism legislation that has been found to be incompatible with the HRA
    E.   Foreign nationals in the UK receive many benefits

409. Which one of the following is stated as a possible cause of draconian counter-terrorism legislation?
    A.   The need to combat the threat posed by foreign nations
    B.   The government felt that in order to prevent mayhem their actions needed to be swift and decisive
    C.   The side-lining of the HRA
    D.   The portrayal of Islam as a religion which fuels terrorism
    E.   Giving policeman in Ulster the right to stop, search, and arrest members of the public

Section B

The ultimate goal of any essay is to convey an argument to the reader. In order to do that, the essay needs to be as clear as possible, follow a logical structure, and develop a coherent argument. Even though you do not get your mark back from the essay, it is read directly by the admissions tutors at the LNAT Universities you have applied to, which will have a weighting on your application's chances of success.

In the exam, you will have 45 minutes to write the essay.

The key to creating a solid essay in the exam is to develop a good, persuasive argument in clear written English. It is **not** about writing as much as you can – indeed, some of the best essays are the shortest; and a rambling essay can attract low marks.

Ultimately, the examiners are testing your **ability to argue** and **not** particularly on your knowledge. That being said, having a good general knowledge will help you create good arguments and will stand you well for the exam. Crucially, it means that you'll be comfortable answering the questions in the exam.

Structuring your Essay

The structure of an essay consists of 3 parts:
1. Introduction
2. Main Body
3. Conclusion

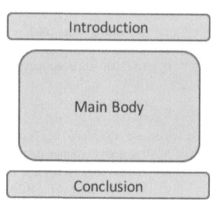

This is a well-known structure and while it is <u>not</u> necessary to give headings or to say that you're writing your introduction, keeping your essay in this format will be more clear and understandable.

A well-known saying is that: In your introduction, say what you're going to say; in the main body, you say it and in your conclusion, say what you've already said by bringing it all together.

The Exam Approach

Most students think that the "writing" component is most important. This is simply not true.

The vast **majority of problems are caused by a lack of planning and essay selection** – usually, because students just want to get writing as they are worried about finishing on time. 45 minutes is long enough to be able to plan your essay well and *still* have time to write it, so don't feel pressured to immediately start writing.

Step 1: Selecting

You will be given a choice of 3 essays to choose from and crucially, you will have no idea of what it could be beforehand. Selecting your essay is crucial- make sure you're comfortable with the topic and ensure you understand the actual question- it sounds silly but about 25% of essays that we mark score poorly because they don't actually answer the question!

Take two minutes to read all the questions. Whilst one essay might originally seem the easiest, if you haven't thought through it, you might quickly find yourself running out of ideas. Likewise, a seemingly difficult essay might actually offer you a good opportunity to make interesting points.

Use this time to carefully select which question you will answer by gauging how accessible and comfortable you are with it given your background knowledge.

It's surprisingly easy to change a question into something similar, but with a different meaning. Thus, you may end up answering a completely different essay title. Once you've decided which question you're going to do, read it very carefully a few times to make sure you fully understand it. Answer all aspects of the question. Keep reading it as you answer to ensure you stay on track!

Step 2: Planning

Why should I plan my essay?

There are multiple reasons you should plan your essay for the first 5 minutes of Section B:

- As you don't have much space to write, make the most of it by writing a very well-organised essay.
- It allows you to get all your thoughts ready before you put pen to paper.
- You'll write faster once you have a plan.
- You run the risk of missing the point of the essay or only answering part of it if you don't plan adequately.

How much time should I plan for?

There is no set period of time that should be dedicated to planning, and everyone will dedicate a different length of time to the planning process. You should spend as long planning your essay as you require, but it is essential that you leave enough time to write the essay. As a rough guide, it is **worth spending about 5-10 minutes to plan** and the remaining time on writing the essay. However, this is not a strict rule, and you are advised to tailor your time management to suit your individual style.

How should I go about the planning process?

There are a variety of methods that can be employed in order to plan essays (e.g. bullet-points, mind-maps etc). If you don't already know what works best, it's a good idea to experiment with different methods.

Generally, the first step is to gather ideas relevant to the question, which will form the basic arguments around which the essay is to be built. You can then begin to structure your essay, including the way that points will be linked. At this stage, it is worth considering the balance of your argument, and confirming that you have considered arguments from both sides of the debate. Once this general structure has been established, it is useful to consider any examples or real world information that may help to support your arguments. Finally, you can begin to assess the plan as a whole and establish what your conclusion will be based on your arguments.

How do I plan my essay?

Different methods work best for different students, but some are as follows:

- ➢ A mind-map
- ➢ Bullet-points
- ➢ A side by side list of PROS and CONS

Step 3: Writing

Introduction

The introduction should explain the statement and define any key terms. Here, you can say what you're going to say and suggest (either affirmatively or tentatively) a response or answer to the question.

It is important not to spend too long on an introduction as that would use up too much time unnecessarily, which could be better spent on other parts of the essay.

Main Body

The main body is where you discuss your arguments, consider counter arguments or consider the pros and cons of a particular statement or policy position.

In particular, while you may have numerous ideas, it is generally better to spend more time developing and evaluating fewer points, rather than listing as many points as possible and not going into much depth on each point.

Just like in GCSE English, using the Point-Evidence-Evaluation technique can help ensure you develop and deploy your ideas more fully.

In particular, using relevant examples where you can will help bolster your argument and provide for a more persuasive essay. However, it is crucial that real world examples are only used if they fit in with your argument – otherwise, it adds nothing and will not gain you marks.

How do I go about making a convincing point?

Each idea that you propose should be supported and justified in order to build a convincing overall argument. A point can be solidified through a basic Point à Evidence à Evaluation process. By following this process, you can be assured each sentence within a paragraph builds upon the last and that all the ideas presented are well solidified.

How do I achieve a logical flow between ideas?

One of the most effective ways of displaying a good understanding of the question is to keep a logical flow throughout your essay. This means linking points effectively between paragraphs and creating a congruent train of thought for the examiner as the argument develops. A good way to generate this flow of ideas is to provide ongoing comparisons of arguments and discussing whether points support or dispute one another.

Conclusion

The conclusion provides an opportunity to emphasise the **overall sentiment of your essay** which readers can then take away. It should summarise what has been discussed during the main body and give a definitive answer to the question. It's not necessary to restate your points but this is where you can weigh up the advantages and disadvantages and explain why you've attached more weight to an advantage or disadvantage.

Some students use the conclusion to **introduce a new idea that hasn't been discussed**. This can be an interesting addition to an essay and can help make you stand out. However, it is by no means, a necessity. In fact, a well-organised, 'standard' conclusion is likely to be more effective than an adventurous but poorly executed one.

Crucially, it is important to give a judgement in the conclusion, or a decisive response to the question posed, based on the arguments you've advanced in the main body. For example, do you agree with the statement?

Worked Example

"Abortion should only be permitted in certain circumstances" Discuss.

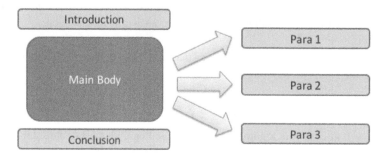

Introduction

In the introduction, it would be useful to present a brief outline of what you're going to discuss. After planning the essay (discussed below), you will know what you're going to talk about in the main body and can give a <u>very brief</u> outline in the introduction.

It is also important to define any key terms in the question here. It is quite clear that 'abortion' could be usefully defined ('the termination of a pregnancy').

If you wish, you can also highlight the key themes that will run through the essay.

Main Body

A key issue is what you write in the main body.

In the planning stage, jot down the ideas that first come to your head. For this question, you should think of possible circumstances where abortion should be permitted and possible circumstances in which it shouldn't be permitted.

Possible circumstances to consider abortion:
➤ When the mother just wants to give up the foetus
➤ In the event of a medical issue
➤ Disability of the child
➤ Sexual Assault
➤ When the mother is too young

Five possible lines of inquiry are listed here but there won't be time in 40 minutes to consider all of them in enough detail. In the exam, it's much better to focus on <u>quality</u> rather than quantity. Accordingly, choose the areas where you have the most knowledge or where you feel like you can make an original contribution and shine.

It is then necessary to choose a structure, and one possible structure is to devote each paragraph to a 'circumstance' and in order to cover fewer points, but in more detail, three circumstances will be considered. In each paragraph, the pros and cons should be considered to produce a balanced essay.

Structure of Main Body:
1. Paragraph 1: Abortion when the mother wants to give up foetus
2. Paragraph 2: Disability of the child
3. Paragraph 3: Medical issue

Detailed plan of Main Body

Abortion when the mother wants to give up foetus

For Allowing Abortion:

1. Some may argue that the mother should be able to give up the foetus should she want to.
2. This is based on her freedom to plan her life as she chooses.
3. Forcing the mother to have a child may not be in the child's best interests –would they be cared for?

Against Allowing Abortion

4. The foetus has a right to life.
5. The mother already made her choice during consummation and exercised her freedom to choose then.
6. Therefore, abortion should not be permitted in this circumstance as the right to life should take greater precedence. The mother should be encouraged to think carefully about having a child before consummation.

Disability of the Child

For Allowing Abortion:

7. The child would have a poor quality of life.
8. Would be more expensive to bring up a child with a disability.

Against Allowing Abortion

9. Again, against the child's right to life.
10. Hard to tell what the child's quality of life would be if there's a known disability.
11. Even if the child will be disabled, disabled people play an important role in society.
12. The rights of a foetus shouldn't be different depending on a disability.
13. Arguably, the right to life of the foetus should prevail here. It would, in any case, be discriminatory to lower the rights of an abnormal foetus when compared to that of a healthy foetus. It's against the law to discriminate against disabled humans and surely the same should be the case for a disabled foetus.

Medical Issue

For Allowing Abortion

14. Health risk to the mother.

Against Allowing Abortion

15. Right to life of the foetus.

On balance, the right to life of a living person should take precedence in this circumstance

This is a detailed plan and your plan in the exam does not need to be this detailed, but it should still cover the main points. Once a plan is written, you can get straight into writing the essay.

Note carefully how alternative points of view are <u>always</u> considered in the detailed plan above. Ultimately, your goal is to write a persuasive and balanced essay. When you consider alternative points of view, it strengthens your main argument. This is because it shows that you have thought about the different sides of the issue.

In the detailed plan above, point (c) is an intermediate (or interim) conclusion at the end of each paragraph. This is simply a statement which concludes a *paragraph*. It is generally desirable to include tentative conclusions where possible here as it makes it easier for the reader to understand your essay.

Conclusion

In the conclusion, the arguments advanced in the main body are brought together. In this question, the interim conclusions on each circumstance went into a lot of depth, so just a basic summary suffices for the main conclusion. An example could be as follows:

"On balance, abortion should only be permitted in certain circumstances. The right to life of the foetus demands that abortion is not allowed at the behest of the mother alone. However, there are certain situations when abortion should be permitted, such as when there is a health risk to the mother as her rights must be considered alongside that of the foetus."

Common Mistakes

1) Ignoring the other side of the argument

Although you're normally required to support one side of the debate, it is important to **consider arguments against your judgement** in order to get the higher marks. A good way to do this is to propose an argument that might be used against you, and then to argue why it doesn't hold true or seem relevant. You may use the format: "some may say that…but this doesn't seem to be important because…" in order to dispel opposition arguments whilst still displaying that you have considered them. For example, "some may say that fox hunting shouldn't be banned because it is a tradition. However, witch hunting was also once a tradition – we must move on with the times".

2) Answering the topic/Answering only part of the question

One of the most common mistakes is to only answer a part of the question whilst ignoring the rest of it as it's inaccessible.

3) Long Introductions

Some students can start rambling and make introductions too long and unfocused. Although background information about the topic can be useful, it is normally not necessary. Instead, the **emphasis should be placed on responding to the question**. Some students also just **rephrase the question** rather than actually explaining it. The examiner knows what the question is, and repeating it in the introduction is simply a waste of space.

4) Not including a Conclusion

An essay that lacks a conclusion is incomplete and can signal that the answer has not been considered carefully or that your organisation skills are lacking. **The conclusion should be a distinct paragraph** in its own right and not just a couple of rushed lines at the end of the essay.

5) Sitting on the Fence

Students sometimes don't reach a clear conclusion. You need to **ensure that you give a decisive answer to the question** and clearly explain how you've reached this judgement. Essays that do not come to a clear conclusion generally have a smaller impact and score lower.

General Advice

✓ Always answer the question clearly – this is the key thing examiners look for in an essay.

✓ Analyse each argument made, justifying or dismissing with logical reasoning.

✓ Keep an eye on the time/space available – an incomplete essay may be taken as a sign of a candidate with poor organisational skills.

✓ Use pre-existing knowledge when possible – examples and real world data can be a great way to strengthen an argument- but don't make up statistics!

✓ Present ideas in a neat, logical fashion (easier for an examiner to absorb).

✓ Complete some practice papers in advance in order to best establish your personal approach to the paper (particularly timings, how you plan etc.).

✓ Attempt to answer a question that you don't fully understand, or ignore part of a question.

✓ Rush or attempt to use too many arguments – it is much better to have fewer, more substantial points.

✓ Attempt to be too clever or present false knowledge to support an argument – a tutor may call out incorrect facts etc.

✓ Panic if you don't know the answer the examiner wants – there is no right answer, the essay is not a test of knowledge but a chance to display reasoning skill.

✓ Leave an essay unfinished – if time/space is short, wrap up the essay early in order to provide a conclusive response to the question.

Annotated Essays

Example Essay 1

"There is a time and place for censorship of the internet" Discuss with particular reference to the right of freedom of expression.

Internet is the main source of connection for people all around the world. It's where we get the latest news and information worldwide effectively and effortlessly. The lack of barrier in this internet world gives easy access to information that we might not want to see and might cause us offence. This essay is about the act of censorship, which filters offensive information of the internet, given there's a time and a place we can do so in this modern era. There's "a place" suggests there's enough room which needs to be censored and there's "a time" suggest it's the time to act on censorship.

Firstly, censorship is necessary to a certain extent as due to freedom of expression, we might be access to information that we found offended by, such as pornography. These might affect viewers mentally and easily cause depression, and affect minds especially the early teens.
However, an age limit could be set on a pornography website and refuse access to such images. It's taking an active role, to click into those websites. Setting age limits can prevent youngsters receiving non-educational information and affect their youth development. Freedom of expression is also offended as some might treat pornography as a form of art. It's very difficult to monitor whether the viewer is really above the age limit.

Also, definition of Art is very blurry, which as an excuse for people to share it through the internet. Parental education is also a key. Instead of setting limits to the children, we should given them advice on what they should go on, and guide them to make the right decision on choosing suitable materials. In this way, not only the children's development is protected, it also trains them to give the right judgements and develop logical thinking.

Another place for censorship is political/religious offensive comments and materials. Religious behaviours should be protected due to the freedom of expression in society and political views in order to keep harmony.
However, in order to achieve this, censorship is not necessary, as it would block the minds and thoughts and might be a chance for the government to brain-wash the citizens with the 'right kind' of political behaviour, which might ironically break harmony in society, and slow down its development.

Examiner's Comments:

Introduction: The introduction rambles on too long – it could have been much shorter and concise. It was also not clear. It's not necessary to say "This essay is about the act of censorship" – this is obvious. It was good to define censorship in the introduction (when the student said 'filters offensive information') although the rest of the introduction was not clear and did not make sense. Accordingly, the student would have lost marks here and wasted time. The student did not address the entire question either – it would have been good to point out how freedom of expression will come into play as well. The introduction only really needs to be 3 or 4 lines.

Main Body: In the main body, the student makes some wild assertions, such as the point regarding depression. The first sentence in the main body doesn't make much sense either. The main issue with the essay is that the student's points are not linked together logically: the essay consists of a large number of separate assertions. The point made about 'art' is not very clear either.

Conclusion: Lastly, the candidate finishes with no conclusion. The final paragraph did not draw together all the relevant information, which is bad practice. Even if a candidate is running out of time, writing a solid conclusion will gain many more marks than not writing one. Indeed, a lack of conclusion would likely lose marks.

Example Essay 2

"There is a time and place for censorship of the internet" Discuss with particular reference to the right of freedom of expression.

Censorship of the internet is not a new concept, nor does it seem like one that will ever cease. I believe that there are a multitude of reasons for its presence and hence in this essay I will argue for the statement provided.

Firstly, censorship of the internet has the capability to protect certain groups. For example, a widely accepted and often promoted form of this is the blocking of certain websites by parents and schools. In this case, the group being protected are children. We accept the fact that children should not be exposed to 'mature' content such as violence and sexual activity and by blocking certain sites, children are free to surf the internet safely. This seemingly innocent form of censorship hardly poses much threat to the right to freedom of expression as no party is being prevented from having their views heard. If this form of censorship was deemed illegal, it may paradoxically push such parties to tone down their content by pressures from parental organisations and child safety bodies in order to create a child friendly internet, which ultimately would be restrictive to free expression.

Secondly, the restriction of content that is deemed extremist or might incite acts of terrorism, whilst going against the right to freedom of expression, could be necessary to uphold the security of a nation. Thus, taking a utilitarian perspective to this issue, as the safety of the nation takes on a greater value than the right to freedom of expression. Although it may seem like a cold approach, it is justified for the well-being of a greater number of individuals and thus, society generally. Truly, the recruitment of ISIS supports is a disturbing example of the necessity of censorship of extremist views and preventative measures must be taken to ensure that the young and susceptible are not taken in by the terrorist propaganda. Accordingly, there must be a point where we deem the safety of a nation as more imperative than the freedom of expression of certain individuals.

Lastly, the freedom of expression is easily abused in the cyber world. The animosity that the internet provides gives the users a sense of invisibility where they see themselves as above the law. Abusive remarks are made without consideration of their consequence or lack of this providing an environment for cyber bullying to grow.

Censorship is not necessarily tyrannical but is a practice that must co-exist with any form of content creation and like all else, there is a time and place for it.

Examiner's Comments:

Introduction: The introduction is very concise, which is good. However, it would have been even better if the student considered freedom of expression in the introduction.

Main Body: This is a good attempt at the question and relates directly to the question. The candidate refers to the relationship between censorship and freedom of expression throughout the main body. She points out when freedom of expression is hindered by censorship and when censorship is justified. Given that it relates to the question throughout, it would yield many more marks than Example 1 (even though both are of a similar length).

However, the candidate did not consider whether censorship might be bad in the first place: why can't the government just censor what they want to? The essay would have been stronger if this were considered (even briefly). It would have been worthwhile to bring in the countervailing principle of 'freedom of expression' at some point in the essay. This would be a reason why the government should *not* be able to censor any information. The penultimate paragraph doesn't add much to the essay though – is censorship acceptable in this situation? Is this the place for censorship? Or should people be expected to put up with other people's views? Other than this paragraph, the rest of the essay is very good.

Conclusion: The conclusion is quite short. While there is nothing inherently wrong with the conclusion being concise, it must answer the question. The candidate's conclusion did not refer to freedom of expression and so she has not fully addressed the question.

Example Essay 3

"There is a time and place for censorship of the internet" Discuss with particular reference to the right of freedom of expression".

In today's day and age it's extremely easy for anyone to access explicit or dangerous content on the internet. There have been talks of censorship on the internet, but is it necessary? One would argue that the censorship of the internet is against our freedom of expression, which is why in this essay I will come with an answer in response to the statement 'There is a time and place for censorship of the internet'.

In our current education system there is a heavy emphasis put on the usage of the internet to aid our learning. However, once children learn how to use the internet, the whole world is just one click away. Children could be easily exposed to indecent images, which is why some say the government should censor the internet for the safety of children. Possible solutions could be only allowing websites with adult material to be accessible at late-night, reducing the chances of indecent exposure to children. Accordingly, in this instance, censorship is justified.

Similarly, one could easily research the internet to find information about illegal activities such as drug or bomb making. This means that the internet could be used as a tool to threaten national security, hence why the internet should have tough censorship in order to prevent criminals from accessing dangerous material, for the benefit of everyone's safety.

On the other hand, blocking certain websites strictly goes against our right of freedom of expression and instead of blocking certain dangerous websites, the government should have a more efficient surveillance strategy in order to track people who are accessing such dangerous websites. This would ensure that our right of freedom of expression isn't breached and at the same time, criminal activity would be prevented.

Furthermore, with regards to the access of sexually explicit websites, more work should be some in order to educate children not to access such websites. Good parent is a better alternative to preventing children accessing such websites, rather than blocking sites which goes against our right of freedom of expression.

In conclusion, there is no time and place for censorship as it goes against our right to freedom of expression. Other alternatives such as internet surveillance would be more effective as it ensures the safety of the general public and at the same time our freedom of expression is not breached.

Examiner's Comments:

Introduction: This is a very good introduction. It highlights the conflict between censorship and freedom of expression, which is a good place to point it out. In the final sentence, though, the student wastes time in saying "which is why in this essay I will come with an answer in response to the statement...." – this is obvious and there's little point in saying it. It just wastes time and prevents one using the time for writing something more useful. Other than this, the introduction is very good and concise.

Main Body: The student considers two main instances of censorship in the main body (indecent images and dangerous websites) and suggests that censorship could be used, but suggests alternatives would be more effective. This is quite a persuasive essay because the student has considered alternative points of view, which makes the essay balanced.

Conclusion: The conclusion is very clear and brings the arguments advanced in the essay to a final judgement. The candidate directly addresses the question and refers to the whole part of the question by considering freedom of expression (unlike in Example Essay 2). On the whole, this is a very impressive essay.

Example Essay 4

"The UK should codify its Constitution" Discuss.

The present constitution of the UK includes sources such as statute law and case law made by the judges. To codify the constitution, the state should come up with one single document that explains and describes the power of the Royal family, the government, the parliament and all the citizens. This essay will critically examine the possibility of codifying the UK's constitution.

Firstly, a codified constitution can make it clear and accessible to the citizens. With a written constitution, the citizens can trace back the rights and obligations of them and also have a better understanding of the relationship between them and the government. This may motivate the UK people to be active in partaking in politics because they can feel that they have a say through voting or other political actions. [This sentence isn't clear – how would codifying the constitution make people feel they can have a say?] It is beneficial for the development of society. [Should use a connective within this – i.e. "It is thus beneficial for…." – as this would make it flow]

Secondly, it helps limit the power of the government. Since the government is formed by the winning party of the election of the House of Commons, the party which is in charge of the government has the strength to implement the policies that they are fond of, even if it opposes the public will. A written constitution which clarifies the duty and the power of a government that when the government does not go in accordance with the public opinion, the constitution will help prevent it from doing so.

However, it is actually unique that the parliament is supreme of all. The parliament should have the ability to change the law or add new rules according to the demand of the public, neither any other organisations nor rules should have a bigger say than parliament, which is elected by the people. If the Constitution is codified, there will be a constitutional court like USA does with judges who are not elected to interpret the law. It actually weakens our democracy.

Furthermore, an uncodified constitution is more flexible. For the UK, getting enough support and votes in the Parliament is the only factor to change the content of a law or add a new act. It only takes a minimal period of time to adopt a new policy which benefits the country. This is totally opposite to the rigid process of changing regulations and rules in a country with a codified constitution such as France, which takes a very long time to pass through all stages to get consent from difference levels. A flexible system helps a country react to international changes faster and diminishes the losses it can possibly come up with.

In addition, the UK does not have a revolution in history to set up the Constitution. Unlike other countries with written constitutions, the UK did not have any revolution (such as in France and the US) to overthrow the old system and set up a democracy. It gradually changes from authoritarian rule of monarch to a state with constitutional monarch. The past has shown that the country actually works well with this system while the democracy of the country does not disappear. It always ranks top 10 in the world, a place where many developing countries would like to learn.

To conclude, it is unnecessary for the UK to set up a codified constitution because the existing system alas makes the country work well. No one can ensure what effects the written one will bring to the UK.

Examiner's Comments:
This is a relatively long essay for 45 minutes, but one which is not that good. It uses a lot of knowledge, but not to good effect.

Introduction: This is concise in just explaining the role of a codified constitution. The issue is that it does not define what a codified constitution is. Having a written constitution is more than just having the constitution written in a single document: It is one which is harder to repeal than other laws.

This is not a big concern to the essay though as the Section B essay is more a test of your persuasive and written skills, rather than your essay. It would also be worth pointing out that an unwritten constitution (like the UK's) is flexible, which means that a government can just change it if it wants. Whereas a written constitution (like in the US) binds the government and Parliament and a special majority is required to alter it. Highlighting such policy tensions can be beneficial in the introduction to essay questions.

Main Body:
Everything you say in the essay must make sense. It is pointless writing the first thing which comes to your mind if it doesn't make sense as this would paint a poor picture with admissions tutors and lead to rejection. In the second paragraph, it is not clear how the sentence "This may motivate the UK people to be active in partaking in politics because they can feel that they have a say through voting or other political actions" relates to the question. It isn't clear that people feel they can have a say through voting by virtue of having a written constitution, and the student does not explain it. The sentence does not actually make sense.

The third paragraph is not entirely clear either – surely a government would reflect the public will. It would be more coherent to suggest that the government can infringe on minority rights.

At the beginning of the fourth paragraph, the candidate starts with the sentence: "However, it is actually unique that the parliament is supreme of all." This particular point is not elaborated on further and it does not add to the essay. So what if Britain's system is unique? Does this mean it's bad? It is important to be clear in essays, to make the reader's job (of understanding what you mean) as easy as possible. The candidate also incorrectly uses the connective 'However' – this connection is used to indicate a contrasting point but the student's points before and after the connective do not contrast! This goes to show that it is important to think before writing and ensure you express yourself clearly.

The candidate has clearly deployed significant knowledge but has not used it well to advance an argument and the essay has been confusing at many points. It is important to think and plan what you're going to write before writing it; planning before writing can significantly help the quality of writing. This essay would have likely led to rejections by admissions tutors.

A crucial suggestion would be to think whether you are addressing the question while answering it.

Conclusion: The conclusion is sufficient as it directly answers the question.

Example Essay 5

"The UK should codify its Constitution" Discuss.

Currently, the UK has an uncodified constitution. This means that there is no single legal document defining and storing all of the laws and regulations in the UK. The UK constitution comprises of many different areas. The ambiguity of the UK constitution has created debate within government itself as well as outside of government.

One main reason as to why the constitution should be codified is due to it being ambiguous. Certain laws become ignored as it may be presumed not legislative despite it being a tradition, or an 'unproven rule'. Theresa May, the Home Secretary, violated one of the constitutional conventions, where she publicly blamed and identified one of the civil servants for a mishap in passport checking in Paris. It is known in government that the head of authority is to take responsibility of everything, whether it is or not their fault. Due to there not being a codified constitution, Theresa May was able slip from it. Therefore, having a codified constitution can ensure full responsibility to government.

Furthermore, a codified constitution provides security to the citizens, they would know that government is bounded by a tight set of rules which has little of no loopholes and can easily be accountable for any mistakes. This reduces confusion among the public as well, as there is a clear definition between right and wrong. A legal and authoritative constitution disregards any form of unfair or unequal treatment within the hierarchy.

However, looking at it from a different perspective, the constitution has been within the UK for a long time and has been fine with no major problems. Codifying the constitution may stir major changes and cause unpopularity among those in power. So if it isn't broken, why change it? The constitution's long standing in the UK and is also a form of historical standing which has become an admiration of many countries. It also symbolises trust in government by the people. The constitution holds more than just rights and rules, but also a devoting history in the country. As long as it has been here, there have been no issues with it.

In addition, an uncodified constitution is said to be flexible and up to date. This is because it can easily be changed and corrected if flawed. The bills of rights in the US takes a lengthy and long process to be altered. However, in the UK, adding or removing the law requires not a great amount of time. The flexibility of the constitution is a greatly appreciated characteristic of it, making it unique.

In conclusion, the constitution is fine as it is and does not require change. It is the government's own responsibility to maintain professionalism within parliament and not the laws. There is still the stance of democracy among the people to hold those of government accountable if there is any distasteful occurrence.

Examiner's Comments:

Introduction: The introduction started on the right approach, which was to explain what is meant by a codified constitution. However, the student gave an incorrect definition. It's not correct that all of a country's laws would be in a constitution.

Main Body: Each of the middle four paragraphs in the main body discuss interesting and relevant points in a lucid and concise way. The candidate presents arguments for and then arguments against, then comes to a final conclusion. To improve the essay, it would have been worth trying to evaluate and weigh up the pros and cons a bit more. For example, considering the final point in the fifth paragraph more would have been better – such as, "the flexibility has received criticisms for governments being able to make significant constitutional change, but it has crucially allowed the UK's constitution to adapt to changes in society, which may not have otherwise been possible". A further relevant point would be that the Human Rights Act may help reach a balance.

Conclusion: The first sentence is fine. It is not clear how the rest of the paragraph addresses the question. It would have been clearer to the reader if the candidate said: "The government is more accountable under an un-codified constitution".

Example Essay 6

"The UK should codify its Constitution" Discuss.

A codified constitution is a formal document where the country's main constitutional laws are set out, which defines the rules governing the political system, the distribution of powers between the different branches of government and the rights and responsibilities of citizens. A particular feature of a codified constitution is that it has a special status over other laws, such that it is harder to amend or repeal. The UK's constitution is not codified but is found in statute, case law and conventions. This essay considers whether a codified constitution would be beneficial.

Constitutional law in the UK is ambiguous and unclear. Accordingly, it tends to lead to greater disputes and uncertainty regarding the true constitutional position. For example, much of the constitution is currently in the form of constitutional conventions, which are non-legal norms. An example of such a rule is the one whereby the Queen always gives assent to bills of Parliament passed by the House of Commons and the House of Lords. While this is widely accepted and understood without controversy, other conventions are unclear. For example, the Salisbury convention requires that the House of Lords shall not oppose any government legislation that is promised in its manifesto. In regard to the recent Tax Credit cuts, the government argued that the public voted for its policies of spending cuts and that the Convention applied. However the opposition sides in the House of Lords argued that the convention did not apply because the government did not promise that specific spending cut. Given that this convention was not in a written constitution meant that it was both unclear and that there was no recourse to a court to resolve the issue. Accordingly, a codified constitution would bring clarity in such areas.

Nonetheless, an uncodified constitution does have its advantages. In particular, an uncodified constitution is easier to alter when compared to a codified one. In the UK, the constitution can just be altered like any other law. Whereas in countries with a codified constitution, it is either not possible to change the constitution or it requires an even larger majority of the legislative assembly (such as 2/3 lawmakers consenting). An advantage of the UK's system is thus that the constitution is flexible and adaptable to the changing nature of society. For example, in 1998, the government could easily establish devolution through the creation of a Scottish Parliament and Welsh Assembly in response to changing public attitudes. However, it is likely that the UK government could have done this even with a codified constitution given that it had direct public support for the devolved assemblies through referendums.

There is indeed a concern though that a codified constitution is inflexible: for example, the US constitution came into force centuries ago and thus, reflected the attitudes of that time. It provided the right for citizens to bear arms and for militias to operate. Nonetheless, this does not have to be an issue for the UK. Firstly, fewer constitutional matters can be placed in the codified constitution (in most countries, including the US, the entire constitutional law is not embodied in the written constitutional document). Secondly, it can require a reasonable, but not impossible, majority to amend the constitution.

The most important reason of all for having a codified constitution though is that it can guarantee fundamental rights of the British people. The UK currently has the Human Rights Act, but that can easily be revoked, just like any other law. Arguably, the time has come to give the public human rights in a codified constitution.

In conclusion, it would be beneficial to have a codified constitution.

Examiner's Comments:

Introduction: The introduction was very good – it defined the key term ('codified constitution') and what the significance of that would be.

Main Body: The essay started well and brings in lots of real world examples to good effect. It would have been good though for the student to use 'sign-posting' language a bit more. In the third paragraph, it appears that the student is implying that a codified constitution would still allow big changes. Writing an intermediate conclusion at the end of the paragraph, such as "Thus, the UK should not hesitate to codify its constitution".

However, the student runs out of time and in the smallest paragraph in the main body, brings in the 'most important reason' to back up his argument. The lack of consideration given to that reason makes it less convincing. Nonetheless, the second and third paragraphs did contain a more thorough discussion.

Conclusion: The conclusion is too short here. It does come to a judgement but does not explicitly weigh up the advantages and disadvantages and say why the advantages outweigh the disadvantages of having a written constitution.

Example Essay 7

"Developed countries have a higher obligation to tackle climate change than developing countries" Discuss the extent to which you agree with this statement

Developed countries, such as the UK, Dubai and Japan have economies and political standings which are considered fairly stable. Developing countries strive to achieve the stable GDP, inflation and political welfare of developed countries. Climate change occurs due to excess emission of gas pollutants, melting of ice caps, global warming and in the case of such serious disasters, is it fair to place greater responsibility towards the more developed first and then the second world countries?

Statistically, it is shown that developed countries have been the main pollutants and indirectly, cause global warming. Due to its higher living standards, citizens of these countries have a higher propensity to consume, thus contributing to greater factory industries. As they would also have the best resources and research and development facilities, these countries would be more capable of producing greater findings to tackle climate change. This can also avoid conflicting interests among developing countries whose main focus would still be on growing. The countries may still be facing radical political changes which could affect their views and focus on the country's wellbeing. The Kyoto protocol, a Japanese originated project to reduce air pollution mainly forced on the more developed countries, as knowing it might create economical downfall as many of these countries on oil refinement and exporting of cheap factory produced goods. Under these circumstances, having more developed and stable countries to hold higher obligations towards issues of climate change would not be too wrong.

However, looking at this from a different perspective, climate change issues should not be delegated only to a few countries, but everyone altogether. It should be notably, responsibility of all, as it would require effort and contribution, as well as a mutual understanding form every country. It would also be considered unfair, as some may take this for granted. Countries at war, Iran and Syria, should be held dully responsible for the countless bombing and air strikes which has greatly contributed to global warming. This should be an issue for all. Other than that, developed countries would be blamed for any mistake or distasteful occurrences, which would have been inevitable, by countries where the cause originated from. Therefore, it should be recognised that combating climate change should be the responsibility of all.

Why should these countries make sacrifices to their economical welfare for others? World organisations like NATO, combinations of many different countries, work as a whole, with more developed countries contributing resources and research equipment required. Tackling of climate change should be the responsibility of all, and every country should feel obligated to it.

Examiner's Comments:

Introduction: The introduction is fine as the student defines the key terms within the quotation. It could have been a bit more specific when questioning whether greater responsibility should be placed on developing countries.

Main Body: The points in the second paragraph are interesting and the student deploys her own knowledge to good effect. The final sentence in the second paragraph could usefully be adjusted to say something along the lines of "Given the greater needs of developing companies and greater resources of developed ones, developed countries may have a greater capacity to take on the higher obligation of climate change". In the third paragraph, some good points are made. The point regarding the war in Iraq and Syria makes little sense and significantly reduces the quality of the essay. It is very important to make sure that the points you make follow from each other.

Conclusion: The student does address the question in the conclusion and is very succinct. It is not clear how NATO specifically fits into the essay at all and it would have been better to leave that out.

Example Essay 8

"Developed countries have a higher obligation to tackle climate change than developing countries" Discuss the extent to which you agree with this statement

Developed countries are the nations with mature economies and modernised environments in their states by the development starting from the 18th century, while developing countries refer to the nations which are investing most of their money into the development for economic growth and are trying to catch up to the developed ones. The recent climate changes are mainly incurred by the industrial and business processes of the economic activities in the world. These led to the emission of greenhouse gases and finally global warming. The obligation that countries should bear for this case is to eliminate the pollution of the world and allow the climate to restore. I will now look into the topic economically, historically and environmentally.

To start with, developed countries should take more responsibilities of damaging the climate because they are the starters of it. Looking back to the history, western countries, especially the UK, experienced Industrial Revolution in the 18th Century. Mass production existed by utilising machines and mining for fuels was one of the main businesses at that time. These activities led to the beginning of pollutants and greenhouse gases. This affected their air quality, such as the dark fog in the UK in the past. The businessmen were so greedy that they wanted to exploit other nations' resources for making profits and invested money for further development of new technologies to enlarge the size of production. These behaviours caused even more pollution to be emitted and started changing the climate and damaging nature. Therefore, they should be the first to respond to global warming.

Furthermore, most of the developed countries caused the huge financial and economic gap between them and the developing ones. This forced the developing nations to further increase their economic activities by ignoring the environmental and climatic impacts. It is no longer news when we hear that a developing state has taken on huge debt from a developing one. With the 'good will' of lending other countries in need money to strengthen their economic power by investment in infrastructure and the like, the developed countries always adopt a high interest rate and the debt seems impossible to return. This makes the poorer ones worse because they risk everything, including neglecting the severe consequence of climate change, to produce their economic goods in the cheapest way, which means without any costs on filtering such pollutants. This is actually not directly caused by the privileged countries.

However, it is argued that the rich nations put efforts to start cutting down the emission such as implementing EU emission quotas and keep investigating in new methods to make the production cycle cleaner. Yet, EU emission quotas are actually a failure that many nations persuaded the EU committees for more quotas and this led to the failure of the plan. In addition, those products from investigations are usually expensive and not many products from companies in developed nation will apply it. It ends up failing because of a lack of effectiveness.

To conclude, it is inevitable that developed countries are recognised as the main cause of today's climatic problems. Thus they have to bear the responsibilities more instead of the developing nations.

Examiner's Comments:

Introduction: The introduction is too long and would thus have taken up valuable time in the exam. Nonetheless, it captures the main themes of the question and sets the scene well for the essay.

Main Body: The second paragraph makes a very good point but waffles a bit much and goes off the point when talking about businessmen being greedy and about the dark fog in the UK – these two points are not relevant to the candidate's line of argument. Accordingly, does not gain marks and weakens the essay.
It would have saved a lot of time if the candidate simply stated the following:

> ➢ Developed countries have already emitted enormous amounts such as in the industrial revolution,
> ➢ So it would only be fair to expect them to do more to clean up the mess which they caused.

In the third paragraph, the student needs to be clearer as to what she means. The implication is that developing countries are hard done by the developed ones, but the student needs to state this clearly. The student also uses a weak link here. The final sentence of the third paragraph didn't make sense and contradicts the rest of the paragraph. The fourth paragraph does not add anything new to the argument. The candidate describes the failure of the EU quotas but it's not clear how this relates to the question. Surely it doesn't matter (for the purposes of this essay) whether the EU have failed with their emission quotas or not. The essay question is asking whether developed countries have a higher obligation to tackle climate change than developing countries. As the paragraph does not advance an argument as to whether developed countries should or should not take on a higher obligation than developing countries, it wastes time and reduces the quality of one's essay

Conclusion: The conclusion is succinct and follows from the main body.

Example Essay 9

"Developed countries have a higher obligation to tackle climate change than developing countries" Discuss the extent to which you agree with this statement

Climate change is a global issue that affects all nations and its peoples, and in light of the newly released global sustainability goals, perhaps we should focus on what actions should be taken to effect a change rather arguing who should take responsibility. Hence, I disagree with this statement and will be presenting my argument in this essay.

Firstly, climate change is a global issue and all nations are obligated to combat it. We must abandon the attitude that developing nations are somehow inferior to developed nations simply because of their global position. With this approach in mind, all nations therefore must be taken as accountable for this global crisis that affects us all. Perhaps the view that combatting climate change is an 'obligation' should be abandoned. Improving the condition of our world and fixing our mistakes should be regarded not as a chore, but as a responsibility to future generations. After we have confronted these issues and changed our perceptions, will a global effort truly be effectively carried forward?

Secondly, whilst it is true that developed nations have a greater capacity financially and structurally to enact a change, efforts to improve the infrastructure of a country to make it more green can be done by developing countries. Rather than seeing sustainability as an expensive undertaking, requiring new carbon capturing machines, knowledge of other ways to lessen our carbon footprint should be made clear. These simple methods such as planting more trees than the number being cut down or effective garbage disposable and recycling to minimise burning of garbage. Such inexpensive methods could easily be undertaken by developing countries and eliminating the idea that climate change is a concern of the rich.

Thirdly, to separate countries into two spheres is damaging. This segregation lead to the belief that 'developing nations' are somehow able to 'get away' with releasing high amount of greenhouse gases or deforestation by simply claiming that they don't have the capacity to make such a change. It is not enough for the developed countries to take the initiative; developing nations are equally obligated to combat climate change.

In conclusion, no country should be viewed as having a higher obligation towards alleviating climate change.

Examiner's Comments:

Introduction: The introduction is excellent. The candidate states her main view concisely and proceeds to continue with the main body. The candidate also adopts a unique take on the question, which is even better and will signal green lights in the heads of admissions tutors.

Main Body: The second paragraph raises interesting points but it is not clear how it relates to the question. A running theme throughout the essay is that every country shares a responsibility to be sustainable and reduce climate change. However, counter-arguments are not readily considered, accordingly the essay is not as persuasive as it might be.

In the third paragraph, the student makes the (very good) point about developing countries still being able to plant trees. This could be framed in a different way and incorporate a counter argument. For example:

➢ Climate change affects every country and, thus, every country should have an obligation to tackle climate change
➢ [**Counter Argument**] Some argue though that richer countries have far more resources than developing countries and, thus, can spend money and develop non-renewable energy sources (e.g. solar panels), whereas poorer countries would not have the finances to do this.
➢ [**Rebuttal of the counter argument**] Nonetheless, poorer countries can still do their part by planting trees and should not feel that they're 'off the hook'. Climate change affects everyone and, therefore, every country should do what they can to tackle it.
➢ [An example of a policy might be that each country pays a certain percentage of GDP to tackle climate change – this way, it's proportionate to each country]

Conclusion: The candidate succinctly presents her final response to the question in the conclusion. This could have been elaborated on a little more but is still fine nonetheless.

Example Essay 10

"Abortion should only be permitted in certain circumstances" Discuss.

Abortion means the mother decides to take off her foetus from her body in the form of taking away the possibility that the foetus can grow and finally become a baby which is ready to be born. With the permission of doing so, it means that the mother is taking off her foetus legally and she should bear no response afterwards. In my opinion, several circumstances should be raised and some of them should be given the chance to the mother to take her foetus off while some should not be allowed.

First of all, the victims of sexual assault or sexual abuse should have the right to abort. This mainly concerns their willingness of having the sexual activity and being pregnant. The victims are not willing to be raped by someone she did not fall in love with or someone other than her partner. Therefore, it is even more unlikely that she would further agree to be pregnant. The abuser, firstly, has this conviction of sexually assaulting someone who disagrees with his actions. Because it is a criminal action, the victim should not bear the consequences that the baby will bring such as nurturing the baby with her economic ability, which has a chance that it is insufficient to do so. Therefore, abortion should be permitted here.

Other than the above reason, a mother who has a baby which is inherently with problems should be allowed to abandon the foetus. Having children with inherent deficiency such as Down syndrome is not a choice of the parents: no one is a wrongdoer in this situation. While the parents have to bear the heavy burden to spend a lot of money on the medical caring of the baby, they may need to take care of them for their whole life. For these accidents, the authority should give them a chance to choose whether to keep their children or not. Abortion should be permitted in this circumstance. Almost all countries in the world give this choice to the parents with this problem.

However, babies which accidentally created during the illegal acts should not be permitted to abort, unless there is a medical issue raised. Increasing numbers of teenagers start to have sex at an early age such as 12. When they attempt to do these activities, they do not consider any possibilities may occur in the future and most of them do not care about the consequences. Not allowing them to have an abortion can prevent the teenagers from acting unconsciously, such as by having sex early or having sex without proper protection, even though special cases such as too young to give birth to a baby should be also under the consideration.

Furthermore, parents who want to abandon their children randomly should not be allowed. A wrong message may be given to the society that having foetus is not a real matter and we can abandon it anyway. It may also raise the moral issues such as selecting a perfect foetus in the laboratory to give birth. A random abortion also easily damages women's bodies which may affect their ability to have a baby in the future. This may lead to a decline in the birth rates. It can hamper our imbalance social structure seriously that our future generation has to bear all the welfare we need. Therefore, no permission on random abortions should be given.

To conclude, circumstances whereby the parents or the mother do not have the responsibility of having a baby inappropriately should be permitted to have abortion, whilst the others should not do so.

Examiner's Comments:

Introduction: The introduction attempts a definition of abortion, which is good. The final sentence of the introduction could have been more concise though, for example: "this essay argues that the mother should be allowed to abort in certain circumstances" – this says exactly the same thing but uses fewer words and looks better.

Main Body: The second paragraph makes a valid point but too long is spent on it. It was also quite basic and one-sided. If the student made reference to the countervailing principle based on the rights of the foetus, it would have been a more convincing paragraph (e.g. explicitly suggest that the right of the mother here should outweigh that of the foetus).

At each paragraph, the student relates it back to the question by considering whether abortion should be permitted in that circumstance. However, the essay is mainly one-sided and does not consider the possible right to life of the foetus. The student also does not consider why abortion should not be allowed in all other circumstances in enough detail.

In the penultimate paragraph, assertions, such as abortions easily damaging women's bodies are made, but this is not true if carried out by medical professionals. It is important not to make up facts and only use genuine knowledge.

Conclusion: The conclusion is not specific enough but the candidate does come to a reasoned judgement. Nonetheless, since the main body was generally one-sided, the conclusion does not feel like it is sufficiently balanced.

Example Essay 11

"Abortion should only be permitted in certain circumstances" Discuss.

Abortion is an act of ending the life of an embryo before birth. In my opinion I think abortion should be allowed when parents have a disability to take care of the child or the child itself is predicted to have a disability after he or she is born.

Firstly, if the child is detected to have a disability such as a cancer or similar life threatening disease, parents should have the choice to choose if they want the baby or not as the child may not be able to survive long. It might also bring the child discrimination which is bad for life development. However, it could be argued that every child is a gift from god and abortion is morally wrong. It's against the human right of the child itself. Moreover, through the difficulties in the process of early stage of the child, such as heavy medical treatment and discrimination from the public might train the child to be a stronger and more determined individual. Instead of a disadvantage, it might be a special advantage and experience to the child.

On the other hand, it might be argued that the inability of the parents to educate the child might be a legitimate reason for abortion. If the mother is too young to educate the child and needs huge financial support, she may be unable to care for the child. It might also be case if the parents are in prison.
In my opinion, parents should take up the responsibility of having a baby. If the mother is too young and still needs to continue her studies, the baby could be put into a foster home, who can be taken care of by a specialist. The government can consider giving out loans to these young mothers until they finish their education and have the ability to earn money themselves. If the parents are in prison, they can also send their children to a foster home and ask for permission to see their child frequently to keep the bond within the family.

In conclusion, the only permission for abortion should be when the child has a permanent disability and the parents can get to choose whether or not to give birth to the baby.

It might be inhuman to kill the embryo, but it is also not virtuous for the parents to see the child suffering. However, if the problem comes from the parent itself, the child as individuals should have the right to live, regardless of the fact that care homes might have less support to theme than normal homes, it's still a chance for them to explore the beauty of life.

Examiner's Comments:

Introduction: A clear and crisp definition in the first sentence and a nice introduction. It appears that the student is only considering a few circumstances from the introduction.

Main Body: The second paragraph involves assumptions. What if the child can live long enough, though?

The essay launches straight into specific and niche circumstances and considers remote points, such as the life development of the child. Given the time available, it would be better to stick to the main points.

The essay does not consider in much detail the alternatives and why abortion shouldn't be allowed generally apart from a brief sentence in the second paragraph. Considering the alternatives is important as that leads to a more balanced and persuasive essay and, thus, higher marks!

Conclusion: The initial conclusion is very good and succinct. However, the student also makes an additional point after the conclusion. Sometimes, this works but here it does not fit in. It's normally best to introduce completely new arguments in the main body.

Example Essay 12

"Abortion should only be permitted in certain circumstances" Discuss.

Abortion occurs when a pregnant woman decides to no longer carry her child and in turn kills the foetus. By law, pregnancy must be under a certain age limit. There are three main circumstances I believe abortion should be permitted: where the parent(s) cannot financially support their child, where the parent(s) would raise the child in an unsafe environment and finally, under the circumstance that the health of the mother is at risk. However, the law preventing abortion after a specific number of weeks should still be implemented at each instance.

Firstly, if there is evidence that parent(s) of the child are not financially capable of supporting their child in the future, then an abortion should be permitted. This is because a lack of food, shelter or clothing – seen as basic necessities – should be provided in order to ensure the child remains health. Although it can be argued that the child could be sent to adoption where these necessities can be provided, psychological impact on the family can be forgotten. The emotional connection of a mother to her unborn foetus must be less than her to her newborn, hence for the health of the mother, abortion should be permitted. It is not to be assumed however that there will not be mental consequences of abortion.

Secondly, the environment which a child is raised in has been scientifically linked to the future behaviour of said child. An abusive environment is undoubtedly unsafe for any child, both for their physical and mental wellbeing. This sort of environment can involve similar behaviour in the child which wants to be just like their parents may eventually lead to a life of crime. However, this cannot be said for all children of course. Where abortion is not an option raised by the parents themselves, should it be enforced by a governing authority? It must be vital for examination of home environments to be conducted on an equal level to ensure justice.

Finally, the health of a mother is important. The life of an unborn foetus should not be considered over the life of the woman. It is said that the right to life is of the utmost importance, hence this should be followed. When considering the health of the women, the physical and mental health must be considered equally. This is to ensure the safety of the unborn child as well as the mother.

I believe these are the circumstances that should be permitted. In turn this could reduce the crime rate, protests by those against abortion and can encourage safe sex, especially by teenagers. Although these circumstances may function in countries such as the UK, China, for instance, would permit abortion outside these guidelines due to the implementation of the 'One Child Policy'. However, I believe such policies reduce the freedoms of citizens which should be considered.

Examiner's Comments:

Introduction: The candidate points out that abortion should be allowed in three specific circumstances but does not consider why abortion shouldn't be allowed in any other circumstances (or be allowed generally). Other than this, the introduction is good, very clear, and sets the scene well for the essay.

Main Body: The third paragraph is unclear and does not consider the issue from enough perspectives to be balanced – what about the right of the child to live? Does the foetus not have that right? Or if it does, does the potential for the child to enter into crime outweigh that right? Consideration of such issues will add a greater depth to the essay and make one's argument more persuasive. It is crucial to consider alternative points of view to have a balanced essay. The candidate then digresses and considers whether abortion should be enforced on people and doesn't actually consider that issue properly – she implies that it's important (by saying that there should be 'home examination' but it's not clear what she is saying). The fourth paragraph could be improved by considering what rights, if any, should be given to the foetus and then whether the mother's right to life outweighs this. As it stands, the paragraph is a bit one-sided.

Conclusion: The conclusion is satisfactory but not great. It introduces a few new assertions (such as reducing protests and encouraging safe sex) which do not follow from the rest of the essay. It also introduces a completely new point regarding the One Child Policy in China: while this is an interesting point, the candidate has not demonstrated how it is/could be relevant to the question and so this point is a waste of space. The conclusion should be used to weigh up the pros/cons of the arguments you've already addressed in the main body and to bring everything together (and generally not to introduce new ones).

Example Essay 13

"The government should legalise the sale of human organs" Discuss.

With the lack of available organs for transplant in hospitals, many patients lost the change of survival for a longer period of time. To a large extent, government should legalise the sale of human organs, in order to boost the amount of organs available for replacement and might help the poor to gain some extra money as well.

Firstly, the sale of human organs involves buying and selling. One can gain money in exchange for given out extra/not necessary organs in their bodies (such as kidneys). This is also a more efficient use of resources while saving lives at the same time. However, it could be argued that the action creates bad incentives especially for the poor to sell as much organs as possible, without looking after their own health. This may cause serious damage to their health and even causing more deaths. On the other hand, monitory controls can be done before organs are sold. Compulsory health checks on the seller should be enforced to see whether he/she is capable of giving away the organs and also whether the organ is suitable for transplant at the same time.

Secondly, the selling of organs of organs generates tax revenue for the Government. This extra amount of revenue could be used on investment in healthcare to cure disease, which could in turn reduce the demand on organ transplants. It could also be used to hire more doctors to meet the increasing demand in the NHS. However, selling organs are morally wrong and our bodies are gifts from God and shouldn't be re-used for commercial transactions. Also, sellers might set high prices and widen the gap between rich and poor, as only rich people can afford the organs, thereby deepening inequality in society. Nonetheless, organs are property of individuals and we should full control on how we use them. We're allowed to engage in other potentially harmful activities, such as smoking and drinking. Furthermore, the government can set up a maximum price for organs to avoid the situation of overpriced organs. Accordingly, strong monitoring should be used in order to ensure everyone has an equal chance of getting suitable organs and organs should be arranged for those who are most in need of them.

In conclusion, the government should legalise the sale of human organs in order to increase the amount of organs available for transplant to save more lives, as people would have more incentive to give off their organs. This can also generate more revenue for the government under tax, which could be used in the NHS to help more people in need. Monitoring is also essential, such as setting up maximum prices and ensuring everyone has the equal chance of getting the organ they need.

Examiner's Comments:

Introduction: This is concise and the student directly addresses the question, which is good.

Main Body: The quality of written communication is not particularly good but otherwise, a number of interesting arguments are made and evaluated.

The point regarding tax was a bit remote (and, thus, did not seem entirely relevant in the context of the debate on human organs). It's not clear that would be taxed in the same way as goods (and would probably be too insignificant to make much of a difference to the government's tax revenues).

That time could have instead been used to delve deeper into evaluating the main pros and cons of legalising human organ sales. For example, the candidate raised the point that "our bodies are gifts from God and shouldn't be re-used for commercial transactions." This is a good point, but it could have been elaborated on further and critiqued on the basis that not all individuals follow a religion. This would have reinforced the candidate's conclusion and showed a greater depth of analysis.

Nonetheless, a variety of interesting and directly relevant arguments and counter-arguments are made, which makes this a very decent essay on the whole.

Conclusion: This is a good conclusion which directly addresses the question and the candidate elaborates on her view. However, the point about tax seems out of place as it is a new piece of information. The monitoring point would have fit well in the conclusion with a linking word, such as 'Nonetheless'. For example:
  ➢ In conclusion, the government should legalise the sale of human organs in order to increase the amount of organs available for transplant to save more lives, as people would have more incentive to give off their organs.
  ➢ Nonetheless, this must be accompanied with an effective monitoring system, including the use of maximum prices, to ensure fairness.

Example Essay 14

"The government should legalise the sale of human organs" Discuss.

The question as to whether the sale of human organs should be legalised in countries has always been a hot debate throughout the centuries. As far as I am concerned, this is illegal in most of the major countries and in this essay I am going to outline the reasons.

First and foremost, moral issues are raised in the sale of human organs because many regard it as inhumane and it is not morally acceptable. As I don't see it as a proper way to earn money in a civilised country, where people earn money by selling their organs. I believe life is sacred and should be treated with great care. The sale of human organs implies that money is more important than the protection of lives.

Personally, I am in favour of organ donations which is completely different from selling organs as this does not involve the transfer of money and the meaning behind it is people really want to help others out of compassion but not for the benefits they can receive afterwards.

In addition, legalising the sale of human organs can exploit the poor, particularly in developing countries because they are vulnerable groups which can be exploited by the rich. In developing countries, the majority are not fully educated and they are poor. There is a potential that the rich in educated countries will exploit their advantages to purchase organs from them. This is unjust and unequal. Allowed to continue, this can lead to dire consequences in the long run which cannot be easily stopped.

In my opinion, the government role is to protect the welfare of citizens, so especially in developing countries, the government should not legalise the sale of human organs.

Examiner's Comments:

Introduction: The introduction appears to miss the point slightly. The quote wants the candidate to consider whether the sale of human organs should or should not be legalised. The candidate, though, says that he is going to outline the 'reasons' why it is illegal in most countries, when actually, the question wants the candidate's view as to whether it should be legal or not (and not an explanation for the present state of affairs).

Main Body: This essay is very basic, considers arguments in little depth, and is only one-sided with no consideration of counter-arguments. The candidate highlights some serious issues about allowing organ sales but does not adequately consider the issues at stake – e.g. what can be done about the current organ shortage? Highlighting this point and then evaluating it, by arguing that you don't (for example) think legalising organ sales would solve it, would make for a more persuasive essay as it shows the reader that you have considered alternative perspectives. Merely stating a bunch of opinions such as "I believe…." and "I think…." is not helpful and does not advance an argument <u>unless</u> they're backed up with reasons.

Conclusion: The conclusion, though, is satisfactory and very succinct but since it does not follow from a balanced or well-argued main body, the essay as a whole is poor.

Example Essay 15

"Sufferers of anorexia nervosa should be force-fed" Do you agree with this statement? If so, evaluate at what point of an individuals' disease this measure should be taken.

Anorexia nervosa is a term used to describe individuals who are unhealthily skinny. Some may say that anorexia is a major issue which needs to be addressed and force feeding seems to be a solution to prevent anorexia. In this essay I will be looking at the points for and against force feeding people suffering with anorexia, and I will come up with a constructive conclusion giving my personal opinion on this controversial topic.

Particular diseases often have a domino effect which results in other diseases, meaning that sufferers from anorexia may have the risk of developing other diseases as well. If an individual's anorexia reaches a limit where their daily lives are impacted, then indeed, sufferers of anorexia should be force-fed

Similarly, sufferers from anorexia may be suffering from social anxiety and insecurity due to their body structure, so in order to help improve their quality of life, sufferers should be force-fed in order to prevent sufferers' life from deteriorating.

On the other hand, one would say that force feeding sufferers of anorexia would be unideal since you would be forcing someone to do something against their own will, even if it was benefiting the sufferers. Instead, other alternatives should be explored, such as psychiatric and medical help, or other programmes which could help track the progress of the sufferer's health, rather than force feeding them. This is why some people believe that force feeding sufferers of anorexia would be unideal.

In addition, it would be difficult to distinguish the severity of anorexia and hence force feeding would be unideal because some sufferers of anorexia would be force-fed, when they really didn't need to be. Finding the cut-off point to where sufferers would need to be force-fed would be unrealistic and so other alternatives such as medical help should be explored. This is why sufferers of anorexia should not be force-fed.

In conclusion, force feeding sufferers of anorexia would be a way to tackle this disease, however I personally believe that there are other better alternatives to tackle this disease such as receiving medical treatment or seeking long term help with a professional, which would benefit sufferers in the long run.

Examiner's Comments:

Introduction: The introduction starts straight off with an attempted definition of anorexia nervosa. It's not entirely correct but given that the LNAT exam is not a test of knowledge, it is a good enough approximation. It's really not necessary for the student to say: "I will be looking at the points for and against force feeding" or that "I will come up with a constructive conclusion giving my personal opinion" – these are both obvious and add zero marks, so the time could be better spent writing something more constructive to the essay.

Main Body: The second and third paragraphs (the beginning of the main body) are certainly valid points and they do indicate force feeding as a treatment, but the student uses the word 'should' as if it was a final conclusion in both of these paragraphs. This contradicts the student's actual conclusion at the end of the essay, which is a poor essay technique. Instead, as interim conclusions, the student should be more tentative within the main body and just say 'therefore, force feeding <u>may</u> be an option to resolve this' or 'some may, thus, argue that sufferers should be force-fed'. The fourth paragraph is very good.

A further point that links with the point on consent is that, since force feeding is against a person's free will, it would not solve the underlying problem of anorexia nervosa and a person may need to be continuously force-fed against their will, which surely does not respect their personal autonomy and capacity to think as a human being. Accordingly, alternatives such as psychiatric help to solve the underlying problem would both be more effective and accord with the patient's free will.

Conclusion: The conclusion is sufficient: it is comprehensive, clear and concise. On the whole, this is a very good essay.

Final Essay Advice

➢ Use linking words.

➢ Do NOT use long introductions – they just waste time.

➢ Answer the Question – a surprisingly common issue is where students don't answer the question. This is by either misinterpreting the question, answering a different question, or only answering part of it. It is absolutely <u>critical</u> to answer the question and it is not something you can just assume you are doing. At the end of each paragraph, it would be useful to ask yourself whether you're answering the question. A good plan would help with this. If you are not answering the question, you are not gaining marks on what you're writing and it will not be impressive to the admissions tutors reading your work.

➢ Give a judgement – if the question asks "Do you agree?" make sure that you say whether you agree or disagree with the quote. Feel free to take a midpoint and say that you agree with it in x, y, z circumstances but disagree with it in other circumstances.

➢ Do not give a rant of opinions – it is important to advance an argument throughout your response.

➢ Do not add completely new arguments in the conclusion – it is ideal to weigh up the pros and cons that you've raised in the conclusion.

➢ Consider counter-arguments – even when you're heavily committed to one side of the debate or argument, considering alternative arguments (and then rebutting the counter arguments) makes your essay more persuasive.

➢ Signposting language – using signposting language helps make your essay clearer and more readable for the admissions tutors. For example, using connectives such as 'however' or 'nonetheless' can be used to highlight contrasting points. Words such as 'therefore' help to indicate either an interim conclusion in the middle of the essay or the final conclusion at the end.

Example Essay Questions:

➢ "There is a time and place for censorship of the internet" Discuss with particular reference to the right of freedom of expression.

➢ "The UK should codify its Constitution" Discuss.

➢ "Developed countries have a higher obligation to tackle climate change than developing countries" Discuss the extent to which you agree with this statement.

➢ "Abortion should only be permitted in certain circumstances" Discuss.

➢ "The government should legalise the sale of human organs" Discuss.

➢ "Sufferers of anorexia nervosa should be force-fed" Do you agree with this statement? If so, evaluate at what point of an individuals' disease this measure should be taken

LNAT Practice Papers

Introduction

The Basics

The Law National Aptitude Test (LNAT) is a 2 hour 15 minutes test that is split into two sections – Section A comprises 42 multiple choice questions based on several passages, and Section B consists of a selection of essay questions from which you will have to attempt one.

Many top law schools use the LNAT, including University of Oxford, University College London and King's College London, hence it is imperative to do well for this test in order to maximise your chances of securing a spot in a top law school.

The only way to improve your LNAT scores, especially in Section A, is to keep practicing and reviewing your answers and examination technique. Hence, we have compiled a few LNAT Mock Papers that have been meticulously written by our expert tutors, designed to resemble the actual test as much as possible.

There is a dearth of information available freely which understandably makes students nervous and unprepared for the test. Our Mock Papers come with expert solutions that aim to let you know where you've gone wrong and prepare you as much as possible for the actual test.

Preparing for the LNAT

Before going any further, it's important that you understand the optimal way to prepare for the LNAT. Rather than jumping straight into doing mock papers, it's essential that you start by understanding the components and the theory behind the LNAT by using a LNAT textbook. Once you've finished the non-timed practice questions, you can progress to using official LNAT papers. These are freely available online at **www.uniadmissions.co.uk/LNAT-past-papers** and serve as excellent practice. Finally, once you've exhausted past papers, move onto the mock papers in this book.

Already seen them all?

So, you've run out of past papers? Well hopefully that is where this book comes in. It contains four unique mock papers; each compiled by Oxbridge law tutors at *UniAdmissions* and available nowhere else.

Having successfully gained a place on their course of choice, our tutors are intimately familiar with the LNAT and its associated admission procedures. So, the novel questions presented to you here are of the correct style and difficulty to continue your revision and stretch you to meet the demands of the LNAT.

Revision Timetable

To help stay organised try filling in the example revision timetable below, remember to factor in enough time for short breaks, and stick to it! Remember to schedule in several breaks throughout the day and actually use them to do something you enjoy e.g. TV, reading, YouTube etc.

	8AM	10AM	12PM	2PM	4PM	6PM	8PM
MONDAY							
TUESDAY							
WEDNESDAY							
THURSDAY							
FRIDAY							
SATURDAY							
SUNDAY							
EXAMPLE DAY		School			Section 1	Essay	Dinner

Getting the most out of Mock Papers

Mock exams can prove invaluable if tackled correctly. Not only do they encourage you to start revision earlier, they also allow you to **practice and perfect your revision technique**. They are often the best way of improving your knowledge base or reinforcing what you have learnt. Probably the best reason for attempting mock papers is to familiarise yourself with the exam conditions of the LNAT as they are particularly tough.

Start Revision Earlier

Thirty five percent of students agree that they procrastinate to a degree that is detrimental to their exam performance. This is partly explained by the fact that they often seem a long way in the future. In the scientific literature this is well recognised, Dr. Piers Steel, an expert on the field of motivation states that *'the further away an event is, the less impact it has on your decisions'*.

Mock exams are therefore a way of giving you a target to work towards and motivate you in the run up to the real thing – every time you do one treat it as the real deal! If you do well then it's a reassuring sign; if you do poorly then it will motivate you to work harder (and earlier!).

Practice and perfect revision techniques

In case you haven't realised already, revision is a skill all to itself, and can take some time to learn. For example, the most common revision techniques including **highlighting and/or re-reading are quite ineffective** ways of committing things to memory. Unless you are thinking critically about something you are much less likely to remember it or indeed understand it.

Mock exams, therefore allow you to test your revision strategies as you go along. Try spacing out your revision sessions so you have time to forget what you have learnt in-between. This may sound counterintuitive but the second time you remember it for longer. Try teaching another student what you have learnt; this forces you to structure the information in a logical way that may aid memory. Always try to question what you have learnt and appraise its validity. Not only does this aid memory but it is also a useful skill for the LNAT, Oxbridge interviews, and beyond.

Improve your knowledge

The act of applying what you have learnt reinforces that piece of knowledge. An essay question may ask you about a fairly simple topic, but if you have a deep understanding of it you are able to write a critical essay that stands out from the crowd. Essay questions in particular provide a lot of room for students who have done their research to stand out, hence you should always aim to improve your knowledge and apply it from time to time. As you go through the mocks or past papers take note of your performance and see if you consistently under-perform in specific areas, thus highlighting areas for future study.

Get familiar with exam conditions

Pressure can cause all sorts of trouble for even the most brilliant students. The LNAT is a particularly time pressured exam with high stakes – your future (without exaggerating) does depend on your result to a great extent. The real key to the LNAT is overcoming this pressure and remaining calm to allow you to think efficiently.

Mock exams are therefore an excellent opportunity to devise and perfect your own exam techniques to beat the pressure and meet the demands of the exam. **Don't treat mock exams like practice questions – it's imperative you do**

Before using this Book

Do the ground work
➢ Understand the format of the LNAT – have a look at the LNAT website and familiarise yourself with it: www.lnat.ac.uk/test-format
➢ Improve your written English by practicing writing and reading frequently.
➢ Try to broaden your reading by learning about different topics that you are unfamiliar with as the essay topics can vary greatly.
➢ Learn how to understand a writer's viewpoint by reading news articles and having a go at summarising what the writer is arguing about.
➢ Be consistent – slot in regular LNAT practice sessions when you have pockets of free time.
➢ Engage in discussion sessions with your friends to give you more ideas about certain essay topics.
➢ Download the LNAT simulator and have a go at doing in online – the actual test is done on a computer so you will want to be familiar with the format – www.uniadmissions.co.uk/lnat-past-papers

Ease in gently
With the ground work laid, there's still no point in adopting exam conditions straight away. Instead invest in a beginner's guide to the LNAT, which will not only describe in detail the background and theory of the exam, but take you through section by section what is expected.

Questions are seldom repeated, so don't rote learn methods or facts. Instead, focus on applying prior knowledge to formulate your own approach. If you're really struggling and have to take a sneak peek at the answers, then practice thinking of alternative solutions, or arguments for essays. It is unlikely that your answer will be more elegant or succinct than the model answer, but it is still a good task for encouraging creativity with your thinking. Get used to thinking outside the box!

Accelerate and Intensify
Start adopting exam conditions after you've done the official mock papers. Remember that **time pressure makes the LNAT hard** – if you had as long as you wanted to sit the exam you would probably get 100%. Doing all the mock papers in this book is a good target for your revision. Choose a paper and proceed with strict exam conditions. Take a short break and then mark your answers before reviewing your progress. For revision purposes, as you go along, keep track of those questions that you guess – these are equally as important to review as those you get wrong.

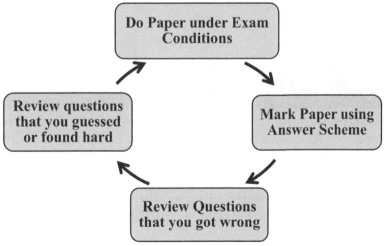

Once you've exhausted all the past papers, move on to tackling the unique mock papers in this book. In general, you should aim to complete mock papers every night in the seven days preceding your exam.

Section A: An Overview

What will you be tested on?	No. of Questions	Duration
Reading comprehension skills, deducing arguments, understanding certain literary tools, questioning assumptions	42 MCQs	95 Minutes

This is the first section of the LNAT, comprising several passages to read and a total of 42 MCQ questions. You have 95 minutes in total to complete the MCQ questions, including reading time for 10-12 passages. In order to keep within the time limit, you realistically have about two minutes per question as you have to factor in the reading time for the passages as well.

Not all the questions are of equal difficulty and so as you work through the past material it is certainly worth learning to recognise quickly which questions you should spend less time on in order to give yourself more time for the trickier questions.

Deducing arguments

Several MCQ questions will be aimed at testing your understand of the writer's argument. It is common to see questions asking you 'what is the writer's view?' or 'what is the writer trying to argue?'. This is arguably an important skill you will have to develop as a law student, and the LNAT is designed to test this ability. You have limited time to read the passage and understand the writer's argument, and the only way to improve your reading comprehension skill is to read several well-written news articles on a daily basis and think about them in a critical manner.

Assumptions

It is important to be able to identify the assumptions that a writer makes in the passage, as several questions might question your understand of what is assumed in the passage. For example, if a writer mentions that 'if all else remains the same, we can expect our economic growth to improve next year', you can identify an assumption being made here – the writer is clearly assuming that all external factors remain the same.

Fact vs. Opinion

It is important to be able to decipher whether the writer is stating a fact or an opinion – the distinction is usually rather subtle and you will have to decide whether the writer is giving his or her own personal opinion, or presenting something as a fact. Section A may contain questions that will test your ability to identify what is presented as a fact and what is presented as an opinion.

Fact	Opinion
'There are 7 billion people in this world...'	'I believe there are more than 7 billion people in this world...'
'She is an Australian...'	'She sounded like an Australian...'
'Trump is the current President...'	'Trump is a horrible President...'
'Vegetables contain a lot of fibre...'	'Vegetables are good for you...'

Section B: An Overview

Section 2 is usually what students are more comfortable with – after all, many GCSE and A Level subjects require you to write essays within timed conditions. It does not require you to have any particular legal knowledge – the questions can be very broad and cover a wide range of topics.

What will you be tested on?	No. of Questions	Duration
Your ability to write an essay under timed conditions, your writing technique and your argumentative abilities	1 out of 3	40 Minutes

Here are some of the topics that might appear in Section 2

> Science
> Politics
> Religion
> Technology
> Ethics

> Morality
> Philosophy
> Education
> History
> Geopolitics

As you can see, this list is very broad and definitely non-exhaustive, and you do not get many choices to choose from (you have to write one essay out of three choices). Many students make the mistake of focusing too narrowly on one or two topics that they are comfortable with – this is a dangerous gamble and if you end up with three questions you are unfamiliar with, this is likely to negatively impact your score.

You should ideally focus on 3-4 topics to prepare from the LNAT, and you can pick and choose which topics from the list above are the ones you would be more interested in. Here are some suggestions:

Science

An essay that is related to science might relate to recent technological advancements and their implications, such as the rise of Bitcoin and the use of blockchain technology and artificial intelligence. This is interrelated to ethical and moral issues, hence you cannot merely just regurgitate what you know about artificial intelligence or blockchain technology. The examiners do not expect you to be an expert in an area of science – what they want to see is how you identify certain moral or ethical issues that might arise due to scientific advancements, and how do we resolve such conundrums as human beings.

Politics

Politics is undeniably always a hot topic and consequently a popular choice amongst students. The danger with writing a politics question is that some students get carried away and make their essay too one-sided or emotive – for example a student may chance upon an essay question related to Brexit and go on a long rant about why the referendum was a bad idea. You should always remember to answer the question and make sure your essay addresses the exact question asked – do not get carried away and end up writing something irrelevant just because you have strong feelings about a certain topic.

Religion

Religion is always a thorny issue and essays on religion provide strong students with a good opportunity to stand out and display their maturity in thought. Questions can range from asking about your opinion with regards to banning the wearing of a headdress to whether children should be exposed to religious practices at a young age. Questions related to religion will require a student to be sensitive and measured in their answers and it is easy to trip up on such questions if a student is not careful.

Education

Education is perhaps always a relatable topic to students, and students can draw from their own experience with the education system in order to form their opinion and write good essays on such topics. Questions can range from whether university places should be reduced, to whether we should be focusing on learning the sciences as opposed to the arts.

Section B: Revision Guide

SCIENCE

Resource	What to read/do
1. **Newspaper Articles**	• The Guardian, The Times, The Economist, The Financial Times, The Telegraph, The New York Times, The Independent
2. **A Levels/IB**	• Look at the content of your science A Levels/IB if you are doing science subjects and critically analyse what are the potential moral/ethical implications • Use your A Levels/IB resources in order to seek out further readings – e.g. links to a scientific journal or blog commentary • Remember that for your LNAT essay you should not focus on the technical issues too much – think more about the ethical and moral issues
3. **Online videos**	• There are plenty of free resources online that provide interesting commentary on science and the moral and ethical conundrums that scientists face on a daily basis • E.g. Documentaries and specialist science channels on YouTube • National Geographic, Animal Planet etc. might also be good if you have access to them
4. **Debates**	• Having a discussion with your friends about topics related to science might also help you formulate some ideas • Attending debate sessions where the topic is related to science might also provide you with excellent arguments and counter-arguments • Some universities might also host information sessions for sixth form students – some might be relevant to ethical and moral issues in science
5. **Museums**	• Certain museums such as the Natural Science Museum might provide some interesting information that you might not have known about
6. **Non-fiction books**	• There are plenty of non-fiction books (non-technical ones) that might discuss moral and ethical issues about science in an easily digestible way

POLITICS

Resource	What to read/do
1. **Newspaper Articles**	• The Guardian, The Times, The Economist, The Financial Times, The Telegraph, The New York Times, The Independent
2. **Television**	• Parliamentary sessions, Prime Minister Questions • Political news
3. **Online videos**	• Documentaries and YouTube Channels
4. **Lectures**	• University introductory lectures • Sixth form information sessions
5. **Debates**	• Debates held in school, Joining a politics club
6. **Podcasts**	• Political podcasts • Listen to both sides to get a more rounded view (e.g. listening to both left and right wing podcasts)

RELIGION

Syllabus Point	What to read/do
1. Newspaper Articles	• The Guardian, The Times, The Economist, The Financial Times, The Telegraph, The New York Times, The Independent
2. Non-fiction books	• Read up about books that explain the origins and beliefs of different types of religion • E.g. Books that talk about the origins of Christianity, Islam or Buddhism, theology books etc.
3. Talking to religious leaders	• Talking to religious leaders may be a good way of understanding different religions more and being able to write an essay on religion with more maturity and nuance • Talking to people from different religious backgrounds may also be a good way of forming a more well-rounded opinion
4. Online videos	• Documentaries on religion • YouTube channels providing informative and educational videos on different religions – e.g. history, background
5. Lectures	• Information sessions • Relevant introductory lectures
6. Opinion articles	• Informative blogs and journals • Read both arguments and counter-arguments and come up with your own viewpoint

EDUCATION

Syllabus Point	What to read/do
1. Newspaper Articles	• The Guardian, The Times, The Economist, The Financial Times, The Telegraph, The New York Times, The Independent
2. A Levels/IB	• Draw inspiration from what you are studying in your A Levels or IB – do you feel like what you are studying is useful and relevant? E.g. Studying arts versus science • Compare the education you are receiving with your friends in different schools or different subjects
3. Educational exchange	• If you have an opportunity to go on an educational exchange, this might be a good opportunity to compare and contrast different educational systems • E.g. the approach to education in Germany versus the UK
4. University applications	• Have a read of how different universities promote themselves – do they claim to provide students with academic enlightenment, or better job prospects, or a good social life? • Why do different universities focus on different things?
5. Online videos	• Documentaries • YouTube Channels
6. Talk to your teachers	• Your teachers have been in the education industry for years and maybe decades – talk to them and ask them for their opinion • Talk to different teachers and compare their opinions regarding how we should approach education

Top Tip! Although you aren't required to have extra knowledge for the LNAT essay, doing so will allow you to make your essay stand out from the crowd. However, you should first **prioritise perfecting your writing style rather than doing extra reading** as the former will have a greater impact on your mark.

How to use this Book

If you have done everything this book has described so far then you should be well equipped to meet the demands of the LNAT, and therefore **the mock papers in this book should ONLY be completed under exam conditions**.

This means:
➢ Absolute silence – no TV or music
➢ Absolute focus – no distractions such as eating your dinner
➢ Strict time constraints – no pausing half way through
➢ No checking the answers as you go
➢ Give yourself a maximum of three minutes between sections – keep the pressure up
➢ Complete the entire paper before marking and mark harshly

This means setting aside 2 hours and 15 minutes to tackle the paper. Completing one mock paper every evening in the week running up to the exam is an ideal target.
➢ Return to mark your answers, but mark harshly if there's any ambiguity.
➢ Highlight any areas of concern and read up on the areas you felt you underperformed
➢ If you inadvertently learnt anything new by muddling through a question, go and tell somebody about it to reinforce what you've discovered.

Finally relax… the LNAT is an exhausting exam, concentrating so hard continually for two hours will take its toll. So, being able to relax and switch off is essential to keep yourself sharp for exam day! Make sure you reward yourself after you finish marking your exam.

Scoring Tables

Use these to keep a record of your scores from past papers – you can then easily see which paper you should attempt next (always the one with the lowest score).

Iock Paper A	

You will not be able to give yourself a score for Section 2 per se as an essay is always marked rather subjectively – the best way to gauge your performance for Section 2 will be to compare your arguments and counter-arguments

Mock Papers

Mock Paper A

1. Disc or Download: A Virtual Energy-Savings Debate

Adapted from 2014©. Copyright Environmental News Network

One of the best ways to spark an energy revolution is through the younger generation — and nothing quite speaks their language like video games.

But this issue has less to do with the content of these addictive games and more with how the younger generation consumes them.

Fantasy and adventure, sci-fi and first-person shooters, strategy and racing — video games today come in all types of genres with thousands of add-ons and customizable features to make each story a virtual reality. And with all of these choices comes two more: buy a copy of the video game on a disc or download the video game straight from the console?

At the crossroads of the decision lies Sony, one of the leading manufacturers of video games including the PlayStation® console. Sony offers the choice between direct download through an Internet connection and purchasing via game stores a Blu-ray copy of the game.

Now, one would think the former option is friendlier on the environment — no carbon footprint from shipping or driving to stores. However, cartridge collectors everywhere are about to rejoice. According to the *Journal of Industrial Ecology*, downloading a game from an Internet server can create a larger carbon footprint than driving to a store to purchase the game on Blu-ray disc.

A study was conducted by the *Journal of Industrial Ecology*, and although the report was based on a series of assumptions, the results were as follows:

"The CF [carbon footprint] of the life cycle of a downloaded 8.80 GB game amounted to 21.9 to 27.5 kg CO_2-eq (for lower and upper bounds of Internet energy intensity), whereas the result for a BD [Blu-ray disc] game was 20.8 kg CO_2-eq. Gameplay (use phase) accounted for 19.5 kg CO_2-eq emissions in both scenarios."

As observed, the carbon footprint from driving to the store and purchasing a Blu-ray disc is less comparatively, but there are some caveats. There is a threshold of 1.3 gigabytes, under which PlayStation® games are still more efficient to download. PlayStation® games are only getting larger, though.

According to the *Journal of Industrial Ecology*, game files have doubled in size between 2010 and 2013, and have increased by 25 percent from PlayStation 3 to PlayStation 4.

In fact, the advantage of discs is their ability to store massive amounts of data, which allows them to become exponentially better than direct downloads as their GB capacity increases. All that's needed is a little more transparency for all video games to be rated "E" — no, not "E" for everyone, but "E" for efficient.

1. Why does the author compare '"E" for everyone' and '"E" for efficient?
A. To illustrate the importance of efficiency
B. To use humour to emphasise the importance of universality
C. To bring together themes of environmentalism and gaming
D. To show that efficiency is not the main concern of video-game makers
E. To argue that the efficiency of video games should be more visible to consumers

2. Which of the following is an assumption that the author has made?
A. The younger generation will be more likely to engage in environmental debate if it surrounds a topic which interests them
B. That transparency regarding the efficiency of video games will encourage younger people to think more carefully about which games they purchase
C. That a higher carbon footprint is more detrimental for the environment
D. That a significant proportion of the younger generation have an interest in video games
E. Discs care better than downloads as they are capable of being shared amongst friends

3. Why does the author use a question as a conclusion to the third paragraph?
A. He wants the reader to engage in the debate through online comment
B. He implies that the issue of efficiency should be as important as the choice between different types of game
C. He wants to show that young people are overwhelmed by questions
D. He attempts to show that there is a real choice to be made between the two options, one which shouldn't be overlooked
E. He is trying to create an informal tone to the piece

2. Why Homework Matters

Adapted from 2014©. Edarticle

As an elementary/middle school teacher, I hear constant complaints about the issue of homework. There are valid points against overdoing it and even studies that suggest, in some cases, it doesn't always help. There's a big difference between busy work and assignments that are meaningful. Some researchers, like Sara Bennett and Nancy Kalish, propose that homework is a hidden cause of childhood obesity. Others, like Alfie Kohn, believe that the quality and quantity of assignments done at home should be addressed, pronto. So, why do students today still have to do this archaic activity?

➢ Although some teachers assign busy work, that is not always the case. Often times, the assignments students bring home are not only continuations of what was done in class, but are activities that help apply the knowledge learned earlier in the day.
➢ Some at-home assignments are enrichment activities that will nurture the creative side of children.
➢ It allows parents to see at firsthand what their child is doing in school, and gives them an opportunity to connect or communicate with their child on a different level, seeing where they stand academically.
➢ Even if it's studying or reviewing what was covered in class, it is an important reinforcement activity. This can eventually lead to information being stored into long-term memory.
➢ Research conducted by Harris Cooper (2006) suggests that students who complete homework tend to have higher scores on content related tests, than do students who have not completed content related homework.
➢ It can teach a child responsibility, which in turn will form into a good work ethic that will be useful for the rest of his/her life.
➢ Time management skills will develop and students will learn a valuable lesson about procrastination.
➢ It will allow students the opportunity to teach themselves, and learn how to manage schoolwork independently.

Although there is an ongoing debate about the merits and necessity of homework, the bottom line is that it doesn't look like it's going away any time soon. Sure, parents should question teachers and make sure that the homework is relevant and not going to last five hours, but the best possible action a student can take at the elementary age is to accept its inevitability. This will lead to time management strategies, allowing them to learn how to juggle a variety of activities on a daily basis, just like parents and all other adults do every day.

Studies at Duke University under Harris Cooper found a positive correlation between test scores and homework. "The results of such studies suggest that homework can improve students' scores on the class tests that come at the end of a topic. Students assigned homework in 2nd grade did better in mathematics, 3rd and 4th graders did better on English skills and vocabulary, 5th graders on social studies, 9th through 12th graders on American history, and 12th graders on Shakespeare." Of course, when it is overloaded, there's the case of diminishing returns.

Life in the working world is more competitive than ever before. Children need to learn the methods that will allow them to be successful in the life that's just ahead. Kohn suggests that homework doesn't help elementary school students, when research states otherwise. He is also a proponent of eliminating grades. Kohn's ideas generate excellent debate topics, but are not practical in a world that uses rankings for everything. So, even if you don't buy into the fact that homework will make a child a higher academic achiever in the short term (even though research states otherwise), realize that it just might create a human being with good habits, a rich work ethic, and success later on in the world outside academia.

4. Which of the following is a statement of fact rather than an assertion of opinion?

A. 'Children need to learn the method that will allow them to be more successful'

B. 'Kohn's ideas generate excellent debate'

C. 'Some researchers, like Sara Bennett and Nancy Kalish propose that homework is a hidden cause of childhood obesity'

D. 'There's a big difference between busy work and assignments that are meaningful'

E. 'it just might create a human being with good habits'

5. Which of the following does the author imply in the piece

A. Nurturing of a child's personality is equally as important as academic success

B. Good habits will help a child throughout his/her adult life

C. That s/he is fed up with hearing complaints about the issue of homework

D. Teachers who assign busy work are failing their students

E. Kohn's research is not trustworthy

6. What, according to the author, do children need to learn?

A. The methods to help them earn money in their later life

B. The mindset required in order to become successful in adulthood

C. To manage their schoolwork independently

D. Techniques that will help them to be successful

E. How to deal with an overload of homework

7. What is the main conclusion of the article?

A. Homework leads to good grades, which is important for children's prospects

B. Too much homework can be overwhelming for children

C. Homework is not helpful for academic achievement, but is instrumental in creating a rounded human being

D. Parents' opinions on how much homework their children have should be taken into consideration

E. Homework is likely to be helpful in a child's development, however important it ends up being from an academic perspective

3. What would independence mean for Scotland's racial minorities?

Adapted from 2014© <u>Nasar Meer</u> for the Guardian

Scotland's 'other' has firmly been the English. What happens when it starts to look more inward?

On Edinburgh's Royal Mile, the ancient road running from its 12th-century castle at the top down to the Enric Miralles-designed parliament at its bottom, you'll find some unlikely proponents of Scottish nationalism. Sporting tartan turbans and proudly brandishing the Saltire, the Sikh small-shop owners are sometimes viewed curiously by tourists and festival goers. The example is symbolic because at a time when Scottish identity is being appropriated in various arenas, it does raise the question of where ethnic and racial minorities fit into a country dominated by myths and legends of an ostensibly "white" nation dating back millennia.

Another way of putting this is to say that whatever else "Britishness" might be, we know that it's not the sole preserve of white Christians, but we're less sure about "Scottishness" (as we might be of "Englishness" if we looked close enough). Of course it depends where you go in Scotland, as it does in Britain as a whole, but surveys tell us that the Scots are no more exclusionary in their attitudes than the English.

With only around 2% ethnic minorities, however, it's a theory of tolerance that is yet to be tested and, as any student of nationalism will tell you, there's a fine line between inclusive and exclusive national identities. Until now though, and sectarian issues aside, it appears that Scotland's "other" has firmly been the English – what happens when it starts to look more inward?

Perhaps with the exception of Herman Rodrigues' 2006 exhibition on Scotland's Asian communities, little is said of the "new Scots". It is nevertheless a matter of enormous pride to the SNP that the only ethnic minority MSPs have been members of their party, the most recent being Humza Yousaf who earlier this month swore his oath of allegiance in Urdu, wearing traditional Pakistani clothes supplemented with a band of tartan. Elsewhere, less "visible" minorities, such as the Italians and now eastern Europeans, have stitched themselves into the fabric of Scotland's major cities, as indeed have Chinese groups and other east Asians. Hence one of Edinburgh's most beloved Scottish folk musicians is the Kirkcaldy-born Andy Chung.

Yet the most fascinating feature of Scottish nationalism is also the least noticed: there's a mighty difference between a nation's identity and people's national identities, which reveals itself in the saying that while England owned the British empire, it was the Scots who ran it. No wonder then that the Indian military has a Scottish tartan in its formal regalia (3rd battalion, the Sikh Regiment, traces its lineage from "Rattray's Sikhs" named after Captain Thomas Rattray of the 64th Regiment of Bengal Infantry). Whatever else people in Scotland think makes up their idea of Scottishness, the identity of Scotland as a historical nation cannot really be understood apart from that of India and other places of empire.

8. Why does the author use the example of Sikh shop owners selling Scottish paraphernalia?
A. To help explain the idea the concept of 'unlikely proponents of Scottish nationalism'
B. To raise the question of where ethnic and racial minorities fit into a historically 'white' country
C. To emphasise the concept that Scottish nationalism is not only for white Christians
D. To suggest that white Scottish people are intolerant of other ethnicities purporting to be Scottish nationalists
E. To ridicule the idea of multi-ethnic nationalism

9. What does the final paragraph of the extract conclude?
A. That there is a difference between people's individual national identities and the identities of a nation.
B. That a nation's identity and people within that country's national identity are entirely different concepts
C. The identity of Scotland is identical to those of other nations all around the world
D. That a nation's identity will have an effect on the national identities of individuals
E. That if Scotland loses the England as 'the other' it may turn inwards on its own minorities

10. All of the following are implied or stated in the piece except:
A. Scotland is becoming more multicultural
B. Scottish people are generally wary of their ethnic minorities
C. There has been very little writing done on ethnic minority Scots
D. Britain is established as a multicultural country
E. 'Englishness' as a concept is much less clear as a nationality than 'Britishness'

4. What price for honesty?

Adapted from 2014© Neil Reaich for BizEd

Pricing seems to be mentioned quite a lot in recent news headlines. The latest is the investigation by the Office of Fair Trading (OFT) into just how genuine the advertised price cuts in some large furniture stores really are. The OFT use the term 'reference pricing'. They've found cases in shops under investigation where not a single product had, in reality, been sold at the (supposedly original) higher price. The argument goes that since 95% of sales were at the lower or 'now' price then the stated original prices cannot really be genuine. Are we really so stupid as to be misled by reference pricing? When my wife buys some clothes in a sale and I ask how much it cost, she always responds by saying how much money she has saved. Well, unless she had the definite intention of buying it in the first place, she hasn't saved anything: in fact, she has spent it. The question is whether or not the reference pricing made a difference to her decision to buy the product. It must do, otherwise why is it so commonly used?

Heuristics suggests that people are not as rational as the standard economic model implies. Instead they use rules of the thumb, educated guesses or short cuts in decision-making. Anchoring refers to people making decisions based upon something they know to start with. For example, the anchor price of a jumper was £40 and there is 20 per cent discount offer making the new price £32 a saving of £8. If I decide to buy I might be thinking of the £8 saved because I didn't have to pay the full price, when I should be thinking logically about the actual price I am paying. In this case, there is too much focus on the anchor price, which is at such a level that I wouldn't have purchased the item anyway. In some cases, reference pricing is part of a strategy of price discrimination over time. Clothes shops attempt to capture consumer surplus by charging a high initial price for a few weeks for the 'must have and will pay' customers. The shops then give a discount which increases over time. New potential customers must now weigh up whether to buy now and get a 25 per cent discount or wait longer for a 5 percent saving and risk losing the deal because they have run out of their size.

Another tactic is to use time-limited offers where notice is given that the offer ends soon: buy now or miss out. Double glazing sales were notorious for using this tactic, encouraging customers to 'sign up today to get an extra 20 percent off'. Supermarkets are always giving volume offers, such as three for the price of two or get the second purchase for half the price. The supermarkets know we will pay more for the first item than we would for a second or third one. This tactic must lead to food waste as we are tempted to buy some products that we cannot possibly use before the sell-by date. The OFT also look at baiting sales where only a very limited number of products are available at the most discounted price. This sounds a little like discounts given for advanced booking train tickets. The OFT always makes mention of portioned 'drip' pricing where price increments drip through the buying process. It's those little add-ons that all add-up. Booking certain airline tickets comes to mind here. Restaurants often make an offer of a free second main course meal after buying one main course at full price. Now add on the fact that we may have full priced deserts, order drinks at a high mark up and then add the tip. Get your calculator out and the deal doesn't seem so good.

All these examples of price framing seek to alter a consumer's perception of the value of the offer. But there is more: have you purchased a printer at a ridiculously low price only to be stung on purchasing printer ink cartridges? Some businesses operate at a loss or low profit margin on certain items, to entice customers to part with more money on higher-profit goods. They bundle items together.

Some advice then:
➢ Use a calculator when shopping
➢ When buying clothes, divide the total by 50 to give you the average cost of wearing something once a week for a year: a shirt costing £25 will work out at 50 pence for every day worn;
➢ Avoid impulse buying on larger-ticket items;
➢ Only purchase multi-buys with long sell-by dates;
➢ Ignore the original price: it's only the current price you need to know about;
➢ Avoid limited-time offers.

11. What is the purpose of the inclusion of the phrase 'Booking *certain airlines* comes to mind here.'?
A. The author wants to leave the name of the airline open to individual inference, to make the statement as relevant as possible to all readers
B. The author wants to create a commonality between himself/herself and the reader, thereby making themselves more legitimate
C. The author wants to refer to a specific budget airline or group of airlines, but is fearful of naming one in particular
D. None of the above
E. All of the above

12. Which of the following is *not* a way that the author suggests retailers attempt to persuade customers into spending money on their products?
A. Stating that the 'original' price of an item is higher that it has ever really been sold, thereby creating the illusion of saving money
B. Splitting up the real cost of a product into a number of small, necessary, purchases
C. Raising the price of product - marketing as a 'luxury product' so that it will become more desirable
D. Giving some items for a reduced price if you buy in bulk
E. Slowly decreasing the price over time, creating a highly-pressured decision between buying now, or gambling that a product will be available at a cheaper price later

13. Why is it suggested that people are not as rational as the standard economic model implies?
A. To further the author's main point that people's decisions are influenced by different pricing models implemented by retailers
B. People make mistakes in their decision making - they aren't as thorough as they could be in the analytical process
C. To emphasise that the standard model is wrong, and that we must look beyond it to explain people's commercial behaviour
D. People are irrational, incapable of making decisions based on evidence
E. People are blind to price-based marketing

14. What is the overall purpose of the extract?
A. To advise consumers on how to make more rational decisions when shopping
B. To persuade retailers to be more honest in their pricing strategies
C. To warn consumers of the dangers of shopping without thinking about marketing
D. To inform readers about the different strategies utilised by retailers
E. To explore the concept of 'honest pricing' and come to a conclusion about the meaning of that term

5. A-level students: if you don't get into a Russell Group university, skip going altogether

Adapted from 2014© Christopher Giles for the Telegraph

There is little point getting into thousands of pounds of debt for a degree in 'Make-up and Hair Design', writes Christopher Giles.

Students who have received your A-level results, please listen up. If you have not – or do not – get into a top UK university, don't go at all. Results day is stressful. Euphoria. Relief. Disappointment. Frustration. Anxiety. It's emotional, but you'll be fine. Just follow my advice: if you aren't going to a Russell Group university or otherwise respected institution, forget about it altogether. Life is about making the right investments and university is pretty big one. It has to be worth every penny.

If, after the emotional roller coaster, you do decide to go to university, you are going to have a lot of debt. It is likely you'll owe £43,000, thanks to recent policies from the suits in Westminster. This will be money taken from your income for years to come (perhaps even for 30 years).

It is not just a financial investment. It is an investment of your time. If you live to 81, the average UK life expectancy, you would have spent 2.4 percent of your life at university. If this is 2.4 per cent of your life studying adventure sports, tourism or homeopathy, it probably won't propel your career. It is true that many have a love for academia – but if that's not you, don't be a sheep. Getting into and going to university used to be a big deal; now it's considered the norm. Don't let that fear of missing out sway you.

Time for some optimism: choosing not to go to university because you haven't got into one of the elite institutions shows good judgment. The appetite for mass university attendance is flagging. In fact, the stigma attached to having a weak degree could be worse than not having one at all. It is a nice sentiment to want everyone to go to university, but in reality it's not such a good idea.

To help you out, I've found some of the worst universities for graduate employability (via The Complete University Guide) and some strange courses to avoid. These universities really are the bottom of the pile and here is a selection of courses they offer, along with annual tuition fees.

The University of Sunderland: Fashion Product and Promotion (£7,800), Glass and Ceramics (£8,500), Sports Journalism (£8,500).

Southampton Solent University: Make-up and Hair Design, Football Studies, Cruise Industry Management. All of these courses cost £8,050.

The University of East London: Songwriting, Computer Game Design (Story Development), Hospitality and International Tourism Management. The University of East London ranks bottom in terms of graduate prospects in the UK and they charge £9,000 a year for full-time undergraduate courses.

Despite the huge debt, and wasted hours of your life, university does provide you with general life skills. Like the ability to work with a hangover, pay rent, and call your mum every weekend. Virtues these may be, but I am certain life experience isn't exclusive to university. It is just a costly way of getting it. Especially if what you're paying for is a tumbleweed degree that's going to float straight past employers.

The UK has some of the world's best universities. It also has quite a few average ones. So when you finally get through UCAS's 10,000-mile internet traffic jam, try to think with a cold rational mind: is it worth it?

15. What assumption does the author make in his discussion of university as 'an investment'?
A.   Going to university will cost a significant amount of money
B.   People will not be supported by family members in their financial commitment to pay university fees
C.   Time spent at university is time wasted
D.   Politicians do not have students' best interests at heart
E.   There is no other reason why someone would choose to go to university except as an investment for the future

16. Which of the following is true according to the author?
A.   University teaches people more than only the subject they study
B.   A decision not to go to university is indicative of shrewdness
C.   Since the rise in fees, university has become too expensive
D.   A degree in 'Glass and Ceramics' is worthless
E.   Only Oxbridge universities are worth attending

17. Which of the following is a possible description of the writer's opinion of the nature of the university system inferred from this passage?
A.   Extortionate
B.   Beneficial
C.   Flawed
D.   Worthless
E.   Good value for money

6. Maria Miller: thank you and goodbye

Adapted from 2014© Simon Tait for 'The Stage' (edited)

She "has done an effective job in making the case for the value of public funding" Bazalgette says in response to the news that the arts are to get a ring-fenced 5% cut when ACE had been told to model for 10% and 15% scenarios for 2015/16.

But it is rather damning with faint praise. He goes on to say: "It is hugely encouraging that the Chancellor and the Treasury have listened to the argument that the arts and culture makes such a valuable contribution to our quality of life and the economy".

The argument made by the secretary of arts and culture, note, not the secretary of state.

DCMS (Department for Culture, Media and Sport) as a whole has got a less generous cut of 8%, and the deal for the arts appears to have been negotiated separately by the likes of Bazalgette and national museum directors like Nicholas Serota who, three weeks ago, went to George Osborne directly with economic arguments for a more lenient treatment of the sector. It was at this point that Osborne and the Treasury finally "got it" and realised how damaging a bigger arts cut would be to the economy, for negligible saving.

It means that Mrs. Miller cannot simply pass on to the arts the 8% cut as she and her predecessor, Jeremy Hunt, have done in the past because there is no fat in the DCMS, having been cut to the bone already, to absorb a new reduction itself. As it is, she will have to find the saving from elsewhere in her budget. Nevertheless, she has hung out and got a better settlement than most other government departments who are suffering at least 10% reductions as the government tries to find more savings, but it seems the knives are out not for culture or the arts but for Miller herself.

The knives appear to be out for her in government for a number of reasons, including not dealing decisively with Leveson. The culture secretary has also been the subject of unprecedented vilification in the Tory press, with the Daily Mail's drama critic Quentin Letts declaring a couple of weeks ago that <u>"Culture is the department where a country can assert its character. If only its Secretary of State had one"</u>. In May, she made her first speech on the arts, calling for the economic argument to be made, Letts conceded, but "Where was the question of morality in Mrs. Miller's approach to the arts? Where was the vision that the arts can civilise us? Where was the idea of the arts as the most meritocratic of gifts, a route which can offer talented and aspiring youngsters a route to self-fulfillment…? There is not even much impression she is an arts lover. It was a speech that could have been given by any one of her departmental officials".

Her desperate attempts to grab a positive headline culminated last week in a damp squib of an announcement about the First World War centenary commemoration, in which nothing new was announced (except that 600-odd streets in England were to be renamed after VC winners from the Great War), and the major news about the cultural element cannot be revealed before August. On Friday, The Times's normally gentle columnist Richard Morrison wrote that "Some (culture secretaries) have been bores; some bluffers. But not one has depressed me as Maria Miller does.

As for the arts, the triumph is substantial and this might be a seachange in the way governments see the sector. The Arts Council, as fuel for the Bonfire of the Quangos, has taken an enormous battering since 2010 and the sector has correctly acknowledged that there is no case for "special treatment" while cuts amounting to 33% have been meted out, and of 50% to ACE itself. But now culture has established the principle that it is a special case after all, and with sense and imagination much of the effect of the new 5% cut might be ameliorated through the National Lottery.

The question now is whether that principle will be accepted by the other great subsidisers of the arts, the local authorities in whose hands the futures of dozens of theatres lie and whose extreme economic pain is even greater than Osborne's.

18. Which of these premises, if true, would most weaken the author's argument for the importance of funding for the arts?

A. Funding for the Emergency services is in decline and has been for years
B. The EU has recently passed legislation mandating that each country must increase state funding for the arts yearly in line with inflation
C. There is no evidence that the cost involved in increasing funding for the arts is justified by the eventual benefits
D. There is little evidence to suggest that funding for the arts would benefit a broad range of people
E. There is a correlation between increased funding for the arts and drop in   the average GCSE grade

19. None of the following show the author's distaste for Maria Miller except:

A. 'not one has depressed me as Maria Miller does'
B. 'her desperate attempts to grab a positive headline'
C. 'it is rather damning with faint praise'
D. 'she has hung out and got a better settlement'
E. 'In May she made her first speech on the arts'

20. Which of the following is not stated in the extract?

A. The culture secretary now cannot pass cuts in budget straight onto the arts
B. Ms Miller is not universally accepted within government
C. The arts have had a better than expected outcome in relation to forecasts
D. The National Lottery's input will make up the whole of the 5% cut in the arts
E. Ms Miller announced a 'First World War centenary commemoration'

21. What, according to the author, is 'the question'?

A. Whether the idea that arts are more worthy than they once were is yet widely accepted
B. Whether the idea that arts are more worthy than they once were can be widely accepted
C. Whether the idea that arts are a 'special case' is correct given the many differing opinions
D. Whether the arts should be considered a 'special case'
E. Whether the idea that the arts should be treated differently will be reflected at a regional level

7. What makes a terrorist?

Adapted from 2014© Simon Allison for the Institute of Security Studies (edited)

Given the extent to which it dominates global news and politics, it is remarkable how little we know about the men – and, very occasionally, women – behind Islamic extremism. How are people drawn into such radical politics? What type of person becomes a terrorist? What is it that forces radicals out of day-to-day politics and into the extreme and often violent margins of society?

To discourage radicalisation we need a proper answer to the question of why and how people are radicalised. Anneli Botha, a senior researcher on terrorism at the Institute for Security Studies, asked members of radical groups themselves, specifically al-Shabaab and the Mombasa Republican Council.

Al-Shabaab is an Islamic extremist group that seeks principally to create an Islamic state in Somalia, although its ambitions and operations don't stop at the Somali border. The MRC, although often associated with al-Shabaab, is a distinct organisation with a very different agenda. It advocates for the secession of Kenya's coastal areas, and emphasises land grievances and economic and political marginalisation. It is a predominantly, but not exclusively, Muslim organisation.

95 interviews were conducted with individuals associated with al-Shabaab and 45 individuals associated with the MRC. Most of the respondents (96%) were male. From these interviews, Botha was able to observe patterns around their education, family background, religion and ethnicity as well as interrogate their motivations for joining and staying with radical groups.

There is no such thing as a typical extremist, however, it is possible to observe certain trends from the data. Almost all respondents grew up in a male-dominated household. Even when the father had passed away, usually a male relative stepped in to assume the patriarchal role. Over 70% of all respondents experienced corporal punishment at home, although most said that this wasn't too severe. Unexpectedly, around 60% of all respondents were middle children. "Middle children are known to experience the greatest sense of 'not belonging'," said Botha, explaining that this makes them especially vulnerable to close-knit radical groups, which can fill this void.

"When asked to clarify [what] finally pushed them over the edge, the majority of both al-Shabaab and MRC respondents referred to injustices at the hands of Kenyan security forces, specifically referring to 'collective punishment'," said Botha.

Respondents complained that "all Muslims are treated as terrorists" and that "government and security forces hate Islam". Some pointed to more specific examples, such as the assassination of Muslim clerics, or even particular incidents, such as an alleged assault by Kenyan police on a group of Muslims.

It is this last point that is most relevant for policymakers who are looking to contain the threat of extremism in Kenya. Simply put, a counter-terrorism strategy that relies on mass arrests, racial profiling and extrajudicial killings is counter-productive. These tactics have radicalised dozens, if not hundreds, of individuals, argues Botha, "ensuring a new wave of radicalism and collective resolve among their members, ultimately indicating that threats of violence or imprisonment are rarely effective deterrents."

Botha's research gives unprecedented insight into the background and motivations of the people who make up extremist organisations in Kenya. This is what a Kenyan radical looks like. Now it is up to Kenya's policymakers to tailor their responses accordingly.

22. According to the article, which of the following statements are accurate?
A.   Extremists tend to share similar childhood experiences
B.   Extremists are mostly male
C.   Recent counter-terrorism strategies have led to increased radicalisation
D.   All of the above
E.   Both B and C

23. Based on the article, which of these pairs of ideas are not shown to be causally related?
A.   Global coverage of radicalism and rise in extremism
B.   Mass arrests and radicalism
C.   Patriarchy and membership of extremist groups
D.   An understanding of radicalisation and a reduction in radicalisation
E.   Generalisation and frustration

24. What can be inferred from Botha's research as the definitive reason why most people become radicalised?
A.   Patriarchal upbringing
B.   Being a 'middle child'
C.   Being overtly religious
D.   None of the above
E.   Impossible to say

25. Which of the following can be said to be the main conclusion of the final two paragraphs?
A.   Radicalism is not always a result of indoctrination, background and childhood circumstances play a part too
B.   The knowledge gained from this study about extremism in Kenya can be used to inform governments throughout the world
C.   Policymakers should realise that most extremists identify unjust government activity as the ultimate reason why they became radicalised
D.   Policymakers should take into consideration that the threat of extremism grows with an aggressive counter-terrorism policy
E.   Policymakers should realise that extremism is best tackled with greater understanding of the reasons behind fanaticism

8. Drugs laws play 'cat and mouse' with creators of legal highs, says senior government adviser

Adapted from 2014© David Barrett for the Telegraph

Drug control legislation is being forced to play "cat and mouse" with the creators of an ever-expanding range of new "legal highs", the government's senior drug adviser has warned.

Professor Les Iversen, chairman of the Advisory Council on the Misuse of Drugs (ACMD), said the category of synthetic drugs – formally known as "novel psychoactive substances" – was growing all the time and legislation was struggling to keep up. He said the council, which advises the Home Secretary on classifying legal and illegal substances, had already reviewed the controls on legal highs in 2009 and 2012 before the launch of its current enquiry earlier this year.

"It looks as though we might have to go on doing this," he said. "It's a cat and mouse game but the cat should not withdraw in defeat."

Chemists who create synthetic drugs, such as mephedrone, to mimic the effects of illegal drugs like cocaine or heroin have been known to tweak the chemical composition of substances to stay one step ahead of the law. A European body which monitors new substances appearing on the black market said 80 different new types of legal high appeared last year.

Prof. Iversen added that the term "legal high" was now inappropriate because many of the substances, including mephedrone, have been made illegal. "The number of serious harms such as deaths emerging from this group of drugs is not all that staggering but it is an issue that we will certainly continue to consider," he said, adding that several announcements on the council's work in this area are due soon once they have secured ministerial approval.

The chairman also said during a meeting of the ACMD in central London that the growth of online pharmacies was leading to far wider abuse of prescription drugs which are legitimately available as painkillers or other types of medication.

"The misuse of medicines has become more of a topic because prescription medicines are so much more easily available on the internet," said the chairman, a retired Oxford University professor of pharmacology. "It's pretty easy to buy most medical products and this gives us a problem that is not easy to solve. We are going to try to evaluate the magnitude of the problem."

Prof. Ray Hill, chairman of the ACMD's technical committee, said it was currently looking at abuse of a drug called pregabalin which is used to alleviate pain and anxiety, and to treat epilepsy. "We are getting a variety of reports now from various parts of the UK that this drug is being misused," said Prof Hill.

"It's a drug that has some similar characteristics to opioid drugs when taken in very high doses. It's probably too soon to have substantive data on its harmful effects."

Prof. Iversen said: "The misuse of pregabalin is a surprise to all of us. This is a relatively new phenomenon and we're looking at how serious a problem this is."

26. Why is it suggested that the misuse of medicines is becoming more prevalent?
A.   Growth of online pharmacies
B.   Widespread use of the internet
C.   The range of 'legal highs' is 'ever-expanding'
D.   Legislation cannot keep up with chemists creating synthetic drugs
E.   It is becoming easier to buy prescription drugs

27. Which of the following can be seen as being implied by Professor Iversen?
A.   Legislators are giving up on the fight against legal highs
B.   There is a labelling issue with so-called 'legal highs'
C.   Pregabalin is being used more widely than before
D.   The problem is becoming more difficult to deal with
E.   The misuse of medicine is a longstanding issue

28. How many individuals are referred to in the extract?
A.   1
B.   2
C.   3
D.   4
E.   5

9. The world needs to talk about child euthanasia: Mercy for all?

Adapted from 2014© David Barrett for the New Scientist

EUTHANISING an infant is not technically difficult. Intravenous sedatives are used to silence the brain, followed by a pain medication such as morphine. This is often enough to trigger respiratory arrest and death, but if not, neuromuscular blockers are added, and the child dies. The process takes 5 to 10 minutes. Belgium has just become the first country to legislate in favour of child euthanasia at any age. However, there is a partial precedent. In 2005, the Netherlands recognised the Groningen protocol, a set of criteria outlining the circumstances under which ending the life of an infant under the age of 1 is permissible. Under those guidelines – which were written by Verhagen – euthanasia can only be undertaken if an infant's diagnosis and prognosis are certain and confirmed by an independent doctor, there is evidence of hopeless and unbearable suffering, both parents give their consent, the procedure follows medical standards, and all details are documented. Dutch children aged between 1 and 12 cannot be euthanised under any circumstances, although Verhagen and others are working to change that. While euthanasia remains technically illegal for infants in the Netherlands, doctors are not prosecuted so long as the protocol's criteria are met.

Opponents argued that this would lead to a slippery slope of infant euthanasia. The opposite happened. Since 2005, there have been only two cases in the Netherlands. Both involved babies with lethal epidermolysis bullosa, a disease of the connective tissues. This decline in euthanasia correlates with an increase in late-term abortions. Previously, most euthanasia cases involved babies born with severe to extreme spina bifida – a congenital disorder in which some of the vertebrae do not fully form. Doctors found that surgery was not possible and that the child would suffer constantly. In 2007, the Netherlands began offering free ultrasound scans at 20 weeks of pregnancy, at which point spina bifida can be detected. Mothers whose babies are diagnosed with the disease can then decide whether to terminate the pregnancy. This is not necessarily the best course for everyone in this situation. Only the most extreme cases of spina bifida are deemed hopeless, and it is impossible for doctors to precisely gauge the severity in utero. Having infant euthanasia as an option allows mothers to be sure that their baby has no chance of survival before ending its life. But, as Verhagen says, in practice most in this situation decide not to take any chances and terminate the pregnancy.

The means of ending a baby's life are subject to debate. Recently, the line between proactive palliative care – applying pain medications that may hasten death – and euthanasia has become more blurred. In some countries, including the US, food and fluid may be withdrawn in some circumstances. But palliative care practices do not necessarily result in a quick death for a terminally ill infant. Death by dehydration and starvation can take days or weeks and it is impossible to guarantee that the child – even heavily medicated – does not suffer. Moreover, no one doubts that death is the outcome of withholding life-sustaining care and support. Rather than draw out the inevitable, would it not be less cruel to swiftly end that life, alleviating all risk of unnecessary suffering? Belgium and the Netherlands have chosen to face this dilemma directly. Of course, not every country is as progressive, there will always be those who – due to religious or personal beliefs – oppose ending a human life. In the US, for example, reaching a federal consensus on the subject of infant euthanasia seems unlikely. On the other hand, progressive states such as Oregon might someday implement their own laws on it, much as they have for assisted suicide in adults. Whether this will ever come to pass remains to be seen. As Verhagen wrote in *The New England Journal of Medicine*, "This approach suits our legal and social culture, but it is unclear to what extent it would be transferable to other countries." For most parts of the world, a refusal to even discuss the subject dominates. As unpleasant as it is, parents, physicians, hospitals and nations need to confront this issue as a matter of responsibility towards both infants born into hopeless circumstances and their families.

29. What assumption does the author make in the last paragraph?
A. That late term abortion has recently become legalised
B. That fathers do not have a say in whether or not a child is to be aborted
C. It is a positive step that infant euthanasia has decreased in frequency
D. Epidermolysis Bullosa will always lead to early death
E. Mothers who terminate a pregnancy without being sure that their foetus' disease will be fatal are negligent

30. What is the main conclusion of the article?
A. Spina bifida is a horrific disease
B. Infant euthanasia should be widely legalised
C. Late term abortion can pre-empt a logical decision on the continuation of an infant's life
D. More of an agreement should be reached about the benefits of legalisation of child euthanasia in certain, narrow, circumstances
E. Infant euthanasia should be discussed more openly

31. Which of the following (if true) strengthens the author's argument?
A. Spina bifida's survival rate for newborns is 10%
B. Upon hearing about the inhumane palliative care treatment, 70% of parents with children with terminal, painful illness would choose euthanasia above practice palliative care options
C. Belgium's child mortality rate has risen by 25% since the legalisation
D. Spina bifida has been proven to cause continuous excruciating pain to infants from the moment of birth
E. The President of the European Union has given a speech suggesting that an investigation should be conducted into the possible benefits and problems of extending palliative care practices to cover child euthanasia

32. What is most likely to be the reason that the author begins the article with the sentence 'Euthanizing a child is not technically difficult.'?
A. Stating that euthanasia is not *technically* difficult implies that the real problems lies in its *emotional* difficulty
B. To grab the readers' attention at the beginning of the extract
C. To inform the reader about the methods used in infant euthanasia
D. None of the above
E. All of the above

10. Brazil releases 'good' mosquitoes to fight dengue fever

Adapted from 2014© Julia Carneiro for the BBC

The hope is they will breed, multiply and become the majority of mosquitoes, thus reducing cases of the disease.

The initiative is part of a programme also taking place in Australia, Vietnam and Indonesia.

The intracellular bacteria, Wolbachia, being introduced cannot be transmitted to humans.

The programme started in 2012 says Luciano Moreira of the Brazilian research institute Fiocruz, who is leading the project in Brazil. "Our teams performed weekly visits to the four neighbourhoods in Rio being targeted. Mosquitoes were analysed after collection in special traps. "Transparency and proper information for the households is a priority."

Ten thousand mosquitoes will be released each month for four months with the first release in Tubiacanga, in the north of Rio.

'Good' bacteria

The bacterium Wolbachia is found in 60% of insects. It acts like a vaccine for the mosquito which carries dengue, Aedes aegypti, stopping the dengue virus multiplying in its body.

Wolbachia also has an effect on reproduction. If a contaminated male fertilises the eggs of a female without the bacteria, these eggs do not turn into larvae. If the male and female are contaminated or if only the female has the bacteria, all future generations of mosquito will carry Wolbachia.

As a result, Aedes mosquitoes with Wolbachia become predominant without researchers having to constantly release more contaminated insects.

In Australia this happened within 10 weeks on average. The research on Wolbachia began at the University of Monash in Australia in 2008. The researchers allowed the mosquitoes to feed on their own arms for five years because of concerns at the time Wolbachia could infect humans and domestic animals.

Three more neighbourhoods will be targeted next, and large scale studies to evaluate the effect of the strategy are planned for 2016.

Dengue re-emerged in Brazil in 1981 after an absence of more than 20 years. Over the next 30 years, seven million cases were reported. Brazil leads the world in the number of dengue cases, with 3.2 million cases and 800 deaths reported in the 2009-14 period.

33. What is the ultimate aim of the project?
A.   To release 'good' mosquitoes
B.   To reduce the number of people suffering from Dengue Fever
C.   To reduce the number of mosquitoes carrying Dengue Fever
D.   To reduce the number of people suffering from diseases carried by mosquitoes
E.   To follow in Australia's footsteps

34. Which of the following statements would be most damning to the decision to release 'good' mosquitoes?
A.   The discovery that Wolbachia has the capacity to mutate such that it can be transmitted to, and will be fatal to, humans
B.   The discovery that Wolbachia does not act as a vaccine for Aedes aegypti
C.   The discovery that Wolbachia can only be transmitted from male to male mosquitoes
D.   The discovery that the evidence from Australia that showed that Wolbachia could not infect humans was tampered with
E.   The discovery that Wolbachia can be transmitted to, and causes infertility in, humans

35. Which of the following is not addressed as a potential (but surmountable) issue when releasing mosquitoes?
A.   Members of the public panicking about the sudden influx of mosquitoes
B.   Wolbachia being transmitted to humans
C.   The whole process taking longer than 10 weeks
D.   The need for more and more mosquitoes to be released periodically
E.   Wolbachia being transmitted to livestock

11. If you're a feminist you'll be called a man-hater. You don't need rebranding

Adapted from 2014© Laurie Penny for the Guardian (edited)

Nobody likes a feminist. At least not according to researchers at the University of Toronto, following a study where it emerged that people still defer to stereotypes about "typical" feminist activists, stereotypes including "man-hating" and "unhygienic". These stereotypes are apparently seriously limiting the appeal of women's liberation as a lifestyle choice. Feminism is a mess, and needs to sort itself out. In order to be "relevant to young women today" it needs to shave its legs and get a haircut.

Elle seems to think so too. The fashion and beauty magazine, not a historically notable manual for gender revolution has weighed in this month with a spread on "rebranding feminism", asking three advertising agencies to give gender politics a nip, tuck and polish. The result is flowcharts and a lot of hot pink equivocation that airbrushes out the ugly, uncomfortable bits of women's liberation. They'd prefer us to consider men's feelings first when we speak about work, pay and sexual violence, to be less threatening, to dress it up; they'd prefer us to talk about "equalism" if we must speak at all. Those with a vested interest in the status quo would prefer young women to act more like they're supposed to – to make everything, including our politics, as pretty and pleasing as possible.

The rebranding of feminism as an aspirational lifestyle choice, a desirable accessory, as easy to adjust to as a detox diet and just as unthreatening, is not a new idea. Nor is ELLE magazine even the first glossy to attempt the task in recent years. But unfortunately there's only so much you can "rebrand" feminism without losing its essential energy, which is difficult, challenging, and full of righteous anger. You can smooth it out and sex it up, but ultimately the reason many people find the word feminism frightening is that it is a fearful thing for anyone invested in male privilege. Feminism asks men to embrace a world where they do not get extra special treats merely because they were born male. Any number of jazzy fonts won't make that easy to swallow.

It is not "young women today" who need to be convinced that feminism remains necessary and "relevant". Changing technology has shaken up a tsunami of activism around gender and politics, from initiatives like the Unslut project and Everyday Sexism to sea changes in culture like the backlash against sexual violence in India. In all of these movements, young women are leading the charge, along with a few fighters from older generations who have not been worn down by decades of mockery and marginalisation. While the fashion press and the beauty industry remain invested in the idea of young women as pliant, affable and terminally anxious about getting boys to like them, real women and girls are fighting back against a culture that persists in trying to present our desires and rebrand our politics as fluffy and marketable.

The stereotype of the ugly, unfuckable feminist exists for a reason – because it's still the last, best line of defence against any woman who is a little too loud, a little too political. Just tell her that if she goes on as she is, nobody will love her. Correct me if I'm wrong, but I've always believed that part of the point of feminist politics – part of the point of any sort of radical politics – is that some principles are more important than being universally adored, particularly by the sort of men who would prefer women to smile quietly and grow our hair out.

In the words of the early suffragist and civil rights campaigner Susan B Anthony: "Cautious, careful people always casting about to preserve their reputation or social standards never can bring about reform. Those who are really in earnest are willing to avow their sympathies with despised ideas and their advocates and bear the consequences."

I am not so very old, but I'm old enough to have noticed that the times in my life when I was most admired by men, the times when I was considered most likeable, were also the times when I was most vulnerable, most powerless and unsure of myself. The times when I've been strongest and most daring, the times when I've been proudest of my own achievements – that's when I've been called a difficult bitch. That's what women get to choose, now as much as at any point in history: how much we are willing to sacrifice to make men like us.

36. All of the following words and phrases suggest disapproval except:
A. 'seriously limiting the appeal of women's liberation'
B. 'has weighed in this week'
C. 'how much we are willing to sacrifice to make men like us'
D. 'the sort of men who would prefer women to smile quietly'
E. 'a lot of hot pink equivocation that airbrushes out the ugly'

37. Why might Laurie Penny uses the word 'jazzy' when describing fonts?
A. To explain to the reader what the fonts look like
B. To draw attention to the illogicality of making feminism more digestible by using the juxtaposition of a seemingly uncomplicated and simple word 'jazzy' against the inherently complex idea of feminism
C. To draw attention to the idea that 'sexing' up the idea of feminism is laughable
D. To belittle what fashion magazines are trying to do
E. All of the above

38. Which of the following phrases can be said to be sarcastic?
A. 'hot pink equivocation that airbrushes out the ugly, uncomfortable bits of women's liberation'
B. '"typical" feminist activitists'
C. 'not a historically notable manual for gender revolution'
D. None of the above
E. All of the above

39. Which of the following is not a description of a feminist found in the extract?
A. Difficult
B. Unsanitary
C. Pliant
D. Relevant
E. Unattractive

12. Complex organic molecule found in interstellar space

Adapted from 2014© Michael Eyre for the BBC (edited)

Scientists have found the beginnings of life-bearing chemistry at the centre of the galaxy.

Iso-propyl cyanide has been detected in a star-forming cloud 27,000 light-years from Earth. Its branched carbon structure is closer to the complex organic molecules of life than any previous finding from interstellar space. The discovery suggests the building blocks of life may be widespread throughout our galaxy.

Various organic molecules have previously been discovered in interstellar space, but i-propyl cyanide is the first with a branched carbon backbone; this is important as it shows that interstellar space could be the origin of more complex branched molecules, such as amino acids, that are necessary for life on Earth.

Dr Arnaud Belloche from the Max Planck Institute for Radio Astronomy is lead author of the research, which appears in the journal Science. "Amino acids on Earth are the building blocks of proteins, and proteins are very important for life as we know it. The question in the background is: is there life somewhere else in the galaxy?"

Watch the skies

The molecule was detected in a giant gas cloud called Sagittarius B2, an active region of ongoing star formation in the centre of the Milky Way. As stars are born in the cloud they heat up microscopic dust grains. Chemical reactions on the surface of the dust allow complex molecules like i-propyl cyanide to form. These molecules emit radiation that was detected as radio waves by twenty 12m telescopes at the Atacama Large Millimeter Array (Alma) in Chile.

Each molecule produces a different "spectral fingerprint" of frequencies. "The game consists in matching these frequencies… to molecules that have been characterised in the laboratory," explained Dr Belloche. "Our goal is to search for new complex organic molecules in the interstellar medium, the ultimate aim is to know whether the elements that are necessary for life to occur… can be found in other places in our galaxy."

Prof. Matt Griffin, head of the school of physics and astronomy at Cardiff University, commented on the discovery. "It's clearly very high-quality data - a very emphatic detection with multiple spectral signatures all seen together.

"There seems to be quite a lot of it, which would indicate that this more complex organic structure is possibly very common, maybe even the norm, when it comes to simple organic molecules in space."

"It's a step closer to discovering molecules that can be regarded as the building blocks or the precursors… of amino acids." The hope is that amino acids will eventually be detected outside our Solar System. "That's what everyone would like to see," said Prof Griffin. If amino acids are widespread throughout the galaxy, life may be also.

"So far we do not have the sensitivity to detect the signals from [amino acids]… in the interstellar medium," explained Dr Belloche. "The interstellar chemistry seems to be able to form these amino acids but at the moment we lack the evidence. […] Alma in the future may be able to do that, once the full capabilities are available."

Prof Griffin agreed this could be the first of many further discoveries from the "fantastically sensitive and powerful" Alma facility.

40. Which of the following best describes Professor Griffin's opinion on the implication of the discovery?
A. It is unremarkable
B. It is a reliable conclusion
C. It is a significant discovery
D. It is unexpected
E. It could be a catalyst for further discovery

41. What is Iso-propyl cyanide?
A. Warmed microscopic dust grains
B. A molecule with a branched carbon backbone
C. Radioactive atoms
D. An organic molecule carrying life
E. A component of an amino acid

42. What can be inferred about amino acids from the extract?
A. In the future, amino acids may be identified outside of the Milky Way
B. The Alma facility could make discoveries at the forefront for molecular science
C. Amino acids are radioactive
D. Amino acids give off signals which are currently undetectable
E. Amino acids are essential for discovering space

END OF SECTION

YOU MUST ANSWER ONLY <u>ONE</u> OF THE FOLLOWING QUESTIONS

Your answer should be a well reasoned argument, showing your understanding of the issue in question and giving the reasons for any opinions you hold.

1. "The government should legalise the sale of human organs" Discuss

2. "Developed countries have a higher obligation to combat climate change than developing countries" Discuss the extent to which you agree with this statement

3. "'Putin is a serious threat to global stability" Discuss

4. "Sufferers of anorexia nervosa should be force fed" Do you agree with this statement? If so, evaluate at what point of an individual's disease this measure should be taken.

END OF TEST

THIS PAGE HAS BEEN
INTENTIONALLY LEFT BLANK

Mock Paper B

1. Water Wars: Egyptians Condemn Ethiopia's Nile Dam Project

Adapted from 2013 © Peter Schwartzstein for National Geographic

As the Grand Ethiopian Renaissance Dam takes shape, tempers rise.

"Ethiopia is killing us," taxi driver Ahmed Hossam said, as he picked his way through Cairo's notoriously traffic-clogged streets. "If they build this dam, there will be no Nile. If there's no Nile, then there's no Egypt."

Projects on the scale of the $4.7 billion, 1.1 mile long (1.7 kilometre long) Grand Ethiopian Renaissance Dam often encounter impassioned resistance, but few inspire the kind of dread and fury with which most Egyptians regard plans to dam the Blue Nile river.

Egypt insists Ethiopia's hydroelectric scheme amounts to a violation of its historic rights, a breach of the 1959 colonial-era agreement that allocated almost three-fourths of the Nile waters to Egypt, and an existential threat to a country largely devoid of alternative freshwater sources.

But what Egyptians regard as a nefarious plot by its historic adversary to control its water supply, Ethiopians see as an intense source of national pride and a symbol of their country's renewal after the debilitating famines of the 1980s and '90s.

"People are enthusiastic. They're excited, because no leader has tried such a project in Ethiopia's history," said Bitania Tadesse, a recent university graduate from the capital, Addis Ababa. "It's a big deal that is going to be beneficial to future generations."

Ethiopia maintains that Egypt and Sudan downstream have no reason to be fearful. The government says it's merely redressing the inequalities of previous water-sharing arrangements, which had left the nine upstream countries largely bereft of access to the Nile.

1. What is implied by the Ethiopian government in the last paragraph?
A. Egypt and Sudan have treated Ethiopia badly in the past
B. Ethiopia has been overlooked in previous water sharing schemes
C. Ethiopia has been purposefully disadvantaged as a result of previous water sharing schemes
D. The Dam is but one retaliation in a long-standing feud between Ethiopia and Egypt
E. Egypt has less need for access to water than Ethiopia does

2. None of the following convey Egypt's distaste for Ethiopia except:
A. 'People are enthusiastic'
B. 'Egypt insists Ethiopia's hydroelectric scheme amounts to a violation of its historic rights'
C. 'Ethiopia is killing us'
D. '(I)mpassioned resistance'
E. 'Egyptians regard as a nefarious plot'

3. Why does Ahmed Hossam use exaggeration in the phrase 'If there is no Nile, there is no Egypt'?
A. He believes that Egypt will cease to exist as a State if the Dam is built
B. He uses stark terms to illustrate how strongly he feels about Ethiopia's decision to build the Dam
C. He hopes to explain how important the Nile is to Egyptian culture
D. He wants to emphasise the severity of the potential problem caused by the Dam
E. He uses sarcasm to belittle the fears of other Egyptians

2. Who is to judge which lives are worth living?

Adapted from 2011 © Barbara Ellen for The Guardian (edited)

The able bodied should never dictate the fates of the ill and weak

There has been some fuss about the forthcoming BBC documentary, *Choosing to Die*, presented by novelist and Alzheimer's sufferer Terry Pratchett, which features a man with motor neurone disease, travelling to Swiss clinic Dignitas and – a first on terrestrial television – dying on screen.

The BBC has been accused of acting "like a cheerleader for legalising assisted suicide", which it denies. Pratchett says: "Everybody possessed of a debilitating and incurable illness should be allowed to pick the hour of their death." Clearly, with him, the dignity of choice is paramount. However, while one has enormous sympathy for Pratchett suffering such a vile disease, the fact remains that he is a rich, powerful man and it is highly unlikely that his wishes would be ignored. With respect, euthanasia laws are not in place to protect people such as him. What of those who may have their "choice" taken away, even if they don't want to die?

There are bigger issues at stake, not least the arrogance of the pro-euthanasia able bodied towards the profoundly ill – the unseemly rush to pronounce the lives of others "not worth living". A recent study discovered that some sufferers of locked-in syndrome – as many as three out of four of the main sample – were happy and did not want to die. Such studies are flawed (some sufferers are unable to articulate either way), but it should still give us pause for thought before blasting off about "lives not worth living".

Likewise the knee-jerk: "They wouldn't have wanted to end up like this." Of course not – who would? – but that might not be the end of the story. How individuals feel when they are fit may change considerably when their health fails. Like those with locked-in syndrome, they may adjust to a life that is very different, often difficult, but just as precious. Who are we to judge?

Bizarrely, the one thing the pro- and anti-euthanasia lobbies have in common is an obsession with God. Sometimes, it's almost as if the antis are tricked into talking about the "sanctity of life" and "God's will", to make the pros look more modern and credible.

Personally, if I ever get something nasty, I'd rather be with a God-botherer than somebody who decides I'm looking peaky, books a Swiss flight and whisks me off to the ghouls at Dignitas. Or maybe I wouldn't – maybe I'd be begging for death. The hope is that I'll choose.

At the moment, assisted death is illegal in Britain, with the caveat that each case is assessed individually, with empathy for the individual and their carers. It could be worse. One reason we don't have the death penalty is that there is no guarantee that mistakes would not be made. Who could guarantee that mistakes wouldn't be made with euthanasia? Not all seriously ill people can communicate their current wishes (not necessarily the same as when they first became ill). And no one else should be deciding for them, in worst-case scenarios "putting them down" against their will.

The phrase: "It's what they would have wanted" belongs after death, not before it. A prolonged, pointlessly agonising end is everyone's nightmare, but that doesn't mean the able bodied should ever get to dictate the fates of the ill and weak. Terry Pratchett should be commended for speaking up for those who wish to die with dignity. However, others who might not want to die, but can't articulate that, need a voice to speak up for them too.
Is this man dead or just dead stupid?

4. What is the main conclusion of the article?

A. The conclusion of Terry Pratchett's documentary is skewed by his own personal circumstances

B. Euthanasia is not yet a precise science and legalising it may give rise to irrevocable mistakes being made

C. Those who have the means to be euthanised should be able to do so if they wish to.

D. Euthanasia should be a personal choice, and legalising euthanasia may give rise to a situation where the seriously ill are robbed of their choice by the healthy

E. Euthanasia could only work if all seriously ill people could articulate their intention to die in such a way that it could not be misunderstood

5. What can be inferred from the discussion of Terry Pratchett's situation?

A. A less wealthy person's wishes are more likely to be ignored

B. Mr Pratchett does not care about those whose choice has been taken away

C. For Mr Pratchett, choice is primarily important

D. A wealthier person's wishes are less likely to be ignored

E. Mr Pratchett is not best placed to discuss any Euthanasia laws

6. Which of the following, if true, would rectify the main problem the author identifies in relation to euthanasia?

A. A law is passed requiring that all people must choose when healthy what they wish to happen to them if they become ill (at varying stages of illness)

B. A new medical procedure is discovered which can analyse the brain's electrical impulses and determine how content a seriously ill patient is

C. Euthanasia is made illegal worldwide, so that even the rich and powerful cannot choose to die

D. A study is published, concluding that 85% of people in a persistent vegetative state will never recover brain function

E. A scientific discovery means that people can be kept alive using life support machine for a fraction of the current cost

7. Why does the author use the phrase 'knee jerk'?

A. To emphasise the importance of evaluating an argument before one makes it

B. To cast doubt upon the value of the argument that follows it

C. To show that, when discussing euthanasia, often people have automatic reactions without truly analysing why they feel this way

D. To suggest that people's reflexive reactions to euthanasia are often wrong

E. To imply that there is no need to make the argument that follows due to it being universally acknowledged to be true

3. Illegal downloaders are one of music industry's biggest customers

Adapted from 2009 © Demos.co.uk

People who use peer-to-peer filesharing websites like Pirate Bay to illegally download music spend over £30 more on music per year than those who do not download illegally.

Internet users who claim to never illegally download music spend an average of £44 per person on music per year, while those who do admit to illegal downloading spend £77, amounting to an estimated £200m in revenue per year.

A new poll commissioned by Demos found that almost one in ten adults (9%) aged 16-50 who have internet access admit downloading music illegally. But this group are also active music buyers, with 8 in 10 buying CDs, vinyl or MP3s in the past year. The poll also found that 42 percent of illegal downloaders agree that they 'like to try things out before I decide whether to buy them.'

The findings suggest that government plans to disconnect repeat illegal downloaders from the internet, announced yesterday by Lord Mandelson, could do the music industry more harm than good by punishing core consumers. The poll showed that the availability of new, appealing legal music provision services is the step most likely to encourage illegal downloaders to stop, above fines or the threat of disconnection.

The research reveals a gap between what consumers are willing to pay for music tracks and current market prices. If official music distribution sites like iTunes lowered the cost of a single track to 45p they could expect prospective buyers to double in number.

Peter Bradwell, a researcher at Demos specialising in digital rights and consumer trends said:

"The latest approach from the government will not help to prop up an ailing music industry. Politicians and music companies need to recognise that the nature of music consumption has changed and consumers are demanding lower prices and easier access to music."

8. Which of the following is true, according to the article?
A. The Government's plans are destined to fail
B. Illegal downloading is not as big a problem as first imagined
C. The poll had surprising results
D. Most people who admit to illegally downloading music also buy music
E. All of the above

9. Which statement (if true) would be most damning to the validity of the poll?
A. Demos' CEO has been charged with tax evasion
B. The participants in the poll were friends of Peter Bradwell
C. Only 25 people were questioned
D. A typo in the questionnaire puts into question how many participants understood exactly what they were being asked
E. Half of the participants were bribed by the government's opposition to lie

10. Which of the following is an assertion of opinion rather than a statement of fact?
A. 'The poll showed that the availability of new, appealing legal music provisions is the step most likely to encourage illegal downloaders to stop
B. 'Politicians and music companies need to recognise that the nature of music consumption has changed'
C. 'The research reveals a gap between what consumers are willing to pay for music trackers and current market prices'
D. '8 in 10 buying CDs, vinyl or MP3s in the last year'
E. 'If official music distribution sites like iTunes lowered the cost of a single track to 45p they could expect prospective buyers to double in number'

4. Heavy Military Recruitment at High Schools Irks Some Parents

Adapted from 2005 © Kelly Beaucar Viahos for <u>FoxNews.com</u> (edited)

A little-known provision in the No Child Left Behind Act that compels public high schools to open their doors and pupil records to military recruiters has some parents, students and anti-war groups up in arms. "We think most people were unaware of it," Amy Hagopian, co-president of the Garfield High School Parent-Teacher-Student Association in Seattle and an active counter-recruiter in the school, said of the provision.

Hagopian said parents are just becoming aware of the policy, which gives recruiters the same access to high school campuses and students' phone numbers and addresses as colleges and businesses have. Districts that don't comply could risk annual federal funding. According to the law, parents must be notified and can refuse to release their children's information. Every school has adopted different notification policies, some being more effective than others, school officials said. Military recruitment issues have been making headlines in recent weeks, as the Army, Marine Corps and National Guard have announced shortfalls in their goals this year. Reports say recruitment pressure is translating into inappropriate tactics by recruiters to the extent that the Army halted recruiting for one day in May to refresh staff with proper protocol in dealing with prospective soldiers.

Paul Rieckoff, an Iraq war veteran and founder of Operation Truth, a veterans' advocacy organization, said parents are now reacting to "major recruiting problems" and bad news coming out of Iraq. "I think it's safe to say there is concern and even the beginning of a movement to combat or to face the recruiters at the high schools," he said. "We don't necessarily endorse that but the critical issue is that the Army has missed their goals again this year." The military has always had access to schools but not all have opened their doors and records equally. Now, the No Child Left Behind Act emboldens efforts to gain "access to the best and brightest this country has to offer," said Department of Defence spokeswoman Lt. Col. Ellen Krenke.

"For some of our students, this may be the best opportunity they have to get a college education," wrote Secretary of Defence Donald Rumsfeld and former Secretary of Education Rod Paige in an October 2002 letter to school superintendents announcing the new law. But some parents and teachers say school is not an appropriate place for the military's message, and complain the hard sell has gotten harder since the Iraq war began and following lacklustre recruitment numbers. "The recruiters really harangue people, and this is what parents are trying to avoid," said Tina Weishaus, president of the Highland Park Middle School/High School Parent Teacher Organization in New Jersey. "Personally, I think the whole thing should be struck from No Child Left Behind," Weishaus said. "I don't think the federal government should be mandating that schools become a recruiting ground for the war."

Army spokesman Doug Smith said the Army has not accelerated recruitment at the schools in the face of missed goals. It is primarily targeting college students, he said, with the average age of 21 for new Army recruits. In 2002, 12,560 out of the 77,000 enlistments were recruited out of high schools. Other critics say they have no problem with military recruiters, but are concerned about students' privacy. According to Montclair school officials, more than 80 percent of the parents who responded to that campaign asked that their records not be given to recruiters this year.
Bill Cala, superintendent of the Fairport Central School District in New York, said his school has been found non-compliant with the law because it doesn't release the names of students to the military unless parents specifically give their consent.

He said about 80 out of the 1,600 students in the school consented this school year, but recruitment among seniors hit 2 percent. "This really, for us, is a privacy issue and doesn't have anything to do with support for the military or for the war," Cala said.

11. Which of the following is a provision of the 'No Child Left Behind' Act?
A. That high schools must give over all the contact information of all their students to military recruiters
B. That high schools are given the option to hand over some students' contact information
C. That high schools must hand over all contact information unless the parents specifically withhold their consent
D. That high schools must give over all students' contact information provided that they have consent from the students to do so
E. That high schools must seek consent from parents before giving over students' contact information to recruiters

12. All of the following are opinions stated in the piece except:
A. The main concern about this provision centres around the students' privacy
B. School is not the right place for military recruitment
C. Recruitment tactics have become inappropriate
D. The recruiters' approach is too aggressive for children in a school environment
E. Most parents are unaware of this specific provision of 'No Child Left Behind'

13. What is the purpose of the inclusion of the phrase 'the hard sell has gotten harder'?
A. The author uses repetition to draw the reader's attention to the idea that recruitment tactics are increasingly rough
B. The author plays on the term 'hard sell' to add variation to the language of the piece
C. The author attempts to be facetious by making fun of the term 'hard sell' to belittle the parents' concerns
D. None of the above
E. All of the above

14. What can be inferred from Bill Cala's figures concerning the number of his students who consented to having their details shared with the military?
A. Most students do not want their details shared with the military
B. Seniors are more likely to consent to having their details shared than younger students
C. Current recruitment strategies are more likely to succeed with older students
D. All of the above
E. Both A and C

5. Found: giant spirals in space that could explain our existence

Adapted from 2015 © Michael Slezak for New Scientist (edited)

Giant magnetic spirals in the sky could explain why there is something rather than nothing in the universe, according to an analysis of data from NASA's Fermi space telescope.
Our best theories of physics imply we shouldn't be here. The Big Bang ought to have produced equal amounts of matter and antimatter particles, which would almost immediately annihilate each other, leaving nothing but light.

So the reality that we are here – and there seems to be very little antimatter around – is one of the biggest unsolved mysteries in physics.

Monopole monopoly

In 2001, Tanmay Vachaspati from Arizona State University offered a purely theoretical solution. Even if matter and antimatter were created in equal amounts, he suggested that as they annihilated each other, they would have briefly created monopoles and antimonopoles – hypothetical particles with just one magnetic pole, north or south.

As the monopoles and antimonopoles in turn annihilated each other, they would produce matter and antimatter. But because of a quirk in nature called CP violation, that process would be biased towards matter, leaving the matter-filled world we see today.
If that happened, Vachaspati showed that there should be a sign of it today: twisted magnetic fields permeating the universe – a fossil of the magnetic monopoles that briefly dominated. And he showed they should look like left-handed screws rather than right-handed screws.

So Vachaspati and his colleagues went looking for them in data from NASA's Fermi Gamma ray Space Telescope. As gamma rays shoot through the cosmos, they should be bent by any magnetic field they pass through, so if there are helical magnetic fields permeating the universe that should leave a visible mark on those gamma rays.

All of a twist

Lo and behold, that's just what they found – well, maybe. "What we found is consistent with them all being left-handed," says Vachaspati. "But we can't be sure." He says there's less than a one per cent chance that what they see in the Fermi data happened by chance. "That's being conservative," he says.

They also found that the twists in the field are a bit bigger than they predicted. "So there is some mystery there," says Vachaspati. He says more data from Fermi, which is expected this year, will help narrow down the odds.

Nicole Bell from the University of Melbourne in Australia warns that magnetic fields could have been caused in other ways, including from inflation. What's more, for CP-violation to provide enough matter in the universe you usually need "new physics" – stuff beyond the standard model of particle physics – which hasn't been confirmed experimentally yet. "But it is a very interesting idea," she says.

15. Which of the following best describes the author's stance in the first paragraph?
A.   Human existence is inexplicable
B.   Current theories suggest we should not be in existence
C.   Our existence can be explained by an analysis of data from NASA's Fermi space telescope
D.   The quantities of matter and antimatter particles created during the Big Bang can best explain our existence.
E.   Physics will never be able to explain why humans exist

16. The presence of which of the following could indicate helical magnetic fields permeating the universe?
A.   Antimatter particles
B.   Magnetic monopoles
C.   Gamma rays which have been bent by any magnetic field they pass through
D.   A mark left on the magnetic field
E.   All of the above

17. What does the author mean by "new physics"?
A.   Theories beyond the normal model of particle physics
B.   Theories which haven't been confirmed by experiments yet
C.   Theories which are currently beyond our understanding
D.   A and C
E.   A and B

6. Making an opinion illegal is not going to stop terrorism

Adapted from 2015 © Jonathan Russell for The Telegraph (edited)

If 'British values' mean anything, they must prevent us from legislating against non-violent views that we find abhorrent

The Government's shift in focus to target extremism and tackle radicalisation rather than counter terrorism is a welcome move, but I question whether the measures set out by the Prime Minister will do this effectively, proportionately and in a way that befits our "British values". Among other things, the proposals are to include "extremism disruption orders", which would not criminalise the act of hate speech or promotion of terrorism, for they are already illegal under the Terrorism Act 2006, but rather the intent to do so, if ministers reasonably believe this to be the case. In essence, these measures target those who operate in what the police have called the "pre-criminal space" and therefore expand the definition of people who could be incarcerated from those who do bad things to those who think bad things. This is problematic for a number of reasons. First, we have the ethical issue of clamping down on freedom of expression, one of our universally accepted human rights. The risk is that the measures are likely to be used in instances when contemplation never graduated to, nor was ever going to graduate to, action. Our courts will have to prosecute entirely on *mens rea* in the absence of *actus reus*, not only an ethically dangerous step towards criminalising thought, but also very difficult in practice to achieve a prosecution in our legal system. The inclusion of "reasonable belief" and a decision by "ministers" likely pre-empts this legal challenge, by having the Home Secretary make judgements rather than the traditional criminal justice system.

Secondly, we risk trying to legislate our way out of the extremist mess our country faces, when we should instead be investing in non-legislative measures to tackle the causes of extremism rather than its symptoms. We have all agreed that "a poisonous ideology" is the root cause and radicalisation is the biggest challenge, yet these measures tackle neither, only serving to disrupt its symptoms – hate preachers on campuses or extremist propaganda disseminators online, for example. A third issue is that we must get beyond simply whacking whichever mole is perceived to be the current nature of the threat. Islamic extremism is like electricity, always seeking to take the path of least resistance. Rather than merely disrupting paths and wasting resources trying to catch up with extremists, we must cut this off at source, and do so without negatively altering the fabric of our nation. The fact that we cannot take water bottles onto planes now is a direct result of the foiled 2006 transatlantic aircraft plot. This will have disrupted future plots and may indeed have made us safer in the last decade, and has done so without clamping down on civil liberties, for carrying water on a plane is not a human right.

But the introduction of Terrorism Investigation and Prevention Measures (TPIMs), which curtail an individual's freedom of movement and freedom of expression without being proven guilty at a fair and public hearing, is problematic because of the failure to uphold the three aforementioned human rights. The measures now being proposed will also fail to strike the correct balance between national security and civil liberties; between counter-terrorism and human rights. Our legislation must be valued and not tinkered with every six months to catch up with the extremist tactics du jour. On top of all of this, it is plain to see that we don't have the capacity within our security services and police force to monitor even more people and enforce an ever-widening set of laws. We increasingly see extremists on the fringes of groups carrying out terrorism-related offences, rather than those at the centre. The new measures might disrupt the communication between those at the centre and those on the fringe, but what we really must do is understand why people are attracted to such groups and their ideologies, and prevent people being vulnerable to radicalisation.

We need a broader, smarter and better financed counter-extremism strategy that engages communities and mobilises civil society to understand and comprehensively challenge the Islamic ideology and refute extremist narratives. The strategy must develop resilience in our institutions and among our young people so we can address their grievances though an alternative secular, democratic, liberal lens. It must also build capacity within our civil society by engaging the private sector and training front line workers, as this is not simply an issue for the state but for all of us. These things combined will truly tackle radicalisation.

18. What is the main conclusion of the article?

A. The Government's proposed measures will be ineffective in tackling radicalisation

B. The Government must pursue a more complex strategy to counter extremism that is flexible enough to change according to the more pressing threat

C. The Government needs to be one step ahead of the terrorists

D. Terrorism can only truly be beaten if the problem is addressed at the source, with measures in place to challenge extremist ideologies in the community

E. The Government's proposed measure would be incompatible with our right to freedom of expression

19. Which of the following is not addressed as a potential problem resulting from the expansion of the law on terrorism?

A. Damaging freedom of expression

B. Extremists moving to the fringes of groups

C. Not confronting a 'poisonous ideology'

D. Difficulties in implementation (lack of resources)

E. Paternalism being taken too far

20. Why does the author use the example of Islamic extremism being 'like electricity'?

A. To scare the reader by emphasising the depth of extremist infiltration of society

B. To explain how information is shared amongst extremist groups

C. To emphasise how flexible extremists can be at finding new methods to spread their message

D. To support his point that the government will fail if they continue to address each new issues as it arises

E. To explain why the new measures will not work

21. Which of the following is a statement of fact, rather than an assertion of opinion?

A. 'the proposals are to include "extremism disruption orders", which would not criminalise the act of hate speech'

B. 'it is plain to see that we don't have the capacity'

C. 'We need a broader, smarter and better financed counter extremism strategy'

D. 'the inclusion of "reasonable belief" and a decision by "ministers" likely pre-empts this legal challenge'

E. 'we should instead be investing in non-legislative measures to tackle the causes of extremism rather than its symptoms'

7. Does Death Penalty Save Lives? A New Debate

Adapted from 2007 © Adam Liptak for The New York Times (edited)

For the first time in a generation, the question of whether the death penalty deters murders has captured the attention of scholars in law and economics, setting off an intense new debate about one of the central justifications for capital punishment. According to roughly a dozen recent studies, executions save lives. For each inmate put to death, the studies say, 3 to 18 murders are prevented. "I personally am opposed to the death penalty," said H. Naci Mocan, an author of a study finding that each execution saves five lives. "But my research shows that there is a deterrent effect." The studies have been the subject of sharp criticism, much of it from legal scholars who say that the theories of economists do not apply to the violent world of crime and punishment. Critics of the studies say they are based on faulty premises, insufficient data and flawed methodologies.

The death penalty "is applied so rarely that the number of homicides it can plausibly have caused or deterred cannot reliably be disentangled from the large year-to-year changes in the homicide rate caused by other factors," John J. Donohue III, a law professor at Yale with a doctorate in economics. But the studies have started to reshape the debate over capital punishment and to influence prominent legal scholars. "The evidence on whether it has a significant deterrent effect seems sufficiently plausible that the moral issue becomes a difficult one," said Cass R. Sunstein, a law professor at the University of Chicago who has frequently taken liberal positions. "I did shift from being against the death penalty to thinking that if it has a significant deterrent effect it's probably justified."

Professor Sunstein and Adrian Vermeule, a law professor at Harvard, wrote in their own Stanford Law Review article that "Capital punishment may well save lives. Those who object to capital punishment, and who do so in the name of protecting life, must come to terms with the possibility that the failure to inflict capital punishment will fail to protect life." There is also a classic economics question lurking in the background, Professor Wolfers said. "Capital punishment is very expensive," he said, "so if you choose to spend money on capital punishment you are choosing not to spend it somewhere else, like policing." A single capital litigation can cost more than $1 million. It is at least possible that devoting that money to crime prevention would prevent more murders than whatever number, if any, an execution would deter.

The available data is admittedly thin, mostly because there are so few executions. In 2003, for instance, there were more than 16,000 homicides but only 153 death sentences and 65 executions. "It seems unlikely," Professor Donohue and Professor Wolfers concluded in their Stanford article, "that any study based only on recent U.S. data can find a reliable link between homicide and execution rates." The two professors offered one particularly compelling comparison. Canada has executed no one since 1962. Yet the murder rates in the United States and Canada have moved in close parallel since then, including before, during and after the four-year death penalty moratorium in the United States in the 1970s.

"Deterrence cannot be achieved with a half-hearted execution program," Professor Shepherd of Emory wrote in the Michigan Law Review in 2005. She found a deterrent effect in only those states that executed at least nine people between 1977 and 1996.

Professor Wolfers said the answer to the question of whether the death penalty deterred was "not unknowable in the abstract," given enough data. "If I was allowed 1,000 executions and 1,000 exonerations, and I was allowed to do it in a random, focused way," he said, "I could probably give you an answer."

22. Which of the following, if true, would be most compelling in support of the idea that the death penalty does save lives?

A. A study concludes that 80% of murderers do not understand the personal implications of their actions

B. A larger study was undertaken showing a higher degree of correlation between dropping murder rates and death sentences

C. A study determines that 70% of murders are premeditated

D. The comparison between Canada and the United States is proven to have been falsified

E. A survey was circulated to hundreds of thousands of free citizens. 59% of them suggested that they would be less likely to kill someone if they knew that there was a possibility they might be executed as a result

23. What can be inferred from Professor Sunstein and Adrian Wolfers' discussion in the Stanford article?

A. Choosing not to inflict capital punishment may not result in lives being saved

B. It is unlikely that those who oppose the death penalty will change their minds following this research

C. Those who support capital punishment will be troubled by this research

D. The authors believe that those who oppose capital punishment may do so without much thought as to why

E. Murderers on death row will want this information to be circulated

24. Why does the author use the word 'admittedly' in the statement 'The available data is *admittedly* thin, mostly because there are so few executions'

A. He wants to show that the studies are worthy of discussion, despite not having many participants

B. He wants to emphasise what his personal opinion is, whilst taking notice of the opposition

C. He knows he must share the opposition to his own argument, but wants to do so whilst maintaining his own opinion

D. He wants to show reluctance in sharing the opposition argument as he knows it has merit and could weaken his point

E. All of the above

25. What, on balance, does the article suggest is the main reason these studies have attracted criticism?

A. Economic theories do not apply in real-life situations

B. The method under which these conclusions were drawn was flawed

C. The studies consider capital punishment in abstraction and do not compare with other options such as crime prevention

D. The authors who have undertaken them have been the subject of professional criticism

E. There is not enough available data to conclude whether or not capital punishment has a deterrent effect

8. The answer to Britain's economic woes? Make pensioners work longer

Adapted from 2015 © Lauren Davidson and Peter Spence for The Telegraph (edited)

The UK could have a stronger economy if it encouraged its older population to stay in the workforce, a new report has found.

The World Economic Forum has ranked the UK 19th out of 124 nations on an index that measures how well countries "nurture, develop and deploy their great asset – its people" with a focus on education, skills and employment. With an overall score of 79.07%, the UK capitalises on its population better than such economic powerhouses as Germany and China, but fails to match up to many of its peers including the US, France and Japan.

The UK slipped down the index because of its weak score in the 65 and over age group, coming 47th in this category, dragged down by a particularly low labour force participation rate of 10.2% that saw 86 other countries fare better than the UK.

An unemployment rate of 3.7%, ranking the UK just 38th on the global index, suggests that the silver population is lacking in job opportunities for those who want to work.

"There's always a lot of focus on youth unemployment and disillusionment which has a long term effect on a country," Saadia Zahidi, head of the WEF's Employment, Skills and Human Capital Initiative and co-author of the report, said. "But the same issue applies to women and older workers as well."

The average age of retirement in the UK is 64.7 for men and 63.1 for women, but ministers believe it should be higher to prevent healthcare and pensions crises caused by the ageing population and have previously indicated that they would like the average retirement age to rise by as much as six months every year. A recent report from the Department for Work and Pensions found that an extra £25bn would be generated if just half of the older people seeking work were given jobs.

"Talent, not capital, will be the key factor linking innovation, competitiveness and growth in the 21st century," said Klaus Schwab, founder and executive chairman of the World Economic Forum. "To make any of the changes necessary to unlock the world's latent talent – and hence its growth potential – we must look beyond campaign cycles and quarterly reports."

Finland topped the ranking, which the WEF claims is the only global report of its kind, with a score of 85.78%, followed by Norway and Switzerland. Canada, which is the world's best at maximising its 15 to 24 age group, took fourth place, while Japan rounded out the top five thanks to its chart-topping deployment of people over the age of 65.

Julio A. Portalatin, president and chief executive officer of Mercer, which produced the report in collaboration with the WEF, called the Index "a critical tool for global employers".
He said: "It allows them to determine the most pressing issues impacting talent availability and suitability around the world today and identify those issues that have the potential to impact business success in the future – invaluable insight for guiding the allocation of workforce development and investments."

26. Which of the following is an assumption that the authors have made?
A.   The UK focuses too much attention on youth unemployment
B.   Most over 65's currently out of work would like to work if the opportunity arose for them to do so
C.   The figures suggest that there are not enough job opportunities available for over 65's (as opposed to over 65's just not wanting to work)
D.   Finland, Norway and Switzerland properly utilise their ageing population
E.   Talent is more important than capital to ensure a stable economy

27. Which of the following most accurately describes the index identified in the extract?
A.   A critical tool for global employers
B.   An indication of how efficiently countries utilise their population within the economy
C.   A method by which a country's education, skills and employment are evaluated
D.   A source of insight to help allocation of investment in countries' workforces
E.   A set of figures which shows how successful countries are at providing opportunities for their ageing workforce

28. What does Klaus Schwab imply in his discussion of the value of talent?
A.   Campaign cycles and quarterly reports are an inefficient use of time
B.   Talent is more important than capital
C.   Those who look only to campaign systoles and quarterly reports are naive
D.   The world will not be able to meet its potential for growth unless unutilised talent is eliminated
E.   Quarterly reports are useless

9. Deadly Frog Fungus Pops Up in Madagascar, An Amphibian Wonderland

Adapted from 2015 © Jane Lee for National Geographic

Madagascar has been spared the scourge of the chytrid fungus, until recently.

Madagascar is home to a mind-boggling array of frogs, 99 percent of which are found nowhere else in the world. But a study released Thursday finds the island nation now also hosts the greatest threat to amphibian biodiversity in modern times—the chytrid fungus.

As many as 7 percent of the world's amphibian species live only in Madagascar, says Molly Bletz, a researcher at the Braunschweig University of Technology in Germany. Chytrid is responsible for the decline or extinction of hundreds of amphibian species around the world. One forest in Panama lost 30 amphibian species to the fungus in about a year, according to a 2010 study.

Why It Matters

Researchers had thought Madagascar was chytrid-free. A 2014 study found chytrid on Madagascar frogs shipped to the U.S. for the pet trade, but researchers weren't sure whether the animals were contaminated en route or infected in Madagascar.

But a new study in the journal *Scientific Reports* finds that chytrid is present in multiple Madagascar frog species. Bletz and colleagues examined skin swabs and tissue samples from 4,155 amphibians tested for chytrid from 2005 to 2014. They found, to their surprise, that the fungus began to appear on frogs starting in 2010. What they haven't found yet is sick frogs. "It could mean we just caught it very early," Bletz says, or it's possible the chytrid strain in Madagascar isn't very lethal.

The Big Picture

"It's the best worst-case scenario," says Jonathan Kolby, a researcher at Queensland's James Cook University, who was not involved in the study. "[Chytrid] is there, but the frogs aren't dying right now."

Scientists need to figure out where the chytrid came from, though, he says. If it was introduced, scientists need to know how it got into the country and how they can prevent another introduction. "Because next time, it could be a strain that's supervirulent," says Kolby, a National Geographic grantee.

What's Next

Meanwhile, experts are working on a multipronged response to the threat. Bletz is working on a possible preventive treatment using frog skin bacteria that may fight off the fungal invader. Other groups around the world—such as in Panama—are setting up breeding facilities for especially vulnerable amphibians just in case, while others in places including Madagascar and Panama are working on long-term amphibian monitoring efforts.

29. Which of the following best describes Jonathan Kolby's approach to the situation in Madagascar?
A.   Optimistic
B.   Cautiously Relieved
C.   Pessimistic
D.   Concerned
E.   Prudent

30. What can be inferred about chytrid from the article?
A.   It takes some time to kill frogs
B.   It is supervirulent
C.   It is a fungus
D.   It is harmful to all amphibians
E.   It is not harmful to frogs

31. Which of the following are mentioned in the article as being worked on as a response to the threat?
A.   A prophylactic treatment
B.   Developing reproduction centres for vulnerable Madagascan frogs
C.   Improving observation procedures
D.   B and C
E.   A and C

10. Why presume a child raised by a gay couple will be emotionally damaged?

Adapted from 2015 © Eleanor Morgan for The Guardian (edited)

In the summer holidays before my 14th birthday, my mum told my siblings and I that my dad would be leaving because she'd begun a relationship with her friend – a woman. I remember the scene clearly; Mum sat on the bottom bunk and told us, calmly, what had happened and what was going to happen now. We cried. We shouted. I ran out of the room screaming at her while my sister gingerly cuddled her.

It's gut-wrenching, whatever the circumstances, to watch your dad pack his things into sports bags and leave the family home. In a fit of utterly emo teenageness I made myself listen to John Lennon's Jealous Guy again and again on my Discman until I could cry out my frustration. I laugh now when I remember writing "I was shivering inside" at the back of my school organiser. But I was. My teenage heart hurt, too, when we'd spend weekends at his temporary bedsit in Chigwell and wake up in the night on our blow-up lilo to see him looking at us with watery eyes. But aren't these experiences par for the course of any parental separation? I had fears and prejudices relating to my mum's new relationship: what will I tell my friends? (I didn't, for ages.) What will the neighbours think? What will people think if we go on holiday together? But looking back, despite my desperate sadness, the gender shift in the house was never really a black cloud. It was the idea of someone else full stop.

This is partly why I have found pieces like Hetty Baynes' in the Daily Mail, detailing how being raised by two mums "ruined her life", and this open letter-style piece that says "Dear gay community: your kids are hurting" so uncomfortable. The layers of irresponsibility in the Daily Mail article are myriad, and while Baynes may well be damaged by her upbringing and feel justified speaking about her family in such a way, the poison dart for me is the line, "So how to explain the bizarre construct which passed for my family?"

When adults are emotionally neglectful of children – as Baynes says her mother and her female lover were of her – it can be harmful and you can end up firing your blame in all sorts of directions. But is being emotionally neglectful a gender-specific thing? I don't think so. If a woman isn't considering the needs of her daughter properly, it's not because she's gay or bisexual – these things are multi-causal and cannot, must not, be hung on the idea of gayness somehow poisoning the ability for normal love and rationality.

I had some really horrible times in the house we shared with my mum's partner, her two children and my brother and sister. Two women, five children. There were, as Baynes experienced, lots of people vying for space and prominence. There was resentment, jealousy and incidents that did cause long-term damage, but they happened as a result of two families being thrown into a building together. When I finally told my friends at school (the excuses for not having them round were wearing thin) the ones whose parents had separated and who had new family dynamics all spoke of similar problems. My problems were not unique because my step-parent figure was a woman.

I experience a crisis whenever I read any of these pieces because for every encouraging study that says the children of gay parents are doing all right we have these singular accounts of pain and turmoil attributed to homosexuality. Everyone is entitled to portray their pain how they wish, but at what point do we say, as a society, that using a public platform to make gross generalisations like Baynes does when she says "roll the dice … you are taking a chance with an innocent life" is harmful?

Although perhaps not to the same degree, I feel the same way today as I did when Dolce and Gabbana made their hateful, dehumanising remarks about "synthetic" IVF children in an interview, because such insular considerations, in such a visible context, are shameful. Talking about your own experience definitely falls under the free speech banner, but making blanket remarks about what is and isn't safe for a child doesn't.

As someone who was co-parented by two women for a while, and who is in a same-sex relationship and planning to have IVF to conceive a child some day, seeing these things splashed across front pages makes my insides curdle. I can laugh at the absurdity, but still know that, however much progress we've made, so many people will agree. That's what's frightening.

32. All of the following suggest disapproval except:
A. 'gross generalisations'
B. 'dehumanising remarks'
C. 'layers of irresponsibility'
D. 'gut-wrenching'
E. 'insular considerations'

33. Which of the following is true according to the author?
A. Her teenage reaction to her parents' news was unjustified
B. None of the problems the author encountered were due to having same sex guardians
C. The problems she encountered as a teenager was due to two families being pushed together in a new and difficult scenario
D. People should be able to speak freely about inferences about the wider world they can draw from their own personal experiences
E. Neglectfulness in a same-sex household is more often than not due to resentment

34. What assumption does the author make in the piece?
A. There is no scenario in which a child might be happy to see a father leave the family home
B. Resentment is a logical response to family upheaval
C. Dolce and Gabbana must feel similarly negatively towards same-sex unions
D. Had her father been around, she would have been happier
E. IVF children are less 'human' than other children

35. Why might Eleanor Morgan use the term 'poison dart' in relation to Baynes' argument?
A. To explain the importance of this particular part of Baynes' argument
B. To draw attention to the fact that Baynes' argument is flawed
C. To draw attention to the fact that Baynes is unjustified in extending her argument so far
D. To draw attention to the fact that the author finds Baynes' article distasteful
E. All of the above

11. Calculated risks: Will algorithms make business boring?

Adapted from 2015 © Zoe Kleinman for BBC News (edited)

Ever found yourself wondering whether your boss is human?

Haven't we all.

Now, with the march of artificial intelligence and robotics, it's becoming an increasingly valid question. Algorithms – problem solving computer programmes – are, to put it bluntly, getting much better at doing our jobs than we are. And it's not just in the tech sector where computers are becoming king, school children in South Korea are being taught English by a machine called Robosem.

But is there a danger that this brave new world is going to be a bit, well, dull?

Movies by numbers
The entertainment industry has already adopted an algorithmic approach to working out what we want to watch. When TV and movie streaming service Netflix decided to start commissioning its own material, it turned not to Hollywood veterans, film critics or media forecasters, but to algorithms and user data.

A data trawl of the most-watched and loved content streamed by Netflix customers revealed three key ingredients - actor Kevin Spacey, director David Fincher and political dramas produced by the BBC. So the firm commissioned a remake of 1990 BBC political thriller House of Cards, starring Kevin Spacey and directed by David Fincher. It became the first ever web series to win a prestigious Emmy award, with the first series receiving nine nominations.

High tech traders
The finance sector is another enjoying the comparative stability of computer control.
"The trading floor was a very exciting place. Now it's more like a software company than a financial organisation," said Dr Juan Pablo Pardo-Guerra, assistant professor at the London School of Economics.

High-frequency financial trader Virtu, an electronic trader with computerised strategies, has only ever recorded one day of losses in nearly six years. On its website the firm claims that its 148 employees are the "secret sauce" of its success but its own proprietary technology is at the heart of all of its trading activity.

Financial thrill-seekers will still enjoy working in the sector, predicts Dr Pardo-Guerra – but there will definitely be a change of scene. "It's probably going to have some impact on how people relate with the market, how they find excitement in trading," he said.

And the good news is there is still room for the human touch – at least for now.

"Even within the most analytical part of finance, meetings and relationships are still quite important and that's something algorithms can't do," he added. "Some aspects can be automated – calculated or quantitative decision making. But others rely more on personal cues and networks."

Algorithms have another potentially fatal flaw – they are unable to recognise when someone is taking the tricks of the trade too far.
"People know how to manipulate prices to get more customers or make their strategies more profitable," said Dr Pardo-Guerra. "I think algorithms present challenges in terms of identifying these processes."

36. Why might the author choose to start the piece in this way?
A.   To grab the audience's attention
B.   To establish an informal register from the beginning
C.   To immediately create a connection with the audience
D.   All of the above
E.   A and C

37. Which of the following is not identified in the piece as an example of artificial intelligence being used to make our lives easier?
A.   Computers being used to teach children
B.   Algorithms being used to predict human behaviour
C.   Automating quantitative decision making
D.   Strategising trading decisions
E.   Algorithms being used to develop products

38. What is the main reason is it suggested that artificially intelligent beings are inferior to human beings in certain roles?
A.   They cannot build relationships and trust with other humans
B.   They are unable to identify and respond to uniquely human reactions
C.   They take tricks of the trade too far
D.   They cannot identify when a human is playing outside of the rules
E.   Their efficacy will result in business becoming dull

39. What does the author imply in the piece about the future of business?
A.   It will be computer-centric
B.   It will be more efficient
C.   It might become tedious
D.   Humans will still be necessary for certain tasks
E.   All of the above

12. U.S. opposes honest labelling of GMO foods

Adapted from 2010 © Ethan Huff for Natural News

The official U.S. position on genetically-modified organisms, also known as GMs or GMOs, is that there is no difference between them and natural organisms. Crafted by the U.S. Food and Drug Administration (FDA) and the Department of Agriculture (USDA), the position set forth to the Codex Alimentarius Commission on the issue goes even further to suggest that no country should be able to require mandatory GMO labelling on food items, even though science shows that GMOs act differently in the body than do natural organisms and are a threat to health.

A group of over 80 food processors, farmers and consumer organizations has sent an official letter to Michael Taylor, deputy commission at the FDA, and Kathleen Merrigan, deputy secretary of agriculture at the USDA, protesting the official U.S. position, citing the fact that it creates "significant problems" for all U.S. food producers that wish to label their products as being GMO-free.

Not only is there no mandatory labelling of products sold in the U.S. that contain GMO ingredients, but the FDA and USDA now want to prohibit the labelling of products that do not contain GMO ingredients. In other words, the FDA and USDA are trying to outlaw truth in labelling and are openly working deceive the public.

Among those opposing the draft U.S. position on GMOs are members of the Consumers Union, the Organic Trade Association, the Union of Concerned Scientists and the Centre for Food Safety.

The FDA and USDA actually had the audacity to include in the draft position that mandatory labelling of GMOs is "false, misleading [and] deceptive" because it implies that there is a difference between GMO ingredients and non-GMO ingredients.

Fortunately, science and pure common sense, which are both lacking at the FDA and USDA, indicate that GMOs are different from non-GMOs, and that the public has a right to know the types of ingredients that are in the products they buy.

Many countries already require food processors and manufacturers to label products that contain GMOs, but the FDA and USDA hope to convince the Codex Committee to outlaw this practice.

Not only are GMOs structurally different than non-GMOs, but GMOs are actually toxic. Several studies have shown that they are harmful to the body.

40. What, according to the article, is the U.S. position on GMOs?
A. They are identical to natural organisms
B. No country should be able to demand labelling on products identifying them as GMOs
C. There is no distinction between them and natural organisms
D. A and B
E. B and C

41. Which of the following would best describe the author's perception of the U.S. position?
A. Evasive
B. Obstructive
C. Impudent
D. Illogical
E. Competent

42. Which of the following, if true, would most undermine the author's position?
A. The 'several studies' he mentions in the final paragraph are proven to have been falsified
B. The U.S position is supported by 90% of MEDC countries around the globe
C. The U.S has used no scientific evidence to support its position
D. The author is proven to be a fanatical eco-terrorist
E. An article published in the New Scientist magazine supports the idea that GMOs act no differently to biological organisms

END OF SECTION

YOU MUST ANSWER <u>ONLY</u> <u>ONE</u> OF THE FOLLOWING QUESTIONS

Your answer should be a well reasoned argument, showing your understanding of the issue in question and giving the reasons for any opinions you hold.

1. Tennessee currently protects teachers who wish to teach children to explore the potential value of following creationism. Do you think that this is correct? Identify and analyse any legal problems that may arise in discussion of this law.

2. "There is a time and a place for censorship of the internet." Discuss with particular reference to the right of freedom of expression

3. "The UK should codify its Constitution' Discuss.

4. "The general trend towards the liberalisation of marriage undermines its religious basis." Discuss this comment with reference to the idea of abolishing marriage as a legal concept.

END OF TEST

THIS PAGE HAS BEEN
INTENTIONALLY LEFT BLANK

Mock Paper C

1. Marriage

Ever since the time, nineteen years ago, when Mrs Mona Caird attacked the institution of matrimony in the Westminster Review and led the way for the great discussion on 'Is Marriage a Failure?' in the Daily Telegraph—marriage has been the hardy perennial of newspaper correspondence, and an unfailing resource to worried sub-editors. When seasons are slack and silly, the humblest member of the staff has but to turn out a column on this subject, and whether it be a serious dissertation on 'The Perfections of Polygamy' or a banal discussion on 'Should Husbands have Tea at Home?', it will inevitably achieve the desired result, and fill the spare columns of the papers with letters for weeks to come. People are always interested in matrimony, whether from the objective or subjective point of view, and that is my excuse for perpetrating yet another book on this well-worn, but ever fertile topic.

Marriage indeed seems to be in the air more than ever in this year of grace; everywhere it is discussed, and very few people seem to have a good word to say for it. The most superficial observer must have noticed that there is being gradually built up in the community a growing dread of the conjugal bond, especially among men; and a condition of discontent and unrest amongst married people, particularly women. What is the matter with this generation that wedlock has come to assume so distasteful an aspect in their eyes? On every side one hears it vilified and its very necessity called into question. From the pulpit, the clergy endeavour to uphold the sanctity of the institution, and unceasingly exhort their congregations to respect it and abide by its laws. But the Divorce Court returns make ominous reading; every family solicitor will tell you his personal experience goes to prove that happy unions are considerably on the decrease, and some of the greatest thinkers of our day join in a chorus of condemnation against latter-day marriage.

Tolstoy says: 'the relations between the sexes are searching for a new form, the old one is falling to pieces.' Among the manuscript 'remains' of Ibsen, the profound student of human nature, the following noteworthy passage occurs: '"free-born men" is a phrase of rhetoric. They do not exist, for marriage, the relation between man and wife, has corrupted the race and impressed the mark of slavery upon all. 'Not long ago, too, our greatest living novelist, George Meredith, created an immense sensation by his suggestion that marriage should become a temporary arrangement, with a minimum lease of, say, ten years.

That the time has not yet come for any such revolutionary change is obvious, but if the signs and portents of the last decade or two do not lie, we may safely assume that the time will come, and that the present legal conditions of wedlock will be altered in some way or other.

1. The writer takes the view that:
A. Marriage needs to undergo a revolutionary change now
B. Marriage will not require a change at all
C. Marriage will undergo a change in the future
D. Marriage will become a temporary arrangement
E. Marriage is in decline

2. What does Tolstoy suggest?
A. Marriage as a concept is outdated
B. People do not believe in marriage anymore
C. Marriage is on a steady decline
D. The concept of adult relationships is evolving
E. Couples no longer wish to get married

3. What does the writer **not** suggest in the first paragraph?
A. Marriage is an important topic for people
B. Marriage is an interesting topic for people
C. Marriage provides good fodder for publications
D. There are both serious and trivial publications on marriage
E. Marriage has been discussed very frequently in publications

4. Which word is **not** used by the writer as a form of criticism?
A. Hardy
B. Silly
C. Banal
D. Superficial
E. Distasteful

2. Nature or nurture?

Galton adopted and popularised Shakespeare's antithesis of nature and nurture to describe a man's inheritance and his surroundings, the two terms including everything that can pertain to a human being. The words are not wholly suitable, particularly since nature has two distinct meanings, — human nature and external nature. The first is the only one considered by Galton. Further, nurture is capable of subdivision into those environmental influences which do not undergo much change, — e.g., soil and climate, — and those forces of civilization and education which might better be described as culture. The evolutionist has really to deal with the three factors of germ-plasm, physical surroundings and culture. But Galton's phrase is so widely current that we shall continue to use it, with the implications that have just been outlined.

The antithesis of nature and nurture is not a new one; it was met long ago by biologists and settled by them to their own satisfaction. The whole body of experimental and observational evidence in biology tends to show that the characters which the individual inherits from his ancestors remain remarkably constant in all ordinary conditions to which they may be subjected. Their constancy is roughly proportionate to the place of the animal in the scale of evolution; lower forms are more easily changed by outside influence, but as one ascends to the higher forms, which are more differentiated, it is found more and more difficult to effect any change in them. Their characters are more definitely fixed at birth.

It is with the highest of all forms, Man, that we have now to deal. The student in biology is not likely to doubt that the differences in men are due much more to inherited nature than to any influences brought to bear after birth, even though these latter influences include such powerful ones as nutrition and education within ordinary limits.

But the biological evidence does not lend itself readily to summary treatment, and we shall therefore examine the question by statistical methods. These have the further advantage of being more easily understood; for facts which can be measured and expressed in numbers are facts whose import the reader can usually decide for himself: he is perfectly able to determine, without any special training, whether twice two does or does not make four. One further preliminary remark: the problem of nature vs. nurture cannot be solved in general terms; a moment's thought will show that it can be understood only by examining one trait at a time. The problem is to decide whether the differences between the people met in everyday life are due more to inheritance or to outside influences, and these differences must naturally be examined separately; they cannot be lumped together.

5. The writer suggests in the first paragraph that:
A. We should no longer use the terms 'nature or nurture'
B. Galton created the terms 'nature or nurture'
C. Galton uses the terms 'nature or nurture' precisely
D. 'Nurture' does not undergo much change
E. 'Nature or nurture' is very widely used

6. Why does the writer capitalise the word 'man' in the third paragraph?
A. It is the official name of a species
B. The writer wants to emphasise its importance
C. The writer is not using the word's ordinary meaning
D. The writer wants to refer to a collective group
E. The writer is referring to a particular person

7. The writer is of the view that:
A. We should analyse 'nature or nurture' as a whole
B. We should only focus on one trait at a time
C. Nature plays a bigger part than nurture as shown by biology
D. It is difficult to change what is inherited by man
E. Statistical methods will provide a more accurate answer than summary methods

8. Which of the following pair of words are **not** used as a contrasting pair?
A. Inheritance and surroundings
B. Human nature and external nature
C. Soil and climate
D. Lower forms and higher forms
E. Inheritance and outside influences

3. Biology

I must at the outset remark that among the many sciences that are occupied with the study of the living world there is no one that may properly lay exclusive claim to the name of Biology. The word does not, in fact, denote any particular science but is a generic term applied to a large group of biological sciences all of which alike are concerned with the phenomena of life. To present in a single address, even in rudimentary outline, the specific results of these sciences is obviously an impossible task, and one that I have no intention of attempting. I shall offer no more than a kind of preface or introduction to those who will speak after me on the biological sciences of physiology, botany and zoology; and I shall confine it to what seem to me the most essential and characteristic of the general problems towards which all lines of biological inquiry must sooner or later converge.

It is the general aim of the biological sciences to learn something of the order of nature in the living world. Perhaps it is not amiss to remark that the biologist may not hope to solve the ultimate problems of life any more than the chemist and physicist may hope to penetrate the final mysteries of existence in the non-living world. What he can do is to observe, compare and experiment with phenomena, to resolve more complex phenomena into simpler components, and to this extent, as he says, to "explain" them; but he knows in advance that his explanations will never be in the full sense of the word final or complete. Investigation can do no more than push forward the limits of knowledge.

The task of the biologist is a double one. His more immediate effort is to inquire into the nature of the existing organism, to ascertain in what measure the complex phenomena of life as they now appear are capable of resolution into simpler factors or components, and to determine as far as he can what is the relation of these factors to other natural phenomena. It is often practically convenient to consider the organism as presenting two different aspects— a structural or morphological one, and a functional or physiological—and biologists often call themselves accordingly morphologists or physiologists. Morphological investigation has in the past largely followed the method of observation and comparison, physiological investigation that of experiment; but it is one of the best signs of progress that in recent years the fact has come clearly into view that morphology and physiology are really inseparable, and in consequence the distinctions between them, in respect both to subject matter and to method, have largely disappeared in a greater community of aim.

If I have the temerity to draw your attention to the fundamental problem towards which all lines of biological inquiry sooner or later lead us, it is not with the delusion that I can contribute anything new to the prolonged discussions and controversies to which it has given rise. I desire only to indicate in what way it affects the practical efforts of biologists to gain a better understanding of the living organism, whether regarded as a group of existing phenomena or as a product of the evolutionary process; and I shall speak of it, not in any abstract or speculative way, but from the standpoint of the working naturalist. The problem of which I speak is that of organic mechanism and its relation to that of organic adaptation. How in general are the phenomena of life related to those of the non-living world? How far can we profitably employ the hypothesis that the living body is essentially an automaton or machine, a configuration of material particles, which, like an engine or a piece of clockwork, owes its mode of operation to its physical and chemical construction? It is not open to doubt that the living body is a machine. It is a complex chemical engine that applies the energy of the food-stuffs to the performance of the work of life. But is it something more than a machine? If we may imagine the physico-chemical analysis of the body to be carried through to the very end, may we expect to find at last an unknown something that transcends such analysis and is neither a form of physical energy nor anything given in the physical or chemical configuration of the body? Shall we find anything corresponding to the usual popular conception—which was also along the view of physiologists—that the body is "animated" by a specific "vital principle," or "vital force," a dominating "archæus" that exists only in the realm of organic nature? If such a principle exists, then the mechanistic hypothesis fails and the fundamental problem of biology becomes a problem sui generis.

9. The writer is of the view that:
A. Biology is more worthy of study than physics or chemistry
B. Biology only consists of physiology, botany and zoology
C. The sciences can be split into the living and non-living
D. Biology is the only study of the living world
E. Biology consists of many different sciences concerned with life

10. Which of the following, according to the passage, is a **fact**?
A. Biology lays exclusive claim to the study of the living world
B. To present in a single address, even in rudimentary outline, the specific results of these sciences is obviously an impossible task
C. It is the general aim of the biological sciences to learn something of the order of nature in the living world
D. The biologist may not hope to solve the ultimate problems of life any more than the chemist and physicist may hope to penetrate the final mysteries of existence in the non-living world
E. It is often practically convenient to consider the organism as presenting two different aspects—a structural or morphological one, and a functional or physiological

11. Which of the following does the writer believe biologists are **not** able to offer a solution?
A. 'The phenomena of life'
B. 'Order of nature in the living world'
C. 'The ultimate problems of life'
D. 'Nature of the existing organism'
E. 'resolve more complex phenomena into simpler components'

12. Which of the following rhetorical questions do **not** refer or relate to a metaphorical analysis?
A. How in general are the phenomena of life related to those of the non-living world?
B. How far can we profitably employ the hypothesis that the living body is essentially an automaton or machine, a configuration of material particles, which, like an engine or a piece of clockwork, owes its mode of operation to its physical and chemical construction?
C. But is it something more than a machine?
D. If we may imagine the physico-chemical analysis of the body to be carried through to the very end, may we expect to find at last an unknown something that transcends such analysis and is neither a form of physical energy nor anything given in the physical or chemical configuration of the body?
E. Shall we find anything corresponding to the usual popular conception—which was also along the view of physiologists—that the body is "animated" by a specific "vital principle," or "vital force," a dominating "archæus" that exists only in the realm of organic nature?

4. Fossils

Fossils are the remains, or even the indications, of animals and plants that have, through natural agencies, been buried in the earth and preserved for long periods of time. This may seem a rather meagre definition, but it is a difficult matter to frame one that will be at once brief, exact, and comprehensive; fossils are not necessarily the remains of extinct animals or plants, neither are they, of necessity, objects that have become petrified or turned into stone.

Bones of the Great Auk and Rytina, which are quite extinct, would hardly be considered as fossils; while the bones of many species of animals, still living, would properly come in that category, having long ago been buried by natural causes and often been changed into stone. And yet it is not essential for a specimen to have had its animal matter replaced by some mineral in order that it may be classed as a fossil, for the Siberian Mammoths, found entombed in ice, are very properly spoken of as fossils, although the flesh of at least one of these animals was so fresh that it was eaten. Likewise, the mammoth tusks brought to market are termed fossil-ivory, although differing but little from the tusks of modern elephants.

Many fossils indeed merit their popular appellation of petrifactions, because they have been changed into stone by the slow removal of the animal or vegetable matter present and its replacement by some mineral, usually silica or some form of lime. But it is necessary to include 'indications of plants or animals' in the above definition because some of the best fossils may be merely impressions of plants or animals and no portion of the objects themselves, and yet, as we shall see, some of our most important information has been gathered from these same imprints.

Nearly all our knowledge of the plants that flourished in the past is based on the impressions of their leaves left on the soft mud or smooth sand that later on hardened into enduring stone. Such, too, are the trails of creeping and crawling things, casts of the burrows of worms and the many footprints of the reptiles, great and small, that crept along the shore or stalked beside the waters of the ancient seas. The creatures themselves have passed away, their massive bones even are lost, but the prints of their feet are as plain to-day as when they were first made.

Many a crustacean, too, is known solely or mostly by the cast of its shell, the hard parts having completely vanished, and the existence of birds in some formations is revealed merely by the casts of their eggs; and these natural casts must be included in the category of fossils.

Impressions of vertebrates may, indeed, be almost as good as actual skeletons, as in the case of some fishes, where the fine mud in which they were buried has become changed to a rock, rivalling porcelain in texture; the bones have either dissolved away or shattered into dust at the splitting of the rock, but the imprint of each little fin-ray and every threadlike bone is as clearly defined as it would have been in a freshly prepared skeleton. So fine, indeed, may have been the mud, and so quiet for the time being the waters of the ancient sea or lake, that not only have prints of bones and leaves been found, but those of feathers and of the skin of some reptiles, and even of such soft and delicate objects as jelly fishes. But for these we should have little positive knowledge of the outward appearance of the creatures of the past, and to them we are occasionally indebted for the solution of some moot point in their anatomy.

The reader may possibly wonder why it is that fossils are not more abundant; why, of the vast majority of animals that have dwelt upon the earth since it became fit for the habitation of living beings, not a trace remains. This, too, when some objects—the tusks of the Mammoth, for example—have been sufficiently well preserved to form staple articles of commerce at the present time, so that the carved handle of my lady's parasol may have formed part of some animal that flourished at the very dawn of the human race, and been gazed upon by her grandfather a thousand times removed. The answer to this query is that, unless the conditions were such as to preserve at least the hard parts of any creature from immediate decay, there was small probability of it becoming fossilized. These conditions are that the objects must be protected from the air, and, practically, the only way that this happens in nature is by having them covered with water, or at least buried in wet ground.

13. Which of the following words (in bold) do **not** advance the writer's argument?
A. '**And** yet it is not essential for a specimen…'
B. '**Although** the flesh of at least one of these animals was so fresh that it was eaten…'
C. '**But** it is necessary to include…'
D. '**Yet**, as we shall see, some of our most important information…'
E. '**But** for these we should have little positive knowledge…'

14. How does the writer define 'fossils' in the first paragraph?
A. Remains, or even the indications, of animals and plants that have, through natural agencies, been buried in the earth and preserved for long periods of time
B. Remains of extinct animals or plants
C. Objects that have become petrified or turned into stone
D. All of the above
E. None of the above

15. Which of the following pair of words are **not** used as a direct comparison by the writer?
A. 'Animals' and 'plants'
B. 'Brief' and 'comprehensive'
C. 'Bones of the Great Auk and Rytina' and 'bones of many species of animals, still living'
D. 'Mammoth tusks' and 'tusks of modern elephants'
E. 'Soft mud or smooth sand' and 'enduring stone'

16. What does the writer believe is the answer to the question of why fossils are not more abundant?
A. They are difficult to discover
B. They require very specific conditions to form
C. They get destroyed easily
D. It is hard to define what a 'fossil' is
E. All of the above

5. Sanskrit

What, then, is it that gives to Sanskrit its claim on our attention, and its supreme importance in the eyes of the historian? First of all, its antiquity—for we know Sanskrit at an earlier period than Greek. But what is far more important than its merely chronological antiquity is the antique state of preservation in which that Aryan language has been handed down to us. The world had known Latin and Greek for centuries, and it was felt, no doubt, that there was some kind of similarity between the two. But how was that similarity to be explained? Sometimes Latin was supposed to give the key to the formation of a Greek word, sometimes Greek seemed to betray the secret of the origin of a Latin word. Afterward, when the ancient Teutonic languages, such as Gothic and Anglo-Saxon, and the ancient Celtic and Slavonic languages too, came to be studied, no one could help seeing a certain family likeness among them all. But how such a likeness between these languages came to be, and how, what is far more difficult to explain, such striking differences too between these languages came to be, remained a mystery, and gave rise to the most gratuitous theories, most of them, as you know, devoid of all scientific foundation. As soon, however, as Sanskrit stepped into the midst of these languages, there came light and warmth and mutual recognition. They all ceased to be strangers, and each fell of its own accord into its right place. Sanskrit was the eldest sister of them all, and could tell of many things which the other members of the family had quite forgotten. Still, the other languages too had each their own tale to tell; and it is out of all their tales together that a chapter in the human mind has been put together which, in some respects, is more important to us than any of the other chapters, the Jewish, the Greek, the Latin, or the Saxon. The process by which that ancient chapter of history was recovered is very simple. Take the words which occur in the same form and with the same meaning in all the seven branches of the Aryan family, and you have in them the most genuine and trustworthy records in which to read the thoughts of our true ancestors, before they had become Hindus, or Persians, or Greeks, or Romans, or Celts, or Teutons, or Slaves. Of course, some of these ancient charters may have been lost in one or other of these seven branches of the Aryan family, but even then, if they are found in six, or five, or four, or three, or even two only of its original branches, the probability remains, unless we can prove a later historical contact between these languages, that these words existed before the great Aryan Separation. If we find agni, meaning fire, in Sanskrit, and ignis, meaning fire, in Latin, we may safely conclude that fire was known to the undivided Aryans, even if no trace of the same name of fire occurred anywhere else. And why? Because there is no indication that Latin remained longer united with Sanskrit than any of the other Aryan languages, or that Latin could have borrowed such a word from Sanskrit, after these two languages had once become distinct. We have, however, the Lithuanian ugnìs, and the Scottish ingle, to show that the Slavonic and possibly the Teutonic languages also, knew the same word for fire, though they replaced it in time by other words. Words, like all other things, will die, and why they should live on in one soil and wither away and perish in another, is not always easy to say. What has become of ignis, for instance, in all the Romance languages? It has withered away and perished, probably because, after losing its final unaccentuated syllable, it became awkward to pronounce; and another word, focus, which in Latin meant fireplace, hearth, altar, has taken its place.

Suppose we wanted to know whether the ancient Aryans before their separation knew the mouse: we should only have to consult the principal Aryan dictionaries, and we should find in Sanskrit mûsh, in Greek μῦς, in Latin mus, in Old Slavonic mýse, in Old High German mûs, enabling us to say that, at a time so distant from us that we feel inclined to measure it by Indian rather than by our own chronology, the mouse was known, that is, was named, was conceived and recognized as a species of its own, not to be confounded with any other vermin. And if we were to ask whether the enemy of the mouse, the cat, was known at the same distant time, we should feel justified in saying decidedly, No. The cat is called in Sanskrit mârgâra and vidâla. In Greek and Latin the words usually given as names of the cat, γαλέη and αἴλουρος, mustella and feles, did not originally signify the tame cat, but the weasel or marten. The name for the real cat in Greek was κάττα, in Latin catus, and these words have supplied the names for cat in all the Teutonic, Slavonic, and Celtic languages. The animal itself, so far as we know at present, came to Europe from Egypt, where it had been worshipped for centuries and tamed; and as this arrival probably dates from the fourth century a.d., we can well understand that no common name for it could have existed when the Aryan nations separated. In this way a more or less complete picture of the state of civilization, previous to the Aryan Separation, can be and has been reconstructed, like a mosaic put together with the fragments of ancient stones; and I doubt whether, in tracing the history of the human mind, we shall ever reach to a lower stratum than that which is revealed to us by the converging rays of the different Aryan languages.

17. Which language does the writer state is the oldest?
A. Sanskrit
B. Greek
C. Latin
D. Gothic
E. Anglo-Saxon

18. Which of the following is **not** used as a metaphor?
A. Strangers
B. Eldest sister
C. Human mind
D. Mosaic
E. Fragments of ancient stone

19. The writer is of the view that:
A. We should only study Sanskrit and not Greek or Latin
B. Sanskrit helps us understand the relationship between the different ancient languages
C. Sanskrit is obsolete as it is the oldest language
D. Not enough attention is given to Sanskrit
E. None of the above

20. How does the writer suggest we understand the history of languages?
A. Consulting the dictionaries
B. Looking at the chronology
C. Comparing words which occur in the same form and same meaning
D. Looking at words like 'fire' or 'cat'
E. All of the above

6. Pre-adolescence

The years from about eight to twelve constitute a unique period of human life. The acute stage of teething is passing, the brain has acquired nearly its adult size and weight, health is almost at its best, activity is greater and more varied than it ever was before or ever will be again, and there is peculiar endurance, vitality, and resistance to fatigue. The child develops a life of its own outside the home circle, and its natural interests are never so independent of adult influence. Perception is very acute, and there is great immunity to exposure, danger, accident, as well as to temptation. Reason, true morality, religion, sympathy, love, and aesthetic enjoyment are but very slightly developed.

Everything, in short, suggests that this period may represent in the individual what was once for a very protracted and relatively stationary period an age of maturity in the remote ancestors of our race, when the young of our species, who were perhaps pygmoid, shifted for themselves independently of further parental aid. The qualities developed during pre-adolescence are, in the evolutionary history of the race, far older than hereditary traits of body and mind which develop later and which may be compared to a new and higher story built upon our primal nature. Heredity is so far both more stable and more secure. The elements of personality are few, but are well organised on a simple, effective plan. The momentum of these traits inherited from our indefinitely remote ancestors is great, and they are often clearly distinguishable from those to be added later. Thus the boy is father of the man in a new sense, in that his qualities are indefinitely older and existed, well compacted, untold ages before the more distinctly human attributes were developed. Indeed, there are a few faint indications of an earlier age node, at about the age of six, as if amid the instabilities of health, we could detect signs that this may have been the age of puberty in remote ages of the past. I have also given reasons that lead me to the conclusion that, despite its dominance, the function of sexual maturity and procreative power is peculiarly mobile up and down the age-line independently of many of the qualities usually so closely associated with it, so that much that sex created in the phylum now precedes it in the individual.

Rousseau would leave prepubescent years to nature and to these primal hereditary impulses and allow the fundamental traits of savagery their fling till twelve. Biological psychology finds many and cogent reasons to confirm this view if only a proper environment could be provided. The child revels in savagery; and if its tribal, predatory, hunting, fishing, fighting, roving, idle, playing proclivities could be indulged in the country and under conditions that now, alas! seem hopelessly ideal, they could conceivably be so organized and directed as to be far more truly humanistic and liberal than all that the best modern school can provide. Rudimentary organs of the soul, now suppressed, perverted, or delayed, to crop out in menacing forms later, would be developed in their season so that we should be immune to them in maturer years, on the principle of the Aristotelian catharsis for which I have tried to suggest a far broader application than the Stagirite could see in his day.

These inborn and more or less savage instincts can and should be allowed some scope. The deep and strong cravings in the individual for those primitive experiences and occupations in which his ancestors became skilful through the pressure of necessity should not be ignored, but can and should be, at least partially, satisfied in a vicarious way, by tales from literature, history, and tradition which present the crude and primitive virtues of the heroes of the world's childhood. In this way, aided by his vivid visual imagination, the child may enter upon his heritage from the past, live out each stage of life to its fullest and realize in himself all its manifold tendencies. Echoes only of the vaster, richer life of the remote past of the race they must remain, but just these are the murmurings of the only muse that can save from the omnipresent dangers of precocity. Thus we not only rescue from the danger of loss, but utilize for further psychic growth the results of the higher heredity, which are the most precious and potential things on earth. So, too, in our urbanized hothouse life, that tends to ripen everything before its time, we must teach nature, although the very phrase is ominous. But we must not, in so doing, wean still more from, but perpetually incite to visit, field, forest, hill, shore, the water, flowers, animals, the true homes of childhood in this wild, undomesticated stage from which modern conditions have kidnapped and transported him. Books and reading are distasteful, for the very soul and body cry out for a more active, objective life, and to know nature and man at first hand. These two staples, stories and nature, by these informal methods of the home and the environment, constitute fundamental education.

21. The writer suggests that:
A. Children require more adult guidance
B. Children are able to be independent
C. Children should not read
D. Children should only play outside
E. Children need to be active outdoors on top of reading

22. Which of the following did the writer **not** suggest happens during years 8 to 12?
A. The start of teething
B. Growth in brain size
C. Improvement in health
D. Greater activity
E. Greater resistance to fatigue

23. Which of the following would the writer and Rosseau agree on?
A. Children should play outdoors more
B. Children should read
C. Children should be left to fend for themselves
D. Children should grow up in the wild
E. None of the above

24. Which sentence comes closest to the writer's views?
A. Reason, true morality, religion, sympathy, love, and aesthetic enjoyment are but very slightly developed
B. Heredity is so far both more stable and more secure
C. The boy is father of the man in a new sense, in that his qualities are indefinitely older and existed, well compacted, untold ages before the more distinctly human attributes were developed
D. The child revels in savagery; and if its tribal, predatory, hunting, fishing, fighting, roving, idle, playing proclivities could be indulged in the country and under conditions that now, alas! Seem hopelessly ideal
E. These two staples, stories and nature, by these informal methods of the home and the environment, constitute fundamental education

7. Life of Chopin

Deeply regretted as he may be by the whole body of artists, lamented by all who have ever known him, we must still be permitted to doubt if the time has even yet arrived in which he, whose loss is so peculiarly deplored by ourselves, can be appreciated in accordance with his just value, or occupy that high rank which in all probability will be assigned him in the future.

If it has been often proved that "no one is a prophet in his own country;" is it not equally true that the prophets, the men of the future, who feel its life in advance, and prefigure it in their works, are never recognized as prophets in their own times? It would be presumptuous to assert that it can ever be otherwise. In vain may the young generations of artists protest against the "Anti-progressives," whose invariable custom it is to assault and beat down the living with the dead: time alone can test the real value, or reveal the hidden beauties, either of musical compositions, or of kindred efforts in the sister arts.

As the manifold forms of art are but different incantations, charged with electricity from the soul of the artist, and destined to evoke the latent emotions and passions in order to render them sensible, intelligible, and, in some degree, tangible; so genius may be manifested in the invention of new forms, adapted, it may be, to the expression of feelings which have not yet surged within the limits of common experience, and are indeed first evoked within the magic circle by the creative power of artistic intuition. In arts in which sensation is linked to emotion, without the intermediate assistance of thought and reflection, the mere introduction of unaccustomed forms, of unused modes, must present an obstacle to the immediate comprehension of any very original composition. The surprise, nay, the fatigue, caused by the novelty of the singular impressions which it awakens, will make it appear to many as if written in a language of which they were ignorant, and which that reason will in itself be sufficient to induce them to pronounce a barbarous dialect. The trouble of accustoming the ear to it will repel many who will, in consequence, refuse to make a study of it. Through the more vivid and youthful organizations, less enthralled by the chains of habit; through the more ardent spirits, won first by curiosity, then filled with passion for the new idiom, must it penetrate and win the resisting and opposing public, which will finally catch the meaning, the aim, the construction, and at last render justice to its qualities, and acknowledge whatever beauty it may contain. Musicians who do not restrict themselves within the limits of conventional routine, have, consequently, more need than other artists of the aid of time. They cannot hope that death will bring that instantaneous plus-value to their works which it gives to those of the painters. No musician could renew, to the profit of his manuscripts, the deception practiced by one of the great Flemish painters, who, wishing in his lifetime to benefit by his future glory, directed his wife to spread abroad the news of his death, in order that the pictures with which he had taken care to cover the walls of his studio, might suddenly increase in value!

Whatever may be the present popularity of any part of the productions of one, broken, by suffering long before taken by death, it is nevertheless to be presumed that posterity will award to his works an estimation of a far higher character, of a much more earnest nature, than has hitherto been awarded them. A high rank must be assigned by the future historians of music to one who distinguished himself in art by a genius for melody so rare, by such graceful and remarkable enlargements of the harmonic tissue; and his triumph will be justly preferred to many of far more extended surface, though the works of such victors may be played and replayed by the greatest number of instruments, and be sung and resung by passing crowds of Prime Donne.

25. What point is the writer trying to make in the first paragraph?
A. Chopin died too young
B. It is doubtful whether Chopin should be appreciated
C. Chopin may be underrated
D. Chopin may be overrated
E. Chopin is too unknown

26. What is the writer's argument in the second paragraph?
A. A lifetime is insufficient to have notable achievements
B. Musicians tend to only be appreciated after they die
C. People focus too much on the living
D. A lifetime is insufficient to judge an artist's work
E. Dead musicians are always better than living musicians

27. Which of the following is **not** being used as a metaphor?
A. Electricity
B. Soul
C. Magic circle
D. Composition
E. Chains

28. The writer's main argument is that
A. Only death will increase a musician's value
B. We should focus more on living musicians than dead musicians
C. Young musicians need time for people to appreciate the work as people need to become accustomed to their unconventional music
D. Musicians should emulate the great Flemish painters
E. Musicians should not adopt unaccustomed forms or modes of composition

8. Religion and morality

It had been formerly asserted by theologians that our moral laws were given to man by a supernatural intuitive process. However, Professor E. A. Westermarck's "Origin and Development of the Moral Ideas," and similar researches, give a comprehensive survey of the moral ideas and practices of all the backward fragments of the human race and conclusively prove the social nature of moral law. The moral laws have evolved much the same as physical man has evolved. There is no indication whatsoever that the moral laws came from any revelation since the sense of moral law was just as strong amongst civilized peoples beyond the range of Christianity, or before the Christian era. Joseph McCabe, commenting on Professor Westermarck's work states, "All the fine theories of the philosophers break down before this vast collection of facts. There is no intuition whatever of an august and eternal law, and the less God is brought into connection with these pitiful blunders and often monstrous perversions of the moral sense, the better. What we see is just man's mind in possession of the idea that his conduct must be regulated by law, and clumsily working out the correct application of that idea as his intelligence grows and his social life becomes more complex. It is not a question of the mind of the savage imperfectly seeing the law. It is a plain case of the ideas of the savage reflecting and changing with his environment and the interest of his priests."

Justice is a fundamental and essential moral law because it is a vital regulation of social life and murder is the greatest crime because it is the greatest social delinquency; and these are inherent in the social nature of moral law. "Moral law slowly dawns in the mind of the human race as a regulation of a man's relation with his fellows in the interest of social life. It is quite independent of religion, since it has entirely different roots in human psychology." (Joseph McCabe: "Human Origin of Morals.")

In the mind of primitive man there is no connection between morality and the belief in a God. "Society is the school in which men learn to distinguish between right and wrong. The headmaster is custom and the lessons are the same for all. The first moral judgments were pronounced by public opinion; public indignation and public approval are the prototypes of the moral emotions." (Edward Westermarck: "Origin and Development of the Moral Ideas.")

Moral ideas and moral energy have their source in social life. It is only in a more advanced society that moral qualities are assumed for the gods. And indeed, it is known that in some primitive tribes, the gods are not necessarily conceived as good, they may have evil qualities also. "If they are, to his mind, good, that is so much the better. But whether they are good or bad they have to be faced as facts. The Gods, in short belong to the region of belief, while morality belongs to that of practice. It is in the nature of morality that it should be implicit in practice long before it is explicit in theory. Morality belongs to the group and is rooted in certain impulses that are a product of the essential conditions of group life. It is as reflection awakens that men are led to speculate upon the nature and origin of the moral feelings. Morality, whether in practice or theory, is thus based upon what is. On the other hand, religion, whether it be true or false, is in the nature of a discovery—one cannot conceive man actually ascribing ethical qualities to his Gods before he becomes sufficiently developed to formulate moral rules for his own guidance, and to create moral laws for his fellowmen. The moralization of the Gods will follow as a matter of course. Man really modifies his Gods in terms of the ideal human being. It is not the Gods who moralize man, it is man who moralizes the Gods." (Chapman Cohen: "Theism or Atheism.")

29. Which tone does the writer adopt?

A. Neutral

B. Investigative

C. Critical

D. Ironic

E. Sarcastic

30. Which philosopher mentioned in the passage disagrees with the viewpoint of the others?

A. Professor E. A. Westermarck

B. Joseph McCabe

C. Edward Westermarck

D. Chapman Cohen

E. They all agree with each other

31. Which sentence does not convey the same argument as the others?

A. Our moral laws were given to man by a supernatural intuitive process

B. Social nature of moral law

C. Moral law slowly dawns in the mind of the human race as a regulation of a man's relation with his fellows in the interest of social life

D. Moral ideas and moral energy have their source in social life

E. It is not the Gods who moralize man, it is man who moralizes the Gods

32. Which phrase is **not** used with a negative connotation in this passage?

A. Backward

B. Pitiful

C. Monstrous

D. Delinquency

E. Evil

9. Australia

To the people who lived four centuries ago in Europe only a very small portion of the earth's surface was known. Their geography was confined to the regions lying immediately around the Mediterranean, and including Europe, the north of Africa, and the west of Asia. Round these there was a margin, obscurely and imperfectly described in the reports of merchants; but by far the greater part of the world was utterly unknown. Great realms of darkness stretched all beyond, and closely hemmed in the little circle of light. In these unknown lands our ancestors loved to picture everything that was strange and mysterious. They believed that the man who could penetrate far enough would find countries where inexhaustible riches were to be gathered without toil from fertile shores, or marvellous valleys; and though wild tales were told of the dangers supposed to fill these regions, yet to the more daring and adventurous these only made the visions of boundless wealth and enchanting loveliness seem more fascinating.

Thus, as the art of navigation improved, and long voyages became possible, courageous seamen were tempted to venture out into the great unknown expanse. Columbus carried his trembling sailors over great tracts of unknown ocean, and discovered the two continents of America; Vasco di Gama penetrated far to the south, and rounded the Cape of Good Hope; Magellan, passing through the straits now called by his name, was the first to enter the Pacific Ocean; and so in the case of a hundred others, courage and skill carried the hardy seaman over many seas and into many lands that had lain unknown for ages.

Australia was the last part of the world to be thus visited and explored. In the year 1600, during the times of Shakespeare, the region to the south of the East Indies was still as little known as ever; the rude maps of those days had only a great blank where the islands of Australia should have been. Most people thought there was nothing but the ocean in that part of the world; and as the voyage was dangerous and very long—requiring several years for its completion—scarcely any one cared to run the risk of exploring it.

33. According to the passage, which region was the last to be explored?
A. Europe
B. Americas
C. North of Africa
D. West Asia
E. Australia

34. What was the reason why explorers were reluctant to explore Australia?
A. It was too expensive
B. There were not enough seamen
C. It was too risky
D. They did not think Australia existed
E. Navigation techniques were not good enough

35. Why did the explorers want to venture out further into the unknown?
A. They wanted to complete the map
B. They wanted passages named after them
C. Long navigation was possible
D. They believed they could get rich
E. They were seeking adventure

36. Which of the following was **not** a factor being the explorers being able to travel so far?
A. Money
B. Courage
C. Skills
D. Improvement in navigation
E. None of the above

10. Dreams

The subject which I have to discuss here is so complex, it raises so many questions of all kinds, difficult, obscure, some psychological, others physiological and metaphysical; in order to be treated in a complete manner it requires such a long development—and we have so little space, that I shall ask your permission to dispense with all preamble, to set aside un-essentials, and to go at once to the heart of the question.

A dream is this. I perceive objects and there is nothing there. I see men; I seem to speak to them and I hear what they answer; there is no one there and I have not spoken. It is all as if real things and real persons were there, then on waking all has disappeared, both persons and things. How does this happen?

But, first, is it true that there is nothing there? I mean, is there not presented a certain sense material to our eyes, to our ears, to our touch, etc., during sleep as well as during waking?

Close the eyes and look attentively at what goes on in the field of our vision. Many persons questioned on this point would say that nothing goes on, that they see nothing. No wonder at this, for a certain amount of practise is necessary to be able to observe oneself satisfactorily. But just give the requisite effort of attention, and you will distinguish, little by little, many things. First, in general, a black background. Upon this black background occasionally brilliant points which come and go, rising and descending, slowly and sedately. More often, spots of many colours, sometimes very dull, sometimes, on the contrary, with certain people, so brilliant that reality cannot compare with it. These spots spread and shrink, changing form and colour, constantly displacing one another. Sometimes the change is slow and gradual, sometimes again it is a whirlwind of vertiginous rapidity. Whence comes all this phantasmagoria? The physiologists and the psychologists have studied this play of colours. "Ocular spectra," "coloured spots," "phosphenes", such are the names that they have given to the phenomenon. They explain it either by the slight modifications which occur ceaselessly in the retinal circulation, or by the pressure that the closed lid exerts upon the eyeball, causing a mechanical excitation of the optic nerve. But the explanation of the phenomenon and the name that is given to it matters little. It occurs universally and it constitutes—I may say at once—the principal material of which we shape our dreams, "such stuff as dreams are made on."

37. What, according to the writer, is the 'heart of the question'?
A. What we see in dreams
B. What we hear in dreams
C. Our field of vision when we close our eyes
D. What are the 'play of colours'
E. What the physiologists and psychologists think

38. What does the writer think we see when we close our eyes?
A. Nothing
B. Perceived objects
C. Brilliant spots
D. Dull spots
E. There is no settled opinion

39. What does the writer agree with the physiologists and psychologists on?
A. We see colours when we close our eyes
B. The colours we see can be named 'ocular spectra', 'coloured spots' or 'phosphenes'
C. The phenomenon is due to slight modifications which occur ceaselessly in the retinal circulation
D. The phenomenon is due to pressure that the closed lid exerts upon the eyeball, causing a mechanical excitation of the optic nerve
E. None of the above

11. Pathological liars

The role played in society by the pathological liar is very striking. The characteristic behaviour in its unreasonableness is quite beyond the ken of the ordinary observer. The fact that here is a type of conduct regularly indulged in without seeming pleasurable results, and frequently militating obviously against the direct interests of the individual, makes a situation inexplicable by the usual canons of inference. To a certain extent the tendencies of each separate case must be viewed in their environmental context to be well understood. For example, the lying and swindling which centre about the assumption of a noble name and a corresponding station or affecting the life of a cloister brother, such as we find in the cases cited by Longard, show great differences from any material obtainable in our country. In interpretation of this, one has to consider the glamour thrown about the socially exalted or the life of the recluse—a glamour which obtains readily among the simple-minded people of rural Europe. Then, too, this very simple-mindedness, with the great differences which exist between peasant and noble, leads in itself to much opportunity for cheating.

With us, especially in the newer work of courts, which are rapidly becoming in their various social endeavours more and more intimately connected with many phases of life, the pathological liar becomes of main interest in the role of accuser of others, self-accuser, witness, and general social disturber.

Here again, we may call attention to the fact, which is of great social importance, namely, that the person who is seemingly normal in all other respects may be a pathological liar. It might be naturally expected that the feebleminded, who frequently have poor discernment of the relation of cause and effect, including the phenomena of conduct, would often lie without normal cause. As a matter of fact, there is surprisingly little of this among them, and one can find numerous mental defectives who are faithful tellers of the truth, while even, as we have found by other studies, some are good testifiers. Exaggerated instances of the type represented by Case 12, where the individual by the virtue of language ability endeavours to maintain a place in the world which his abilities do not otherwise justify, and where the very contradiction between abilities and disabilities leads to the development of an excessive habit of lying, are known in considerable number by us. Many of these mentally defective verbalists do not even grade high enough to come in our border-line cases, and yet frequently, by virtue of their gift of language, the world in general considers them fairly normal. They are really on a constant social strain by virtue of this, and while they are not purely pathological liars they often indulge in pathological lying, a distinction we have endeavoured to make clear in our introduction.

It stands out very clearly, both in previous studies of this subject and in viewing our own material, that pathological lying is very rarely the single offense of the pathological liar. The characteristics of this lying show that it arises from a tendency which might easily express itself in other forms of misrepresentation. Swindling, sometimes stealing, sometimes running away from home (assuming another character and perhaps another name) may be the results of the same general causes in the individual. The extent to which these other delinquencies are carried on by a pathological liar depends again largely upon environmental conditions—for instance, truancy is very difficult in German cities; a long career of thieving, under the better police surveillance of some European countries, is less possible than with us; while swindling, for the reason given above, seems easier there.

40. The writer is of the view that:
A. People can often tell who a pathological liar is
B. Pathological liars can easily integrate into society
C. Pathological liars often commit related offences
D. Pathological liars are good verbalists
E. Pathological liars are mentally-impaired

41. What does the writer want to show by using 'case 12'?
A. Pathological liars are often forced to lie
B. Pathological liars can easily integrate into society
C. Pathological liars lack the skills to contribute to society
D. Some pathological liars may be lying to fit into society
E. Pathological liars suffer from lying

42. What tone does the writer adopt?
A. Sarcastic
B. Critical
C. Analytical
D. One-sided
E. Ironic

END OF SECTION

YOU MUST ANSWER <u>ONLY</u> <u>ONE</u> OF THE FOLLOWING QUESTIONS

1. Is social media damaging for teenagers?

2. To what extent should journalists be responsible for 'fake news'?

3. Limitations should be put in place for scientific discovery. Discuss.

4. Too many students are going to University. Discuss

END OF TEST

THIS PAGE HAS BEEN
INTENTIONALLY LEFT BLANK

Mock Paper D

1. The fallacy of protection

All laws made for the purpose of protecting the interests of individuals or classes must mean, if they mean anything, to render the articles which such classes deal in or produce dearer than they would otherwise be if the public was left at liberty to supply itself with such commodities in the manner which their own interests and choice would dictate. In order to make them dearer it is absolutely necessary to make them scarcer; for quantity being large or small in proportion to demand, alone can regulate the price; —protection, therefore, to any commodity simply means that the quantity supplied to the community shall be less than circumstances would naturally provide, but that for the smaller quantity supplied under the restriction of law the same sum shall be paid as the larger quantity would command without such restriction.

Time was when the Sovereigns of England relied chiefly on the granting of patents to individuals for the exclusive exercise of certain trades or occupations in particular places, as the means of rewarding the services of some, and as a provision for others of their adherents, followers, and favourites, who either held the exclusive supply in their own hands on their own terms, or who again granted to others under them that privilege, receiving from them a portion of the gains. In the course of time, however, the public began to discover that these monopolies acted upon them directly as a tax of a most odious description; that the privileged person found it needful always to keep the supply short to obtain his high price (for as soon as he admitted plenty he had no command of price)—that, in short, the sovereign, in conferring a mark of regard on a favourite, gave not that which he himself possessed, but only invested him with the power of imposing a contribution on the public.

The public once awake to the true operation of such privileges, and severely suffering under the injuries which they inflicted, perseveringly struggled against these odious monopolies, until the system was entirely abandoned, and the crown was deprived of the power of granting patents of this class. But though the public saw clearly enough that these privileges granted by the sovereign to individuals operated thus prejudicially on the community, they did not see with equal clearness that the same power transferred to, and exercised by, Parliament, to confer similar privileges on classes; to do for a number of men what the sovereign had before done for single men, would, to the remaining portion of the community, be just as prejudicial as the abuses against which they had struggled. That like the sovereign, the Parliament, in protecting or giving privileges to a class, gave nothing which they possessed themselves, but granted only the power to such classes of raising a contribution from the remaining portion of the community, by levying a higher price for their commodity than it would otherwise command. As with individuals, it was equally necessary to make scarcity to secure price, and that could only be done by restricting the sources of supply by prohibiting, or by imposing high duties on, foreign importations. Many circumstances, however, combined to render the use of this power by Parliament less obvious than it had been when exercised by the sovereign, but chiefly the fact that protection was usually granted by imposing high duties, often in their effect quite prohibitory, under the plea of providing revenue for the state. Many other more modern excuses have been urged, such as those of encouraging native industry, and countervailing peculiar burthens, in order to reconcile public opinion to the exactions arising out of the system, all of which we shall, on future occasions, carefully consider separately. But, above all, the great reason why these evils have been so long endured has been, that the public have believed that all classes and interests, though perhaps not exactly to the same extent, have shared in protection. We propose at present to confine our consideration to the effects of protection, —first, on the community generally; and secondly, on the individual classes protected.

1. The writer takes the view that:
A. Laws were implemented for the benefit of the public
B. Laws were implied indiscriminately
C. The public did not have a say in the laws being implemented
D. Only certain classes of people benefited from the laws at the expense of the rest of the public
E. None of the above

2. The writer suggests in the first paragraph that:
A. Laws are meant for protecting the interests of individuals or classes
B. Commodities should remain scarce
C. Laws result in the price of commodities being increased
D. Laws cannot affect the demand and supply of commodities
E. All of the above

3. According to the second paragraph, what was the original intention behind patents?
A. Exclusive exercise of certain trades or occupations in particular places
B. Taxation
C. Keep the supply short
D. Increase prices
E. None of the above

4. With regards to the imposition of high duties, which reasoning would the writer agree with?
A. Providing revenue for the state
B. Encouraging native industry
C. Countervailing peculiar burthens
D. All of the above
E. None of the above

2. Camouflage

The art of concealment or camouflage is one of the newest and most highly developed techniques of modern warfare. But the animals have been masters of it for ages. The lives of most of them are passed in constant conflict. Those which have enemies from which they cannot escape by rapidity of motion must be able to hide or disguise themselves. Those which hunt for a living must be able to approach their prey without unnecessary noise or attention to themselves. It is very remarkable how Nature helps the wild creatures to disguise themselves by colouring them with various shades and tints best calculated to enable them to escape enemies or to entrap prey.

The animals of each locality are usually coloured according to their habitat, but good reasons make some exceptions advisable. Many of the most striking examples of this protective resemblance among animals are the result of their very intimate association with the surrounding flora and natural scenery. There is no part of a tree, including flowers, fruits, bark and roots, that is not in some way copied and imitated by these clever creatures. Often this imitation is astonishing in its faithfulness of detail. Bunches of cocoanuts are portrayed by sleeping monkeys, while even the leaves are copied by certain tree-toads, and many flowers are represented by monkeys and lizards. The winding roots of huge trees are copied by snakes that twist themselves together at the foot of the tree.

In the art of camouflage—an art which affects the form, colour, and attitude of animals—Nature has worked along two different roads. One is easy and direct, the other circuitous and difficult. The easy way is that of protective resemblance pure and simple, where the animal's colour, form, or attitude becomes like that of its habitat. In which case the animal becomes one with its environment and thus is enabled to go about unnoticed by its enemies or by its prey. The other way is that of bluff, and it includes all inoffensive animals which are capable of assuming attitudes and colours that terrify and frighten. The colours in some cases are really of warning pattern, yet they cannot be considered mimetic unless they are thought to resemble the patterns of some extinct model of which we know nothing; and since they are not found in present-day animals with unpleasant qualities, they are not, strictly speaking, warning colours.

Desert animals are in most cases desert-coloured. The lion, for example, is almost invisible when crouched among the rocks and streams of the African wastes. Antelopes are tinted like the landscape over which they roam, while the camel seems actually to blend with the desert sands. The kangaroos of Australia at a little distance seem to disappear into the soil of their respective localities, while the cat of the Pampas accurately reflects his surroundings in his fur.

The tiger is made so invisible by his wonderful colour that, when he crouches in the bright sunlight amid the tall brown grass, it is almost impossible to see him. But the zebra and the giraffe are the kings of all camouflagers! So deceptive are the large blotch-spots of the giraffe and his weird head and horns, like scrubby limbs, that his concealment is perfect. Even the cleverest natives often mistake a herd of giraffes for a clump of trees. The camouflage of zebras is equally deceptive. Drummond says that he once found himself in a forest, looking at what he thought to be a lone zebra, when to his astonishment he suddenly realised that he was facing an entire herd which were invisible until they became frightened and moved. Evidently the zebra is well aware that the black-and-white stripes of his coat take away the sense of solid body, and that the two colours blend into a light grey, and thus at close range the effect is that of rays of sunlight passing through bushes.

5. Which of the following is **not** a reason put forward by the writer behind the purpose of camouflage for animals?
A. Modern warfare
B. Hide or disguise themselves from enemies
C. Approach their prey without unnecessary noise or attention to themselves
D. All of the above
E. None of the above

6. 'Nature has worked along two different roads' – what are the two different roads mentioned by the writer?
A. Two different ways of camouflage
B. Two different functions of camouflage
C. Two levels of complexity behind camouflage
D. None of the above
E. All of the above

7. Which animal mentioned below has a distinctively different method of camouflage from the others?
A. Desert animals
B. Tiger
C. Zebra
D. Giraffe
E. They all adopt similar methods

8. What tone does the writer adopt throughout the passage?
A. Ironic
B. Sarcastic
C. Analytic
D. Critical
E. Inquisitive

3. History of Nature

We live in and form part of a system of things of immense diversity and perplexity, which we call Nature; and it is a matter of the deepest interest to all of us that we should form just conceptions of the constitution of that system and of its past history. With relation to this universe, man is, in extent, little more than a mathematical point; in duration but a fleeting shadow; he is a mere reed shaken in the winds of force. But as Pascal long ago remarked, although a mere reed, he is a thinking reed; and in virtue of that wonderful capacity of thought, he has the power of framing for himself a symbolic conception of the universe, which, although doubtless highly imperfect and inadequate as a picture of the great whole, is yet sufficient to serve him as a chart for the guidance of his practical affairs. It has taken long ages of toilsome and often fruitless labour to enable man to look steadily at the shifting scenes of the phantasmagoria of Nature, to notice what is fixed among her fluctuations, and what is regular among her apparent irregularities; and it is only comparatively lately, within the last few centuries, that the conception of a universal order and of a definite course of things, which we term the course of Nature, has emerged.

But, once originated, the conception of the constancy of the order of Nature has become the dominant idea of modern thought. To any person who is familiar with the facts upon which that conception is based, and is competent to estimate their significance, it has ceased to be conceivable that chance should have any place in the universe, or that events should depend upon any but the natural sequence of cause and effect. We have come to look upon the present as the child of the past and as the parent of the future; and, as we have excluded chance from a place in the universe, so we ignore, even as a possibility, the notion of any interference with the order of Nature. Whatever may be men's speculative doctrines, it is quite certain that every intelligent person guides his life and risks his fortune upon the belief that the order of Nature is constant, and that the chain of natural causation is never broken.

In fact, no belief which we entertain has so complete a logical basis as that to which I have just referred. It tacitly underlies every process of reasoning; it is the foundation of every act of the will. It is based upon the broadest induction, and it is verified by the most constant, regular, and universal of deductive processes. But we must recollect that any human belief, however broad its basis, however defensible it may seem, is, after all, only a probable belief, and that our widest and safest generalisations are simply statements of the highest degree of probability. Though we are quite clear about the constancy of the order of Nature, at the present time, and in the present state of things, it by no means necessarily follows that we are justified in expanding this generalisation into the infinite past, and in denying, absolutely, that there may have been a time when Nature did not follow a fixed order, when the relations of cause and effect were not definite, and when extra-natural agencies interfered with the general course of Nature. Cautious men will allow that a universe so different from that which we know may have existed; just as a very candid thinker may admit that a world in which two and two do not make four, and in which two straight lines do enclose a space, may exist. But the same caution which forces the admission of such possibilities demands a great deal of evidence before it recognises them to be anything more substantial. And when it is asserted that, so many thousand years ago, events occurred in a manner utterly foreign to and inconsistent with the existing laws of Nature, men, who without being particularly cautious, are simply honest thinkers, unwilling to deceive themselves or delude others, ask for trustworthy evidence of the fact.

9. Why does the writer capitalise the word 'Nature'?
A. To emphasise its importance
B. The writer is not using the word in its natural meaning
C. To refer to 'nature' as an entity
D. The writer is using it as a name
E. The writer think it is customary to capitalise the word 'Nature'

10. Which of the following is **not** used as a metaphor?
A. Universe
B. Mathematical point
C. Shadow
D. Reed
E. None of the above

11. Which of the following is presented as a fact rather than an opinion?
A. 'We live in and form part of a system of things of immense diversity and perplexity'
B. 'But as Pascal long ago remarked, although a mere reed, he is a thinking reed'
C. 'Whatever may be men's speculative doctrines, it is quite certain that every intelligent person guides his life and risks his fortune upon the belief that the order of Nature is constant, and that the chain of natural causation is never broken'
D. 'In fact, no belief which we entertain has so complete a logical basis as that to which I have just referred'
E. 'But we must recollect that any human belief, however broad its basis, however defensible it may seem, is, after all, only a probable belief'

12. Which of the following statements would the author most agree with?
A. Men are always capable of understanding Nature
B. Men can only estimate to the highest probability what they understand about Nature
C. Nature cannot be understood as it is too complex
D. Men are not intelligent enough to understand Nature
E. Men are too insignificant in Nature

4. The selection theory

In artificial selection the breeder chooses out for pairing only such individuals as possess the character desired by him in a somewhat higher degree than the rest of the race. Some of the descendants inherit this character, often in a still higher degree, and if this method be pursued throughout several generations, the race is transformed in respect of that particular character.

Natural selection depends on the same three factors as artificial selection: on variability, inheritance, and selection for breeding, but this last is here carried out not by a breeder but by what Darwin called the "struggle for existence." This last factor is one of the special features of the Darwinian conception of nature. That there are carnivorous animals which take heavy toll in every generation of the progeny of the animals on which they prey, and that there are herbivores which decimate the plants in every generation had long been known, but it is only since Darwin's time that sufficient attention has been paid to the facts that, in addition to this regular destruction, there exists between the members of a species a keen competition for space and food, which limits multiplication, and that numerous individuals of each species perish because of unfavourable climatic conditions. The "struggle for existence," which Darwin regarded as taking the place of the human breeder in free nature, is not a direct struggle between carnivores and their prey, but is the assumed competition for survival between individuals of the same species, of which, on an average, only those survive to reproduce which have the greatest power of resistance, while the others, less favourably constituted, perish early. This struggle is so keen, that, within a limited area, where the conditions of life have long remained unchanged, of every species, whatever be the degree of fertility, only two, on an average, of the descendants of each pair survive; the others succumb either to enemies, or to disadvantages of climate, or to accident. A high degree of fertility is thus not an indication of the special success of a species, but of the numerous dangers that have attended its evolution. Of the six young brought forth by a pair of elephants in the course of their lives only two survive in a given area; similarly, of the millions of eggs which two thread-worms leave behind them only two survive. It is thus possible to estimate the dangers which threaten a species by its ratio of elimination, or, since this cannot be done directly, by its fertility.

Although a great number of the descendants of each generation fall victims to accident, among those that remain it is still the greater or less fitness of the organism that determines the "selection for breeding purposes," and it would be incomprehensible if, in this competition, it was not ultimately, that is, on an average, the best equipped which survive, in the sense of living long enough to reproduce.

Thus the principle of natural selection is the selection of the best for reproduction, whether the "best" refers to the whole constitution, to one or more parts of the organism, or to one or more stages of development. Every organ, every part, every character of an animal, fertility and intelligence included, must be improved in this manner, and be gradually brought up in the course of generations to its highest attainable state of perfection. And not only may improvement of parts be brought about in this way, but new parts and organs may arise, since, through the slow and minute steps of individual or "fluctuating" variations, a part may be added here or dropped out there, and thus something new is produced.

13. Which of the following is a difference between artificial selection and natural selection?
A. Variability
B. Inheritance
C. Selection for breeding
D. None of the above
E. All of the above

14. What contributes to the 'struggle for existence'?
A. Carnivorous animals which take heavy toll in every generation of the progeny of the animals on which they prey
B. Herbivores which decimate the plants in every generation
C. Keen competition for space and food
D. Unfavourable climatic conditions
E. All of the above

15. 'Thus the principle of natural selection is the selection of the best for reproduction' – which of the following does not fall under the writer's definition of 'best'?
A. Organs of an animal
B. Fitness of an animal
C. Avoiding accidents
D. Fertility
E. Intelligence

16. Which of the following is **not** an example of natural selection?
A. The breeder chooses out for pairing only such individuals as possess the character desired by him in a somewhat higher degree than the rest of the race
B. Keen competition for space and food which limits multiplication
C. Numerous individuals of each species perish because of unfavourable climatic conditions
D. Of every species, whatever be the degree of fertility, only two, on an average, of the descendants of each pair survive
E. The others succumb either to enemies, or to disadvantages of climate, or to accident

5. Heredity

One of the best attested single characters in human heredity is brachydactyly, "short-fingerness," which results in a reduction in the length of the fingers by the dropping out of one joint. If one lumps together all the cases where any effect of this sort is found, it is evident that normals never transmit it to their posterity, that affected persons always do, and that in a mating between a normal and an affected person, all the offspring will show the abnormality. It is a good example of a unit character.

But its effect is by no means confined to the fingers. It tends to affect the entire skeleton, and in a family where one child is markedly brachydactylous, that child is generally shorter than the others. The factor for brachydactyly evidently produces its primary effect on the bones of the hand, but it also produces a secondary effect on all the bones of the body.

Moreover, it will be found, if a number of brachydactylous persons are examined, that no two of them are affected to exactly the same degree. In some cases, only one finger will be abnormal; in other cases there will be a slight effect in all the fingers; in other cases all the fingers will be highly affected. Why is there such variation in the results produced by a unit character? Because, presumably, in each individual there is a different set of modifying factors or else a variation in the factor. It has been found that an abnormality quite like brachydactyly is produced by abnormality in the pituitary gland. It is then fair to suppose that the factor which produces brachydactyly does so by affecting the pituitary gland in some way. But there must be many other factors which also affect the pituitary and in some cases probably favour its development, rather than hindering it. Then if the factor for brachydactyly is depressing the pituitary, but if some other factors are at the same time stimulating that gland, the effect shown in the subject's fingers will be much less marked than if a group of modifying factors were present which acted in the same direction as the brachydactyly factor, —to perturb the action of the pituitary gland.

This illustration is largely hypothetical; but there is no room for doubt that every factor produces more than a single effect. A white blaze in the hair, for example, is a well-proved unit factor in man; the factor not only produces a white streak in the hair, but affects the pigmentation of the skin as well, usually resulting in one or more white spots on some part of the body. It is really a factor for "piebaldism."

For the sake of clear thinking, then, the idea of a unit character due to some unit determiner or factor in the germ-plasm must be given up, and it must be recognized that every visible character of an individual is the result of numerous factors, or differences in the germ-plasm. Ordinarily one of these produces a more notable contribution to the end-product than do the others; but there are cases where this statement does not appear to hold good. This leads to the conception of multiple factors.

17. Which of the following is not a possible side-effect of brachydactyly?
A. Reduction in the length of one finger
B. Reduction in the length of all fingers
C. Decrease in height
D. Dropping out of one joint
E. All of the above

18. Based on the information provided in the passage, which set of parents will **not** produce an offspring with brachydactyly?
A. Two normal parents
B. Two affected parents
C. One affected parent and one normal parent
D. A and B only
E. None of the above

19. What explanation does the writer provide for the difference in seriousness of the brachydactyly in affected individuals?
A. Individuals that are more affected have two parents who were affected
B. Individuals that are more affected have only one parent who was affected
C. The more affected individuals had a weak constitution to begin with
D. The more affected individuals had other factors that contributed to the seriousness of the brachydactyly
E. These individuals have weak pituitary glands

20. What is the writer's overall conclusion?
A. The genes determining brachydactyly are solely responsible for producing short fingers
B. Different traits of a person are the result of numerous factors
C. One affected parent is enough for a child to be affected with the illness
D. Brachydactyly and piebaldism are similar
E. Brachydactyly and piebaldism have wide-ranging effects on the body

6. Buddhism in China

Buddhism was not an indigenous religion of China. Its founder was Gautama of India in the sixth century B.C. Some centuries later it found its way into China by way of central Asia. There is a tradition that as early as 142 B.C. Chang Ch'ien, an ambassador of the Chinese emperor, Wu Ti, visited the countries of central Asia, where he first learned about the new religion which was making such headway and reported concerning it to his master. A few years later the generals of Wu Ti captured a gold image of the Buddha which the emperor set up in his palace and worshiped, but he took no further steps.

According to Chinese historians' Buddhism was officially recognized in China about 67 A.D. A few years before that date, the emperor, Ming-Ti, saw in a dream a large golden image with a halo hovering above his palace. His advisers, some of whom were no doubt already favourable to the new religion, interpreted the image of the dream to be that of Buddha, the great sage of India, who was inviting his adhesion. Following their advice the emperor sent an embassy to study into Buddhism. It brought back two Indian monks and a quantity of Buddhist classics. These were carried on a white horse and so the monastery which the emperor built for the monks and those who came after them was called the White Horse Monastery. Its tablet is said to have survived to this day.

This dream story is worth repeating because it goes to show that Buddhism was not only known at an early date, but was favoured at the court of China. In fact, the same history which relates the dream contains the biography of an official who became an adherent of Buddhism a few years before the dream took place. This is not at all surprising, because an acquaintance with Buddhism was the inevitable concomitant of the military campaigning, the many embassies and the wide-ranging trade of those centuries. But the introduction of Buddhism into China was especially promoted by reason of the current policy of the Chinese government of moving conquered populations in countries west of China into China proper, the vanquished peoples brought their own religion along with them. At one time what is now the province of Shansi was populated in this way by the Hsiung-nu, many of whom were Buddhists.

The introduction and spread of Buddhism were hastened by the decline of Confucianism and Taoism. The Han dynasty (206 B. C.-221 A. D.) established a government founded on Confucianism. It reproduced the classics destroyed in the previous dynasty and encouraged their study; it established the state worship of Confucius; it based its laws and regulations upon the ideals and principles advocated by Confucius. The great increase of wealth and power under this dynasty led to a gradual deterioration in the character of the rulers and officials. The rigid Confucian regulations became burdensome to the people who ceased to respect their leaders. Confucianism lost its hold as the complete solution of the problems of life. At the same time Taoism had become a veritable jumble of meaningless and superstitious rites which served to support a horde of ignorant, selfish priests. The high religious ideals of the earlier Taoist mystics were abandoned for a search after the elixir of life during fruitless journeys to the isles of the Immortals which were supposed to be in the Eastern Sea.

21. According to the writer, Buddhism originated in:
A. India
B. China
C. Central Asia
D. West of China
E. Shansi

22. Which of the following is **not** a reason given for the rise of Buddhism in China?
A. The generals of Wu Ti captured a gold image of the Buddha which the emperor set up in his palace and worshiped
B. The emperor sent an embassy to study into Buddhism
C. Buddhism was favoured at the court of China
D. Decline of Confucianism and Taoism
E. None of the above

23. Which of the following is a reason why Confucianism started to decline?
A. The Han Dynasty established the state worship of Confucius
B. The Han Dynasty based its laws and regulations upon the ideals and principles advocated by Confucius
C. The great increase of wealth and power under the Han Dynasty
D. Gradual deterioration in the character of the rulers and officials
E. All of the above

24. What is the writer's purpose behind introducing the 'dream story'?
A. Showing how superstitious the Chinese were
B. The exact date of the introduction of Buddhism was disputed
C. The introduction of Buddhism had questionable origins
D. Buddhism was introduced at an early date
E. Buddhism was not introduced because of the decline of Confucianism and Taoism

7. Science in England

That the state of knowledge in any country will exert a directive influence on the general system of instruction adopted in it, is a principle too obvious to require investigation. And it is equally certain that the tastes and pursuits of our manhood will bear on them the traces of the earlier impressions of our education. It is therefore not unreasonable to suppose that some portion of the neglect of science in England, may be attributed to the system of education we pursue. A young man passes from our public schools to the universities, ignorant almost of the elements of every branch of useful knowledge; and at these latter establishments, formed originally for instructing those who are intended for the clerical profession, classical and mathematical pursuits are nearly the sole objects proposed to the student's ambition.

Much has been done at one of our universities during the last fifteen years, to improve the system of study; and I am confident that there is no one connected with that body, who will not do me the justice to believe that, whatever suggestions I may venture to offer, are prompted by the warmest feelings for the honour and the increasing prosperity of its institutions. The ties which connect me with Cambridge are indeed of no ordinary kind.

Taking it then for granted that our system of academical education ought to be adapted to nearly the whole of the aristocracy of the country, I am inclined to believe that whilst the modifications I should propose would not be great innovations on the spirit of our institutions, they would contribute materially to that important object.

It will be readily admitted, that a degree conferred by an university, ought to be a pledge to the public that he who holds it possesses a certain quantity of knowledge. The progress of society has rendered knowledge far more various in its kinds than it used to be; and to meet this variety in the tastes and inclinations of those who come to us for instruction, we have, besides the regular lectures to which all must attend, other sources of information from whence the students may acquire sound and varied knowledge in the numerous lectures on chemistry, geology, botany, history, etc. It is at present a matter of option with the student, which, and how many of these courses he shall attend, and such it should still remain. All that it would be necessary to add would be, that previously to taking his degree, each person should be examined by those Professors, whose lectures he had attended. The pupils should then be arranged in two classes, according to their merits, and the names included in these classes should be printed. I would then propose that no young man, except his name was found amongst the "List of Honours", should be allowed to take his degree, unless he had been placed in the first class of someone at least of the courses given by the professors. But it should still be imperative upon the student to possess such mathematical knowledge as we usually require. If he had attained the first rank in several of these examinations, it is obvious that we should run no hazard in a little relaxing: the strictness of his mathematical trial.

25. Which of the following is presented as a **fact** by the writer?
A. That the state of knowledge in any country will exert a directive influence on the general system of instruction adopted in it, is a principle too obvious to require investigation
B. And it is equally certain that the tastes and pursuits of our manhood will bear on them the traces of the earlier impressions of our education
C. Much has been done at one of our universities during the last fifteen years, to improve the system of study
D. I am confident that there is no one connected with that body, who will not do me the justice to believe that, whatever suggestions I may venture to offer, are prompted by the warmest feelings for the honour and the increasing prosperity of its institutions
E. The ties which connect me with Cambridge are indeed of no ordinary kind

26. Who is the passage addressed to?
A. Students
B. Professors
C. Universities
D. Scientists
E. Young people

27. What is the writer's main argument?
A. Students are graded too leniently
B. The clerical profession, classical and mathematical pursuits are the sole objects of education
C. Students do not know enough about science
D. Students should aspire to get a First Class Honours
E. Universities are not providing good education

28. What tone does the writer adopt in this passage?
A. Analytical
B. Critical
C. Sarcastic
D. Ironic
E. Neutral

8. Women's suffrage

The prolonged slavery of woman is the darkest page in human history. A survey of the condition of the race through those barbarous periods, when physical force governed the world, when the motto, "might makes right," was the law, enables one to account, for the origin of woman's subjection to man without referring the fact to the general inferiority of the sex, or Nature's law.

Writers on this question differ as to the cause of the universal degradation of woman in all periods and nations. One of the greatest minds of the century has thrown a ray of light on this gloomy picture by tracing the origin of woman's slavery to the same principle of selfishness and love of power in man that has thus far dominated all weaker nations and classes. This brings hope of final emancipation, for as all nations and classes are gradually, one after another, asserting and maintaining their independence, the path is clear for woman to follow. The slavish instinct of an oppressed class has led her to toil patiently through the ages, giving all and asking little, cheerfully sharing with man all perils and privations by land and sea, that husband and sons might attain honour and success. Justice and freedom for herself is her latest and highest demand.

Another writer asserts that the tyranny of man over woman has its roots, after all, in his nobler feelings; his love, his chivalry, and his desire to protect woman in the barbarous periods of pillage, lust, and war. But wherever the roots may be traced, the results at this hour are equally disastrous to woman. Her best interests and happiness do not seem to have been consulted in the arrangements made for her protection. She has been bought and sold, caressed and crucified at the will and pleasure of her master. But if a chivalrous desire to protect woman has always been the mainspring of man's dominion over her, it should have prompted him to place in her hands the same weapons of defence he has found to be most effective against wrong and oppression.

It is often asserted that as woman has always been man's slave—subject—inferior—dependent, under all forms of government and religion, slavery must be her normal condition. This might have some weight had not the vast majority of men also been enslaved for centuries to kings and popes, and orders of nobility, who, in the progress of civilization, have reached complete equality. And did we not also see the great changes in woman's condition, the marvellous transformation in her character, from a toy in the Turkish harem, or a drudge in the German fields, to a leader of thought in the literary circles of France, England, and America!

In an age when the wrongs of society are adjusted in the courts and at the ballot-box, material force yields to reason and majorities.

Woman's steady march onward, and her growing desire for a broader outlook, prove that she has not reached her normal condition, and that society has not yet conceded all that is necessary for its attainment.

29. Which of the following is presented as a **fact**?
A. The prolonged slavery of woman is the darkest page in human history
B. A survey of the condition of the race through those barbarous periods, when physical force governed the world, when the motto, "might makes right," was the law, enables one to account, for the origin of woman's subjection to man without referring the fact to the general inferiority of the sex, or Nature's law
C. Writers on this question differ as to the cause of the universal degradation of woman in all periods and nations
D. The slavish instinct of an oppressed class has led her to toil patiently through the ages, giving all and asking little
E. Justice and freedom for herself is her latest and highest demand

30. Which word is **not** being used with a negative connotation?
A. Darkest
B. Barbarous
C. Inferiority
D. Gloomy
E. None of the above

31. Which tone does the writer adopt?
A. Inquisitive
B. Sceptical
C. Critical
D. Opinionated
E. Factual

32. What is the writer's overall argument?
A. Women are unable to acquire better rights than men
B. Women are constantly oppressed
C. Women are still in the journey of fighting for the rights they deserve
D. Men have a tendency to harm women instead of protecting them
E. Men can also be oppressed

9. Christianity and Islam

A comparison of Christianity with Muhammedanism or with any other religion must be preceded by a statement of the objects with which such comparison is undertaken, for the possibilities which lie in this direction are numerous. The missionary, for instance, may consider that a knowledge of the similarities of these religions would increase the efficacy of his proselytising work: his purpose would thus be wholly practical. The ecclesiastically minded Christian, already convinced of the superiority of his own religion, will be chiefly anxious to secure scientific proof of the fact: the study of comparative religion from this point of view was once a popular branch of apologetics and is by no means out of favour at the present day. Again, the inquirer whose historical perspective is undisturbed by ecclesiastical considerations, will approach the subject with somewhat different interests. He will expect the comparison to provide him with a clear view of the influence which Christianity has exerted upon other religions or has itself received from them: or he may hope by comparing the general development of special religious systems to gain a clearer insight into the growth of Christianity. Hence the object of such comparisons is to trace the course of analogous developments and the interaction of influence and so to increase the knowledge of religion in general or of our own religion in particular.

A world-religion, such as Christianity, is a highly complex structure and the evolution of such a system of belief is best understood by examining a religion to which we have not been bound by a thousand ties from the earliest days of our lives. If we take an alien religion as our subject of investigation, we shall not shrink from the consequences of the historical method: whereas, when we criticise Christianity, we are often unable to see the falsity of the pre-suppositions which we necessarily bring to the task of inquiry: our minds follow the doctrines of Christianity, even as our bodies perform their functions—in complete unconsciousness. At the same time, we possess a very considerable knowledge of the development of Christianity, and this we owe largely to the help of analogy. Especially instructive is the comparison between Christianity and Buddhism. No less interesting are the discoveries to be attained by an inquiry into the development of Muhammedanism: here we can see the growth of tradition proceeding in the full light of historical criticism. We see the plain man, Muhammed, expressly declaring in the Qoran that he cannot perform miracles, yet gradually becoming a miracle worker and indeed the greatest of his class: he professes to be nothing more than a mortal man: he becomes the chief mediator between man and God. The scanty memorials of the man become voluminous biographies of the saint and increase from generation to generation.

Yet more remarkable is the fact that his utterances, his logia, if we may use the term, some few of which are certainly genuine, increase from year to year and form a large collection which is critically sifted and expounded. The aspirations of mankind attribute to him such words of the New Testament and of Greek philosophers as were especially popular or seemed worthy of Muhammed; the teaching also of the new ecclesiastical schools was invariably expressed in the form of proverbial utterances attributed to Muhammed, and these are now without exception regarded as authentic by the modern Moslem. In this way opinions often contradictory are covered by Muhummed's authority.

33. What does the writer think is the purpose of a comparison of religions?
A. Increase the efficacy of proselytising work
B. Secure scientific proof to determine the superiority of a religion
C. Provide a clear view of the of religions upon each other
D. Gain a clearer insight into the growth of a religion
E. None of the above

34. What is the writer's views with regards to Muhammedanism?
A. It is inferior to Christianity
B. It is similar to Buddhism
C. It criticises Christianity
D. It helps us view Christianity more critically
E. It is similar to Christianity

35. Who is more likely to **not** have a biased opinion towards their religion according to the passage?
A. Missionaries
B. Ecclesiastically-minded Christian
C. Inquirer whose historical perspective is undisturbed by ecclesiastical considerations
D. Muhammed
E. Greek philosophers

36. The writer argues in the second paragraph that:
A. Christians are blinded by their religion
B. Muhammedanism provides a false comparison to Christianity
C. We need to compare Christianity to similar religions
D. We need to compare Christianity to dissimilar religions
E. Muhammedanism is critical of Christianity

10. Yeast

I have selected tonight the particular subject of yeast for two reasons—or, rather, I should say for three. In the first place, because it is one of the simplest and the most familiar objects with which we are acquainted. In the second place, because the facts and phenomena which I have to describe are so simple that it is possible to put them before you without the help of any of those pictures or diagrams which are needed when matters are more complicated, and which, if I had to refer to them here, would involve the necessity of my turning away from you now and then, and thereby increasing very largely my difficulty (already sufficiently great) in making myself heard. And thirdly, I have chosen this subject because I know of no familiar substance forming part of our every-day knowledge and experience, the examination of which, with a little care, tends to open up such very considerable issues as does this substance—yeast.

In the first place, I should like to call your attention to a fact with which the whole of you are, to begin with, perfectly acquainted, I mean the fact that any liquid containing sugar, any liquid which is formed by pressing out the succulent parts of the fruits of plants, or a mixture of honey and water, if left to itself for a short time, begins to undergo a peculiar change. No matter how clear it might be at starting, yet after a few hours, or at most a few days, if the temperature is high, this liquid begins to be turbid, and by-and-by bubbles make their appearance in it, and a sort of dirty-looking yellowish foam or scum collects at the surface; while at the same time, by degrees, a similar kind of matter, which we call the "lees," sinks to the bottom.

The quantity of this dirty-looking stuff, that we call the scum and the lees, goes on increasing until it reaches a certain amount, and then it stops; and by the time it stops, you find the liquid in which this matter has been formed has become altered in its quality. To begin with it was a mere sweetish substance, having the flavour of whatever might be the plant from which it was expressed, or having merely the taste and the absence of smell of a solution of sugar; but by the time that this change that I have been briefly describing to you is accomplished the liquid has become completely altered, it has acquired a peculiar smell, and, what is still more remarkable, it has gained the property of intoxicating the person who drinks it. Nothing can be more innocent than a solution of sugar; nothing can be less innocent, if taken in excess, as you all know, than those fermented matters which are produced from sugar. Well, again, if you notice that bubbling, or, as it were, seething of the liquid, which has accompanied the whole of this process, you will find that it is produced by the evolution of little bubbles of air-like substance out of the liquid; and I dare say you all know this air-like substance is not like common air; it is not a substance which a man can breathe with impunity. You often hear of accidents which take place in brewers' vats when men go in carelessly, and get suffocated there without knowing that there was anything evil awaiting them. And if you tried the experiment with this liquid I am telling of while it was fermenting, you would find that any small animal let down into the vessel would be similarly stifled; and you would discover that a light lowered down into it would go out. Well, then, lastly, if after this liquid has been thus altered you expose it to that process which is called distillation; that is to say, if you put it into a still, and collect the matters which are sent over, you obtain, when you first heat it, a clear transparent liquid, which, however, is something totally different from water; it is much lighter; it has a strong smell, and it has an acrid taste; and it possesses the same intoxicating power as the original liquid, but in a much more intense degree. If you put a light to it, it burns with a bright flame, and it is that substance which we know as spirits of wine.

37. Which of the following is **not** a reason why the writer chose yeast as a topic for discussion?
A. It is one of the simplest and the most familiar objects with which we are acquainted
B. The facts and phenomena are simple
C. Pictures or diagrams can be used
D. It forms our every-day knowledge and experience
E. There are considerable issues to be discussed

38. Which of the following is **not** used to describe the formation of yeast?
A. Succulent
B. Turbid
C. Dirty-looking
D. Yellowish
E. Peculiar

39. Which of the following pair of words is **not** being used as a comparison?
A. 'Facts' and 'phenomena'
B. 'Clear' and 'turbid'
C. 'Sweetish' and 'peculiar'
D. 'Innocent' and 'intoxicating'
E. 'Clear' and 'intense'

11. Falling in love

Falling in Love, as modern biology teaches us to believe, is nothing more than the latest, highest, and most involved exemplification, in the human race, of that almost universal selective process which Mr. Darwin has enabled us to recognise throughout the whole long series of the animal kingdom. The butterfly that circles and eddies in his aërial dance around his observant mate is endeavouring to charm her by the delicacy of his colouring, and to overcome her coyness by the display of his skill. The peacock that struts about in imperial pride under the eyes of his attentive hens, is really contributing to the future beauty and strength of his race by collecting to himself a harem through whom he hands down to posterity the valuable qualities which have gained the admiration of his mates in his own person. Mr. Wallace has shown that to be beautiful is to be efficient; and sexual selection is thus, as it were, a mere lateral form of natural selection—a survival of the fittest in the guise of mutual attractiveness and mutual adaptability, producing on the average a maximum of the best properties of the race in the resulting offspring. I need not dwell here upon this aspect of the case, because it is one with which, since the publication of the 'Descent of Man,' all the world has been sufficiently familiar.

In our own species, the selective process is marked by all the features common to selection throughout the whole animal kingdom; but it is also, as might be expected, far more specialised, far more individualised, far more cognisant of personal traits and minor peculiarities. It is furthermore exerted to a far greater extent upon mental and moral as well as physical peculiarities in the individual.

We cannot fall in love with everybody alike. Some of us fall in love with one person, some with another. This instinctive and deep-seated differential feeling we may regard as the outcome of complementary features, mental, moral, or physical, in the two persons concerned; and experience shows us that, in nine cases out of ten, it is a reciprocal affection, that is to say, in other words, an affection roused in unison by varying qualities in the respective individuals.

Of its eminently conservative and even upward tendency very little doubt can be reasonably entertained. We do fall in love, taking us in the lump, with the young, the beautiful, the strong, and the healthy; we do not fall in love, taking us in the lump, with the aged, the ugly, the feeble, and the sickly. The prohibition of the Church is scarcely needed to prevent a man from marrying his grandmother. Moralists have always borne a special grudge to pretty faces; but, as Mr. Herbert Spencer admirably put it (long before the appearance of Darwin's selective theory), 'the saying that beauty is but skin-deep is itself but a skin-deep saying.' In reality, beauty is one of the very best guides we can possibly have to the desirability, so far as race-preservation is concerned, of any man or any woman as a partner in marriage. A fine form, a good figure, a beautiful bust, a round arm and neck, a fresh complexion, a lovely face, are all outward and visible signs of the physical qualities that on the whole conspire to make up a healthy and vigorous wife and mother; they imply soundness, fertility, a good circulation, a good digestion. Conversely, sallowness and paleness are roughly indicative of dyspepsia and anæmia; a flat chest is a symptom of deficient maternity; and what we call a bad figure is really, in one way or another, an unhealthy departure from the central norma and standard of the race. Good teeth mean good deglutition; a clear eye means an active liver; scrubbiness and undersizedness mean feeble virility. Nor are indications of mental and moral efficiency by any means wanting as recognised elements in personal beauty. A good-humoured face is in itself almost pretty. A pleasant smile half redeems unattractive features. Low, receding foreheads strike us unfavourably. Heavy, stolid, half-idiotic countenances can never be beautiful, however regular their lines and contours. Intelligence and goodness are almost as necessary as health and vigour in order to make up our perfect ideal of a beautiful human face and figure. The Apollo Belvedere is no fool; the murderers in the Chamber of Horrors at Madame Tussaud's are for the most part no beauties.

40. The writer uses the examples of the butterfly and peacock to illustrate:
A. Modern biologists' theory of Falling in Love
B. Darwin's theory
C. The fallacy of A and B
D. The workings of A and B
E. The differences between men and animals

41. According to the writer, which of the following is **not** a difference between love in mankind compared to the animal kingdom?
A. Degree of specialisation
B. Degree of individualisation
C. Cognisance of personal traits
D. Cognisance of minor peculiarities
E. All of the above

42. 'The saying that beauty is but skin-deep is itself but a skin-deep saying.' – what does the writer mean by this?
A. Only good-looking people find love
B. A natural attraction towards good-looking people is in line with the selection theory
C. It is superficial to focus on looks
D. Looks do not last
E. Good-looking people tend to be healthier

END OF SECTION

YOU MUST ANSWER <u>ONLY</u> <u>ONE</u> OF THE FOLLOWING QUESTIONS

1. Should the voting age be reduced?

2. Should tuition fees be reduced?

3. Tourism does more harm than good. Discuss.

4. Banning the wearing of a headscarf in the public sector is discriminatory. Discuss.

END OF SECTION

ULTIMATE GUIDE
ANSWERS

Q	A	Q	A	Q	A	Q	A	Q	A
1	B	51	B	101	D	151	C	201	C
2	B	52	A	102	B	152	E	202	B
3	E	53	B	103	E	153	D	203	D
4	C	54	D	104	C	154	C	204	C
5	A	55	D	105	A	155	A	205	D
6	D	56	A	106	B	156	B	206	A
7	A	57	D	107	C	157	A	207	A
8	D	58	A	108	B	158	B	208	E
9	C	59	B	109	C	159	C	209	D
10	E	60	C	110	A	160	D	210	B
11	B	61	D	111	E	161	A	211	A
12	E	62	D	112	D	162	A	212	C
13	C	63	D	113	C	163	E	213	D
14	A	64	E	114	C	164	B	214	C
15	C	65	D	115	E	165	A	215	D
16	A	66	C	116	A	166	D	216	B
17	C	67	C	117	D	167	A	217	A
18	C	68	E	118	E	168	A	218	B
19	E	69	A	119	C	169	C	219	D
20	C	70	D	120	B	170	A	220	A
21	D	71	C	121	A	171	C	221	C
22	C	72	D	122	B	172	D	222	D
23	D	73	C	123	D	173	A	223	B
24	D	74	B	124	C	174	A	224	A
25	B	75	E	125	E	175	A	225	A
26	B	76	A	126	D	176	B	226	D
27	B	77	D	127	B	177	C	227	B
28	E	78	E	128	E	178	B	228	C
29	C	79	C	129	C	179	E	229	B
30	C	80	D	130	A	180	C	230	B
31	C	81	E	131	E	181	D	231	A
32	B	82	A	132	C	182	B	232	D
33	E	83	E	133	A	183	D	233	C
34	D	84	D	134	C	184	A	234	A
35	D	85	D	135	C	185	C	235	D
36	B	86	C	136	C	186	C	236	A
37	A	87	D	137	A	187	D	237	B
38	E	88	A	138	A	188	D	238	B
39	A	89	E	139	B	189	D	239	B
40	E	90	D	140	A	190	D	240	C
41	A	91	C	141	E	191	C	241	A
42	C	92	B	142	D	192	D	242	B
43	B	93	B	143	E	193	A	243	C
44	B	94	B	144	D	194	A	244	A
45	A	95	E	145	B	195	D	245	B
46	B	96	D	146	C	196	E	246	C
47	A	97	D	147	E	197	D	247	D
48	C	98	A	148	D	198	E	248	A
49	E	99	C	149	B	199	E	249	A
50	C	100	B	150	C	200	D	250	C

Q	A	Q	A	Q	A	Q	A
251	A	301	E	351	B	401	B
252	B	302	D	352	E	402	C
253	D	303	E	353	D	403	A
254	A	304	C	354	A	404	B
255	C	305	A	355	B	405	B
256	B	306	D	356	A	406	A
257	A	307	C	357	A	407	E
258	D	308	A	358	C	408	D
259	C	309	A	359	A	409	B
260	C	310	A	360	A		
261	D	311	C	361	C		
262	D	312	B	362	A		
263	A	313	A	363	A		
264	B	314	C	364	B		
265	B	315	C	365	C		
266	B	316	A	366	A		
267	A	317	D	367	B		
268	C	318	B	368	B		
269	B	319	E	369	D		
270	A	320	E	370	D		
271	C	321	A	371	B		
272	D	322	B	372	B		
273	B	323	D	373	D		
274	A	324	A	374	C		
275	C	325	A	375	D		
276	B	326	A	376	C		
277	A	327	C	377	B		
278	B	328	A	378	A		
279	D	329	E	379	E		
280	C	330	A	380	B		
281	B	331	B	381	D		
282	A	332	A	382	B		
283	D	333	C	383	A		
284	C	334	A	384	C		
285	B	335	D	385	E		
286	C	336	A	386	A		
287	A	337	C	387	E		
288	D	338	C	388	C		
289	D	339	A	389	A		
290	E	340	B	390	B		
291	B	341	D	391	A		
292	A	342	E	392	D		
293	B	343	B	393	A		
294	D	344	C	394	C		
295	A	345	D	395	D		
296	C	346	B	396	A		
297	A	347	B	397	B		
298	D	348	E	398	B		
299	C	349	A	399	B		
300	E	350	C	400	E		

Passage 1

Question 1: B

The opponents said that drug consumption is criminogenic because a drug addiction 'can lead to other crimes'. Option A is incorrect as the passage already states that consuming drugs is in itself a criminal offence. Thus, it would be odd and circular if opponents were to argue this.

Question 2: B

It is close between A and B. In regard to A, the author merely points out a 'contrast', which may not necessarily indicate an inconsistency, whereas the author says, in the context of option B in the passage that there is 'incoherence', which clearly means there is an inconsistency. Therefore, the author *presented* B as being paradoxical.

Question 3: E

The entire article is based on the idea of controlled drugs being legalised. Each sub-argument provides an intermediate conclusion from which the main argument is inferred.

Question 4: C

The author does not argue that drug consumption would fall (as distinct from highlighting that it fell for one age group in Portugal) or increase, so A and B are incorrect. The author implied that drug consumption should be seen as a *present* public health problem in the passage – not that it would be a health concern once drugs were legalised so E is wrong. D is not a practical effect which the author believes would happen. C is explicitly stated in the passage.

Question 5: A

The author has based his argument on the fact that there aren't third party effects to taking drugs (e.g. no third party health effects & that the fact of criminalising increases secondary offences). However, if drug consumption increases one's propensity for violence, it contradicts that argument.

Passage 2

Question 6: D

The passage attacks a generalisation and shows an example that refutes one given to the 'musical' genre. Nothing is mentioned of Sondheim's talents, or what his role was in creating the musical, nor are there claims made in relation to Wheeler's literary tastes (he may just like ONE penny dreadful). This musical may deal with morbid themes, but that's not to say that most do - it could be only a select few that do.

Question 7: A

The pies make the crimes 'culinary' in nature, the mention of revenge shows Todd's illegal acts to be 'vengeful', the word 'macabre' indicates it's disturbing and the judge's rape is a 'sexual' crime. There is nothing explicitly suggesting the crimes of any party are funny, or to be considered funny.

Question 8: D

This option is essentially synonymous in the quoted belief, 'we all deserve to die', which includes both bad and good people and makes no significant reference to gender exclusion/inclusion.

Question 9: C

There are four mentioned themes, but that does not mean there are only four themes, nor does 'legal corruption' get named as the central theme. As the entirety of Sweeney Todd is not discussed in the passage, only a central plot line, one could not exclude the potential of something positive happening in the play - even a minor incident. Sadness in itself, therefore, cannot be considered the focus of the play. The themes mentioned are, however, indeed macabre.

Question 10: E

Though the original title 'A String of Pearls: A Romance' may appear to suggest a romantic relationship within the narrative, nothing in the passage states the two are a couple so A is incorrect. The passage only suggested that *some* of the songs were removed, not all of the songs, so B is incorrect. We can't be sure about D as the passage does not refer to the critical acclaim of the film. In regard to C, while some changes were made to the film, we aren't told whether the main storyline changes. Finally, E is explicitly stated in the passage, so that must be true.

Passage 3

Question 11: B

This is mentioned as an impact on the country in the passage and not mentioned as having a personal impact, whereas all others are mentioned as personal impacts. Also, as is evident from the term itself, an increase in government spending affects the government. The focus of the point was on the government's spending (rather than on any welfare benefits derived).

Question 12: E

This option is explicitly stated in the passage.

Question 13: C

This is implied as the author states the numerous benefits of high-quality education provision but also notes how 'more emphasis' should be placed on it. Crucially, it says that vocational education is 'second rate' and that it must change. A clear implication of these sentences is that more should be done to encourage vocational courses. E is already stated in the passage (as the author highlights the negative consequences). The author also explicitly states A, B, and D. Therefore, since they are all stated in the passage, they are incorrect.

Question 14: A

While the author does argue for B, it is only part of the main argument and is in the context of increasing high-quality vocational provision for young people. The skills shortage was used more as a reason to argue for A. Given the emphasis on raising vocational education, it is clear that A is the correct answer. The author does not argue for C and actually points out that the jobs are already there, it's just that young people need to be trained more. D was not argued for and E was an assertion (and, thus, not an argument, let alone a main argument).

Question 15: C

This is explicitly stated in the passage. A is not stated in the passage. While Dulux was increasing young people's skills, the author did not advocate this for businesses generally and, thus, B is incorrect. The author suggested that D and E should be done by the government, rather than by businesses.

Passage 4

Question 16: A

B, C, and D are all stated as influences in the passage but option A is stated as underlying all those reasons and thus, A is the 'ultimate' influence.

Question 17: C

The passage mentions that a son was required to secure the Tudor dynasty. Therefore, there is something familial relating to all of this. Therefore, B and E are incorrect. D doesn't follow from the sentence so it has to be either A or C. A doesn't make sense in this context as it's not clear how having a son would secure his family but C fits in well.

Question 18: C

Royal Supremacy was merely the process of getting control over the church and was not a reason for getting a son (it was the means to getting the son) so A is wrong.

B is incorrect as it should be the other way round – he wanted to divorce Catherine to get a son (and not get a son to divorce Catherine). C fits in with the passage and, thus, is correct.

Question 19: E

A and D were stated in the passage, so are incorrect. Even if B and C were true, they would not weaken any of the conclusions, so they are not correct. However, E is correct as if he believed that he could get a divorce from Rome, he would have tried to get that as it was the 'straightforward' option. The author also stated that without Henry's desire for a break from Rome, it wouldn't have happened. Therefore, point E would be a necessary assumption to support the author's points.

Question 20: C

The passage never suggests that the Act of Supremacy is the royal supremacy, so E is incorrect. The passage states that the Royal Supremacy was established once Henry became the only head of the church, so C is, thus, correct as it means the King had become the leader of the church. D is, therefore, obviously incorrect as the Pope isn't the head of the English church. B is too vague and while option A had to occur for the Royal Supremacy to take place, the passage pointed out that C was, in fact, the royal supremacy.

Passage 5

Question 21: D

The passage explicitly states that charities may not need to pay corporation tax. A is not true based on the passage – as you can provide a public benefit but have not registered to be a charity. It is only after registration as a charity that an institution gets the fiscal benefits; this does not happen automatically. B and C are not proved in the passage, and E is an opinion, which can't be true or false. In regard to C specifically, it cannot be assumed that just because event X is required for event Y to occur that if Y does not occur, X does not occur – in the context of the passage, you can still provide a public benefit and not be a charity. Therefore, C is incorrect.

Question 22: C

The passage explicitly states that it's "necessary" for an institution to demonstrate that its purposes are for the public benefit.

Question 23: D

Both C and E are easily eliminated as neither of these institutions were part of the trial, which is where the argument was advanced. We're told that tribunal judges didn't consider the argument and it is, thus, clear by implication that they didn't bring it forward (therefore, B is incorrect), so we're left with either A or D. It's not A as the argument weakens their position (being independent schools) so it's unlikely they would advance it. By elimination, therefore, it must be the Education Review Group, D.

Question 24: D

Whether the law should or should not allow a free-for-all is not a point that can be tested – it is a person's opinion and cannot be true or false. All the other options can be tested.

Question 25: B

In the passage, it is stated that the tribunal said that children who couldn't afford the fees of private schools must benefit from them for there to be a 'public benefit'. Accordingly, providing scholarships to others who can't afford the fees would help with this. A is insufficient to meet the tribunal's definition. C, D, and E do not support the argument that there's a public benefit.

Passage 6

Question 26: B

Amazon is more powerful. The author said that one is more powerful than the other at the beginning of the passage and it later becomes clear that Amazon is more powerful as they were able to restrict the sales of Hatchette's books in negotiations, thus, indicating a stronger bargaining position.

Question 27: B

There is irony because Hatchette have done aggressive things, just like Amazon have and yet, they are now complaining about Amazon.

Question 28: E

This option draws all of the points expressed in the passage together. Even though Amazon engaged in aggressive tactics, others have done it too and Amazon have done beneficial things. It gets to the heart of the passage, which is considering the actions of Amazon.

Question 29: C

The word 'monstrous' implies not only that there is a big task, but that the task is too big as the word implies some dissatisfaction with all the work that authors have to do.

Question 30: C

If Hachette had offered the same services as Amazon had, it would directly contradict what the author has said – Amazon's services wouldn't be an 'innovation', Hachette (one of the 'big publishers') would not have been acting unfairly in the past and Hachette would have been acting to the benefit of small authors: this would all be contrary to the author's points in the final paragraph; therefore, point C most undermines the author's argument.

Question 31: C

The author has been positive about what Amazon has done throughout the article. Therefore, the author would most likely disagree with the suggestion that Amazon has abused their market position. The author has acknowledged A, B, and E in the article. The author isn't likely to agree with D because Hachette had engaged in a conspiracy.

Passage 7

Question 32: B

The answer here is between A and B, both undermine the author's argument that the court system limits access to justice. B *most* undermines the author's argument because if these cases are normally settled out of court, the court system's high cost does not limit access to justice as people would already be getting justice, albeit outside the court system. Option A doesn't undermine it as much because it just says that there are different options for small claimants, whereas B directly cuts at the option. C, D, and E don't necessarily undermine the author's argument.

Question 33: E

This question necessitates being able to deduce the correct information from the passage and shows whether you've understood the passage. 1 hour of legal advice is £200 and the court fee for each is £35. Therefore, the cost for each side is £235. Therefore, the cost for both sides together is £470. As the passage stated that the losing side pays the winning side's costs (as well as their own), the losing side would have to pay a total of £470. N.B. there have been similar questions to this in recent LNATs.

Question 34: D

This is a statement that cannot be tested as true or false and is a normative statement. Therefore, it is an opinion.

Question 35: D

A and B are true but they are explicitly stated in the passage, so are incorrect (the question explicitly asks for an option which is *not stated*). Neither C nor E can be implied. The passage states that the TV show 'resembles' a courtroom, thereby implying that it is not actually a real courtroom but some kind of mock courtroom that looks similar.

Question 36: B

Despite discussing the Judge Rinder show, the author never advances an argument that an online court should be modelled on it – just that the TV show shows that lawyers aren't needed.

Question 37: A

B doesn't even relate to whether the justice system is inefficient or not, so that's incorrect. Again, D and E don't relate to whether the *current* justice system is efficient. On the other hand, option A points out that the author has only referred to a subset of claims (i.e. small claims) and not considered the justice system generally. In particular, no reference was made to the criminal justice system. It doesn't follow that just because one part of it is inefficient, that the rest of it is inefficient. Therefore, A is correct as it highlights why that argument was weak.

Passage 8

Question 38: E

The author made the analogy in order to demonstrate that even though the driverless car may seem strange, people will still adapt to it. Therefore, E is correct. A is wrong as the author only says that at the first part of the analogy. C doesn't necessarily follow. While D is, in essence, what the author is saying, he's applied it to driverless cars. Therefore, his *main* point is that the public will adapt to the driverless car. Since the author was not arguing that the public *should* take up the driverless car with the analogy, B is incorrect.

Question 39: A

The assumption made is that people adapt to new things.

Question 40: E

The Top Gear test doesn't suggest any of A, B, C or D in respect of the use of autonomous cars *on roads*. A, B, C are obviously not demonstrated by the passage. D is not relevant to the question and would be a generalisation as well.

Question 41: A

Throughout, the author describes different aspects of the take-up of driverless cars by car companies and governments and only considers the benefits of it to drivers in the final paragraph. The author does not point out arguments for and against, so E is incorrect. Indeed, the author's view is that the public will want to drive such a car, but this does not underlie the entire piece, so C is incorrect. The same is true with option B. Again, while D is true, it is not the underlying theme of the article.

Question 42: C

E is incorrect as it doesn't relate to the author's main argument in the final paragraph – indeed, the author even acknowledged point E in respect of luxury cars. In regard to B and D, which relate to how cars are currently (B) or the current views on it (D), the author argues that there will be an uptake of it *in the future*. B and D just relate to the present and, thus, don't undermine the author's argument. If C were true, though, this would undermine the author's argument as the public may not take it up in that case. A doesn't necessarily show that people won't take up driverless cars. Therefore, A is incorrect.

Passage 9

Question 43: B

The Society suggest that one's autonomy involves a condition to act ethically and, therefore, if one restricts euthanasia/assisted suicide (an intentional killing), it wouldn't infringe autonomy. Hence, it is assumed that euthanasia/assisted suicide is unethical so that one is not infringing another's autonomy by restricting the unethical act. Looking at it from the reverse view – if we said that an intentional killing is ethical, the conclusion above wouldn't necessarily follow from the premises.

C and E are not necessarily assumed in the argument – i.e. the argument would hold even if they are untrue. Again, in contrast to A and D, the Society actually states that acting ethically is a part of being autonomous. Therefore, it does not infringe autonomy to stop an unethical act. Hence, A and D are incorrect.

Question 44: B

A clear difference, as stated in the first paragraph, between assisted suicide/euthanasia and suicide is that the former options involve a third party, whereas the latter only involves one person. The Society just argued that euthanasia/assisted suicide should be restricted and not necessarily suicide in itself. While D is true, it does not explain the gap in the logic (which is what the question is asking for). C and E are irrelevant as the author suggested the Society argued that suicide *should* be illegal, which is suggesting an opinion, so it's not relevant that it's in fact currently legal.

Question 45: A

It's implicit that assisted suicide and euthanasia should not be allowed – the society already assumed that it's unethical and said not allowing it wouldn't restrict one's autonomy. B is an assumed assertion, but not an implied *argument*. C and D are incorrect as the Society isn't considering what the current state of the law is. E is incorrect as it did not follow that the Society made this and the author explicitly suggested this.

Question 46: B

This is a question asking for a factual response. The author has pointed out that the court has given more weight to non-maleficence and, thus, that clearly has priority. Although the arguments initially suggest the author may be in favour of autonomy, the question is not asking what the author's opinions are, but which principle, in fact, has priority. C and D are wrong as they're not principles in this context and E is wrong as it has already been shown that non-maleficence has priority.

Question 47: A

In the final paragraph, the case of Miss B was explained on the basis that the doctors were not actually harming the patient, which satisfies the definition of non-maleficence given in the passage. The author also states that non-maleficence was given a significant weighting in the case. B, C, D, and E may well have been relevant to the court's decision but the author only referred to A when discussing the case of Miss B.

Passage 10

Question 48: C

The author never suggested that the IMF should be replaced throughout the passage.

Question 49: E

Benefits of giving the loan to Ireland was explicitly stated.

Question 50: C

Some of the other options were indeed factors in the Greek debt crisis but those other factors were, according to the author, caused by the government spending more money than what it had.

Question 51: B

The author would not agree with the view that the IMF has *only* been beneficial in the world because of the issues the author has stated that they've caused. D and E follow directly from the passage, so the author would agree with those. Given the issues of the IMF in the past, the author would likely agree with A. Even though the author didn't explicitly suggest C in the passage, he is less likely to agree with B compared to C because there have been issues with the IMF, so it is more plausible to agree with C, when compared to B (in particular because B directly contradicts the passage, whereas C does not).

Question 52: A

Incompetence suggests that something was not done *as it should have been done* – accordingly, option A fits in well with the passage and more so than options B and C. Although point E was stated in the passage, it was mentioned in a different context. D is explicitly pointed out, so is incorrect.

Passage 11

Question 53: B

This was not explicitly stated but is quite clearly true as the passage states that since 1953, a UK citizen could take the UK government to the European Court.

Question 54: D

In the third paragraph, the author explicitly stated that UK citizens could sue from 2000 onwards, when the Human Rights Act came into force. While the ECHR was in force since 1950, the author never said that UK citizens could sue in domestic courts then. It was only after the UK Human Rights Act was passed that, as the author said, they could sue in domestic courts.

Question 55: D

The author makes an argument in the final paragraph, citing examples of 'positive cases' and argues that they're some of the examples of the Convention giving rights to people. Therefore, a clear argument is that the Convention is benefiting UK citizens. The author doesn't argue for A and takes the opposite position to B and C. While E is true, it is merely an assertion as opposed to an argument.

Question 56: A

C and E are clearly incorrect as they aren't stated or implied at all in the passage. D was a reason for the European Court of Human Right's decision to block the deportation but the UK still wanted to deport Qatada – they just couldn't do so legally – it was the European Court that blocked the deportation and thus, A is the correct answer.

Question 57: D

The author makes the point that the benefits of the Convention outweigh the negatives as it is pointed out that there are more positive cases than negative cases. The author acknowledges that there are sometimes bad outcomes (so B is incorrect) but that the good ones outweigh this.

Question 58: A

In the final paragraph, the author was suggesting that being a signatory to the ECHR was beneficial to the UK (as there are many more 'good' cases). Accordingly, B doesn't follow. While the author may agree with D or E, they would be inconsistent as a *main* conclusion given that the author's primary contention was that there are many more positive aspects of the system than the negative and doesn't touch on altering it (so C is incorrect).

Passage 12

Question 59: B

C is incorrect as there's no mention or implication of an organisation (and it's not a literary technique), D is incorrect as no irony (when what *appears* to be the case is not *actually* the case) is present and the author does draw a similarity between anything so E isn't correct. The author does not suggest that the invisible hand was 'like' anything, so it can't be a simile (A is incorrect). It is in fact a metaphor, option B.

Question 60: C

A is wrong as while employees are stakeholders, it was stated in the passage (and the question related to an unstated implication). People generally are not stakeholders – the passage defined stakeholders as only those who are significantly affected by a business' decision. Again, this was not implicitly suggested in the passage, so B is incorrect. Neither D nor E is implied either. Shareholders are a stakeholder that is implicitly mentioned in the passage. In paragraph 5, the author discusses moving call centres abroad when considering the impact of business decisions on 'stakeholders'. The author says that the increase in profits benefits shareholders. As shareholders are significantly affected by this (in getting more profit), they are stakeholders.

Further, immediately after the profit point, the author states: "However, it negatively harms other stakeholders"; using the word 'other' also implies that shareholders are stakeholders.

Question 61: D

The author brings up the call centre example to show the effects of it but also to show that it is based on profit maximisation and what the impact of profit maximisation is (i.e. that it can lead to call centres going abroad but it might not, as in the case of BT). The author does not argue for either A, B or C and it is not clear that the author takes a position against moving call centres abroad. Therefore, E is incorrect.

Question 62: D

The author considers throughout that profit maximisation is a relevant objective but that other objectives should also be taken into account and explicitly states this. Accordingly, A and B are incorrect: the author never suggests those points. While C may be broadly true, the author points out that it isn't always beneficial. The author does not argue for or consider point E in much detail.

Question 63: D

As stated in the passage (2nd paragraph), if an individual tries to maximise profits, that will maximise the welfare of society because of the market (or the invisible hand of the market). Accordingly, Smith's view is that profit maximisation maximises the welfare of society. A is in the wrong order (profit maximisation occurs *before* the invisible hand), B and C are wrong because the invisible hand appears to discuss the automatic mechanism of the market and these options do not espouse Smith's ultimate view, which is point D. D more precisely equates with the view in the passage than E, so D is correct.

Passage 13

Question 64: E

A and C are clearly wrong as those groups are harmed. The passage doesn't mention B so that is incorrect. The choice is between D and E, both of whom receive benefits. However, according to the passage, law firms will lose out on fewer oil exploration projects, whereas the consumers have no immediate loss – just the benefit of lower prices. Therefore, E is the correct answer.

Question 65: D

A number of the options are potential reasons but the loss of £4.5 billion was cited as the *immediate* reason in the passage. Low revenues don't necessarily mean a loss is incurred and the passage cites losses as the reason for job cuts generally. Therefore, E is incorrect. Both A and B are factors but not the immediate factors and C just describes the way in which the workers were made unemployed.

Question 66: C

Obviously, to claim unemployment benefits, one cannot have a job. Therefore, it is assumed that those 7,000 individuals would not have a job if they're to seek unemployment benefits. Hence, C is correct. B is not a necessary assumption as the passage just says that the individuals would *seek* unemployment benefits – this does not require that the government actually do have benefits to give. This is a very fine distinction and you're doing very well if you have understood this. Hence, C is a necessary assumption but B is not necessary to support the statement. A, D and E are not relevant as they merely explain or describe the unemployment that has already occurred.

Question 67: C

The first paragraph explicitly attributes the reduction in oil price to the shale boom in the US, which increased supply and ultimately lowered oil prices. A and B are not cited as reasons. E is just a descriptor for the hard times that oil companies are experiencing and not a reason. The whole part of D was not suggested in the passage.

Question 68: E

The author simply points out that the oil companies will go bust if they continue to make losses, which is happening to *some* due to the oil price. Low oil prices are not a reason in and of itself to make redundancies – it is when a company is making a loss that it may need to sack workers.

Passage 14

Question 69: A

All other options are stated in the passage and this is explicitly not true based on the passage – as nuclear power is stated as a non-renewable energy source but the passage also states that it doesn't emit carbon. Therefore, A is correct as not *all* non-renewables emit carbon.

Question 70: D

The author explicitly argues for this at the end of paragraphs one and two. Option A is an assertion and not an argument. B, C and E are plainly not argued for in the passage. The author also appears to disagree with E.

Question 71: C

The author has already argued for a focus on renewable energies and has also expressed concern and doubt as to whether nuclear power is economical given the dangerous waste it produces. Therefore, C is the answer. E is incorrect as the author said that nuclear power is now mostly safe. A and B are not considered. While the author would agree with D, he would be more likely to accede to C as he said at the end of the final paragraph that nuclear power is not economical for the UK.

Question 72: D

This is clearly correct. Earlier in the passage, the author brought up the importance of effective government planning and then the High Court decision showed that the government hadn't undertaken effective planning.

Question 73: C

As the passage highlights the fact that there are issues with fossil fuels emitting carbon, just because a form of energy is cheaper in itself does not mean it will be more beneficial than nuclear power (if the cheaper form pollutes). Therefore, A doesn't necessarily weaken the author's argument based on the passage. B is too vague. While D is true, the author states that there are new safeguards now, so it doesn't necessarily apply in the present day nuclear power stations, so D is incorrect. E is incorrect as all it shows is that the government didn't do enough consideration of waste disposal – i.e. it does not indicate that nuclear power is *not* or is *less* beneficial. Either way, option C weakens the government's argument the most as other options may well be more beneficial if C is true.

Passage 15

Question 74: B

Neither A nor D is suggested in the passage. The author does suggest C, but only if B doesn't work. The author explicitly endorsed option B in the passage and believes that this would stimulate the culture change required in the final paragraph. The author did suggest E at the end of the second paragraph, but this only applies to direct wage discrimination, which the passage says does not account for the entirety of the gender pay gap. The author clearly believes that a reporting obligation would be effective to help the gender pay gap and help with direct discrimination (as stated in the final paragraph). Therefore, it is clear that option B is the preferred option of the author.

Question 75: E

The author clearly argues for this at the end of the second paragraph. B, C and D are not argued for. A is stated in the passage as an *assertion* and not an argument.

Question 76: A

In contrasting both a reporting obligation and the mandatory minimum, the author states that the mandatory minimum may not lead to a culture change and may cause resentment whereas the author states that the reporting obligation will lead to a culture change. So it is, thus, clear that this is the reason for the author's choice. D is not relevant here and E is not the main reason why the author preferred the reporting obligation – it was the culture change. The author doesn't suggest in the passage that B or C will occur from choosing the reporting obligation.

Question 77: D

Each of A, B, and C are cited as reasons for differences in the pay between genders. E is not stated as a reason for a difference.

Question 78: E

The high cost is the author's concern. While the author would agree with B, which is not the author's *concern* but merely the way in which the author would *address his concern*. A and D are not argued in the passage.

Passage 16

Question 79: C

"Oddly, even in spite of a weakening currency, the country is nowhere near to having a trade balance". The use of the words 'oddly' and 'in spite' suggest that the opposite should normally be happening. Since the country is nowhere near to having a trade balance, the opposite would be that the country would be nearer to (or have) a trade balance. Therefore, C is the correct answer as that fits in with this idea the most. E is wrong as that is the current position. A and B are wrong as they fit in with the current position as well. D is not suggested in the passage.

Question 80: D

This is explicitly explained in the passage (fourth paragraph – "As long as Britain exports to the rest of the world as much as it imports…").

Question 81: E

In the final paragraph, the author points out that de-industrialisation and a poor skills base are the wider problems linked with being a net importer. Based on the definition within the passage, being a net importer means the country doesn't have a trade balance. Thus, it is clear that the author is saying that de-industrialisation and poor skills are the reasons for the lack of exports (which mean the UK doesn't have a trade balance). B is incorrect as it means the same thing as not having a trade balance – it does <u>not</u> explain why there is no trade balance.

Question 82: A

This is most consistent with the entire passage. The author explicitly states that our import levels would be OK, but we just need to export more. The final paragraph also indicates that the UK's problem is too few exports as opposed to too many imports. Therefore, B is incorrect. C, D, and E do not underlie the entire passage either.

Question 83: E

A, B and C are irrelevant as they are not referred to in the last paragraph. While both D and E are true, an individual would choose the German goods because they're cheaper (as stated in the last paragraph). They wouldn't have a reason to base their choice on the lack of investment in the UK. Therefore, D is incorrect.

Passage 17

Question 84: D

The author explicitly said this is due to concerns about ESPN, a different part of the Disney Company, so D is obviously the correct answer.

Question 85: D

Ultimately, whether something is 'successful' can be a matter of opinion – some cinema goers did not particularly like the film, even though most did. All other options were facts.

Question 86: C

A prediction is, of course, an estimation of a future result. A is wrong as other companies already have produced such products and the author does not argue for D or E. While B is mentioned in the passage, the author did not predict that it would happen – the author simply stated that some analysts said it would happen. Therefore, B is incorrect. On the contrary, the author expressly predicted C in the passage.

Question 87: D

The author talks about how success does 'not stop' at the success of the box office alone but how merchandising has a role to play. Throughout the passage, the author also pointed out how the merchandising was successful. Therefore, D is the correct answer. E does not follow. Although the author asserts that an important part of *most* successful films is merchandising, that does <u>not</u> mean all – therefore, C is incorrect. The author did not argue for either A or B as well.

Question 88: A

This option is definitely false because the fifth paragraph explicitly states that Disney created products. It also said that Disney did not create all of the products themselves, thus implying that they created at least *some* of the products themselves and that others created some of the products.

Passage 18

Question 89: E

The author does not give a belief as to whether society should adopt same-sex or opposite-sex marriage. The author just argues (by the end) that marriage is a social construct and, arguably, implies that adopting same-sex marriage is a reflection of society. This does <u>not</u> mean that the author argues for same-sex marriage – he could still disagree with society's position. The passage just doesn't refer to his personal view.

Question 90: D

The author explicitly states that the argument of the opponents (in the second paragraph) was circular and only criticises their argument, but does not take a view as to either side of the debate in the process.

Question 91: C

The author was arguing that while marriage was originally a religious construct, it was still a social construct in the 12th Century as much as now. It's just that religion had a greater role in society then.

Question 92: B

The basis for saying that marriage did not require a religious ceremony since 1837 would be that there had to be a religious ceremony before 1837.

Question 93: B

The author states that marriage has continued to evolve, even 'now' to reflect society, thereby suggesting that the recent reform which allowed for same-sex marriage (mentioned at the beginning of the passage), was a reflection of society. It is important in such questions to take a 'global' look at the whole passage while also referring to specific details when required.

Passage 19

Question 94: B

The author argues throughout that sugar should be taxed and considers potential counter-arguments but rebuts them. E is incorrect because, while the author does consider the pros and cons, the author argues for a sugar tax throughout and thus, C is also incorrect. Indeed, the author does point out that D should be done, but that it should be done as well as a sugar tax and this was an incidental point at the end of the article, rather than underlying the entire article. Although the author does allude to point A, that is not the main point of the article, which is to reduce sugar consumption by a sugar tax – the author just deployed point A to suggest that a sugar tax is fair.

Question 95: E

If people were not to reduce consumption of sugary products, it would directly undermine the author's argument because imposing the sugar tax (which would increase prices) would not reduce sugar consumption, which the author highlighted as the reason for imposing the tax in the first place. D does not contradict the argument – it is, in fact, intended that prices would increase. A and C don't undermine the author's argument. While B may be an issue, the author potentially reconciles with this on the basis that those who consume it should bear the consequences of it. Therefore, E is the correct answer.

Question 96: D

This question was limited to the sixth paragraph only. If people don't get health problems from consuming sugar, they're not causing problems for society (i.e. costs to NHS); therefore, it would not be fair to tax them extra given that they aren't causing additional health costs to society. Although the author suggests that people are ignorant of the health consequences, it's not an intrinsic part of the argument that a sugar tax is fair – a sugar tax is meant to be fair (according to the author) because the people who cause the social costs would pay for it. Therefore, E is incorrect. Again, A, B, and C don't take issue with the argument that the people who cause health costs should pay for them. Accordingly, they're incorrect. While B would undermine the author's main argument, it does not undermine the argument he advanced in the sixth paragraph.

Question 97: D

While the passage implies that excess sugar consumption can cause obesity, it does <u>not</u> suggest that all cases of obesity are due to excess sugar consumption. Therefore, we cannot say that option D is true based on the passage. All other options are stated in the passage.

Question 98: A

A is not explicitly stated but is clearly assumed as the link between excess sugar consumption and the health consequences that the author cites (if we said option A was false, then the author's argument would be undermined - option A is a necessary assumption). None of the other options are unstated assumptions.

Question 99: C

The final sentence effectively says that prevention is better than cure, and option C is the only one which illustrates this. E is incorrect because the final sentence does not suggest that anyone is falling.

Passage 20

Question 100: B

Whether something is the 'toughest' sporting challenge is open to debate and can't be tested as being true or false.

Question 101: D

The team leader has the support of his teammates (according to the passage) and, thus, is more likely to win the race compared to if he was just one of the supporting teammates. The team leader also benefits from saving energy. This is also supported by the example stated in the passage, where the leader of Team Sky won in 2012. A, C, and E are incorrect as they don't denote an additional benefit – in particular, A and C are pre-conditions to becoming a team leader. While B is true, it's not correct as it's not necessarily a benefit that the team leader alone gets – in the passage, it was stated that the supporting rider in Team Sky came second in the race (and, by implication, beat all the other team leaders).

Question 102: B

When discussing the slipstreaming effect, the passage explicitly says that this is why cyclists tend to stay together. C is wrong because the passage says that the fact of the cyclists being in a bunch influences team strategy (not the other way round).

Question 103: E

A and B are clearly not implied in the passage. Team Sky's choice was correct for them because their rider won the race. C is stated in the passage ('it would not be possible to say that the result would have been the same') and D does not follow from the passage – the passage never said that the slipstreaming effect would make someone a better rider (just that it's a natural phenomenon that increases one's speed). Therefore, by elimination, E is the only one left. The author seems to believe this based on the fact that the supporting rider did something historic (by coming 2nd) and something 'unheard of' and that the author doubts the result if they were both equals. Hence, E is the correct answer.

Question 104: C

This is the only option that follows from the passage. In the final paragraph, it is explicitly stated that the supporting cyclist used extra energy to help his team leader. As the passage states, helping out a team leader involves keeping the team leader behind the supporting cyclists so that the team leader uses *less* energy – or conversely, the result is that the supporting cyclist may use *more* energy.

Passage 21

Question 105: A

Whether something requires *urgent* attention cannot be tested as true or false and is, therefore, an opinion as opposed to an assertion of fact. B, C, D, and E are all assertions of fact.

Question 106: B

Option A doesn't necessarily follow. The fact that there is a greater proportion of ethnic minorities in prison when compared to their proportion in the population indicates that there's a disproportionate number of ethnic minorities in prison. The government does indeed need to find out about it, but this was explicitly stated in the second paragraph (rather than being implicit) and so was the fact the statistics aren't ideal. Therefore, C and D are incorrect. E does not follow from the statistics so it is incorrect.

Question 107: C

The passage explicitly states that ethnic minorities are more likely to receive longer sentences. The passage doesn't indicate whether there is conscious discrimination, so D is incorrect. Judges don't have discretion for the guilty plea discount and it's not argued that the higher sentences are a breach of the rule of law. Finally, E is incorrect as the passage indicates that socio-economic factors are relevant.

Question 108: B

The passage states that there are a number of factors that judges consider when deciding their sentences and these form part of their discretion when deciding the sentence. The author uses this to explain why there may be different sentences for the same offence so B is the correct option. C and E show specific instances of different sentences but option B explains all the circumstances.

Finally, option A does not, according to the author, appear to always explain the differences in sentences, although it sometimes might. Option D does not have an impact on sentencing so is not relevant.

Question 109: C

This option fits in best with the final paragraph and explains the higher number of ethnic minorities in the prison population. A doesn't follow from the paragraph. D doesn't explain the increase in prison population and E is just a policy suggestion (it's not an explanation) so it's not relevant.

Question 110: A

This is implicit in the final paragraph as the author cites researchers who have found that racial stereotypes do exist in the police force. B and C are explicit (and not implicit). While D is noted in the passage, it is not implied in the final paragraph and the author, if anything, would likely argue against point E and there's nothing to imply that in the paragraph, so E is incorrect as well.

Passage 22

Question 111: E

The definition of a tax inversion, according to the passage, is that a company buys another company in a different country (different jurisdiction). Therefore, the two companies have to be in different countries.

Question 112: D

The passage explicitly states that the purpose of a tax inversion is for a company to move its headquarters to get a lower tax rate. Therefore, D is the essential element. A and B are just a means to an end (the end being D). C doesn't necessarily follow (although it may be the aim of the company) as the defining feature and E doesn't follow either.

Question 113: C

Simply, the aim is to lower their tax liability, as stated in the passage.

Question 114: C

A is wrong as that isn't a benefit to society. D doesn't follow and E was not stated in the passage. However, C was explicitly stated as a potential benefit.

Question 115: E

Buffett was explicitly stating the benefits, with the implication being that businesses should pay their taxes. Therefore, C is incorrect as that is explicit. D does not follow (he was saying that there are benefits from *existing* tax rates). Neither A nor B was argued either.

Passage 23

Question 116: A

The third paragraph was saying that the Black Death transformed the position of peasants *because* ('as') they could now move to different manors. Therefore, a 'transformation' necessitates the assumption that the peasants could not move to different manors before the Black Death. Therefore, A is correct. The passage doesn't say that peasants transformed society, so B is incorrect. D is contradicted and E does not follow. While C is not contradicted, it does not fit in with the discussion in the 3rd paragraph.

Question 117: D

The passage refers to the shortage of the workforce when explaining why women's wages increased. None of the other options follow from the explanation.

Question 118: E

Although peasants moved around, that doesn't necessarily mean that they had permission to do so. E is the only option which is explicitly stated.

Question 119: C

This is a necessary assumption since the author points out that the population started decreasing before the Black Death due to the Great Famine.

Question 120: B

Both views agree on the point that the Black Death had a *role*.

Question 121: A

This would mean that manorial discipline did not weaken and the author's intermediate conclusion would be false.

Passage 24

Question 122: B

The passage explicitly makes the point that legal rights and obligations are lower for a cohabitation than for a marriage. While A may sometimes be true, it doesn't always hold and the passage doesn't argue for this.

Question 123: D

The fourth paragraph explicitly states that the cohabitee wouldn't receive such property "unless it was stated in the will" and the author says it's easy for a cohabiting couple to get round it by creating a will.

Question 124: C

This is the likely argument of the campaigners as their statistic was a 'retort' to the author's point on the infringing couples freedom and the statistic does provide some support (albeit weak) to point C because it shows that a number of couples would consent to the additional legal rights and responsibilities.

Question 125: E

The author expressly disagrees with A and B. C doesn't make sense as the author already stated that it is not a legal concept, so it can't really be banned. The author does not argue for D either. In the final paragraph, though, the author refers to improving the public's understanding, so it is likely that the author would agree with E.

Question 126: D

The author expressly disagrees with giving additional legal rights and responsibilities to cohabiting couples on the basis of absence of consent.

Passage 25

Question 127: B

This naturally fits in the context. Higher costs are not a benefit so point A doesn't make sense. D and E do not easily fit in the sentence either. B makes the most sense as the author is trying to say that the higher costs are resultant from the change in regulations.

Question 128: E

It is open to debate whether Mercedes should or should not be criticised and it, thus, cannot be a fact as it is not something that is provable to be either true or false.

Question 129: C

This is the best answer as the passage explicitly discusses the impact on smaller teams who have been hit by the increase in engine costs. The author does not believe A is the case. B, D and E do not directly relate to whether competition has been hampered so they are incorrect.

Question 130: A

The author simply states that Mercedes built the best car and should get the credit for that. It was not the change in the rules that caused it, but they simply exploited the new rules best.

Question 131: E

The author was not referring to Mercedes specifically, but to other teams dominating previous seasons as well. Therefore, B is wrong and so is C, as that's a recent phenomenon. Ultimately, the author points out that rule changes can increase overtaking (and, thus, competition) and explicitly states that the nature of some rules and regulations is a reason for the dominance.

Passage 26

Question 132: C

Germany paid the victims of the Nazi war crimes themselves whereas the author brings up the contention that the victims of the slave trade are no longer with us. Hence, option C is the relevant distinction between the two.

Question 133: A

This option best fits in with the argument. The company can be equated to the country and the employee being the individual victim. While the individual victims should be compensated, the passage's argument doubts whether the country as the overarching institution should benefit.

Question 134: C

The author considers arguments both for and against giving reparations throughout. It does not appear that the author takes up positions A or B.

Question 135: C

The author does not disagree with Hartley-Brewer's point but merely states that the responsibility of one of the main antagonists (i.e. the UK) should hold. D is an assumption but does not highlight the author's view.

Passage 27

Question 136: C

The passage explicitly states that the government considers the costs and benefits of animal research when deciding whether to give a license to allow it. A and E are wrong as the author was actually saying that no reasonable scientist would make animals suffer *unnecessarily*.

Question 137: A

The author seems to take the position that animal testing is beneficial by highlighting many of the positive examples of it. The author does indeed point out issues and says that a blanket allowance should not be given but that's consistent with allowing animal testing in certain circumstances.

Question 138: A

What counts as reasonable is not something that can be definitively proven to be true or false. Accordingly, it is an opinion – different people can validly disagree on the issue.

Question 139: B

This seems to be the main concern of the author.

Question 140: A

Option A undermines the author's view as, if it is true, this would mean that the government's licenses are given too frequently and that the harm to animals is not minimised.

Passage 28

Question 141: E

This was explicitly stated by the writer.

Question 142: D

This was implied by the writer when stating that unauthorised use would increase.

Question 143: E

As everyone has a natural right in what they produce, it shouldn't be used without their permission.

Question 144: D

Incentivising the production of intellectual works was explicitly stated by the author as a justification for copyright laws.

Question 145: B

This is implied in Hettinger's argument.

Question 146: C

Locke argues for natural rights, which would require copyright protection and Hettinger wants copyright protection to encourage the production of intellectual works.

Passage 29

Question 147: E

The passage explicitly states that tax evasion is illegal and that tax avoidance is legal.
It is true that Google/Starbucks have engaged in tax avoidance in the past, but that point does not indicate the difference between avoidance and evasion.

Question 148: D

The passage says that Starbucks paid zero tax. It also says that the tax rate is 20% on all declared profits. This means that Starbucks both enjoyed a declared profit and <u>evaded</u> tax <u>or</u> it had zero declared profits and may have <u>avoided</u> tax. Given that the writer explicitly stated that Starbucks were within the law, they can't have evaded tax. Therefore, they must have declared zero profits in the years they paid zero tax.

Question 149: B

Starbucks engages in tax avoidance and the author explicitly stated that this 'led' to boycotts.

Question 150: C

It is simply that companies should pay tax because they receive a benefit. The writer also points out an example of a company's use of a public service, thereby implying that companies should pay tax.

Question 151: C

Tax avoidance is where not as much tax is being paid as what should be paid. The author says this can be subjective as well. Accordingly, whether a tax payment is as it *should* be cannot be tested to be true or false. Indeed, most people may think that 20% is the correct amount to be paid but others think that as long as the law is being followed, the company is fine (and those individuals would not call it tax avoidance).

Question 152: E

The author discusses both the responsibilities of the government and the companies.

Question 153: D

This question refers to the tax system *generally* as opposed to just Starbucks' tax arrangement and the author refers to this at the final paragraph when comparing the different groups of taxpayers (large business, small business, and individuals).

Passage 30

Question 154: C

The passage explicitly states that 14 women were arrested and so were the 3 inspectors, thereby making it to a total of 17 arrests.

Question 155: A

It is clear that Anthony was brought to trial but it is not clear whether any other women or whether the 3 inspectors went to trial.

Question 156: B

Hall dissented, so it's clear that the election officers were not united but were divided, so B is the correct response. Neither, C, D nor E follow from the passage.

Question 157: A

It was explicitly discussed in the third paragraph that Anthony believed she had a lawful right to vote due to the advice from her counsel.

Question 158: B

This follows from what is stated in the passage. The passage does not say that the judge believed the jury were biased and there is nothing to suggest that either A or D follows.

Passage 31

Question 159: C

Women in Illinois, not across the USA, were subject to the law, thus A is incorrect, and the passage does not state either a change in fashion or actual arrests, only the potential for arrests. Therefore, B and D are incorrect.

Question 160: D

The pulling out of feathers from live birds was seen as the negative alternative to using ostrich feathers.

Question 161: A

They could be possessed only 'in their proper season'.

Question 162: A

The problem cited is that the article was already in use in the clothing of numerous military men. The authority of the princess/sexist politics does not feature in the passage, and 'D' is patently false.

Question 163: E

None of those are precluded, as only 'harmless' and 'dead' birds (in their entirety) were prohibited. Wearing a living bird was not explicitly banned.

Passage 32

Question 164: B

He thought it was 'a pity that only rich people could own books', and from this, he 'finally determined to contrive' of a new way of printing. The passage does not state that he wished to make money, found books too expensive to get a hold of, or was impatient himself when it came to the production of books.

Question 165: A

The need to be careful is mentioned, as is the fact that the process takes a long time both in creating the block and due to the fact one block can only print one page. That it may tire a carver to make the block is possible, but it is not cited in the passage.

Question 166: D

The statement says it is 'very likely' he was taught to read but is not definite. The fact that his father comes from a 'good family' does not mean he is a member of the aristocracy, necessarily. Though block printing was used as the boy grew up, it does not state this was the most popular process. The mention of Gutenberg's family's 'wealthy friends' indicates that they were sociable.

Question 167: A

The paper was laid on top of the block, not underneath.

Question 168: A

There is nothing written in the passage praising the craftsmanship of manuscripts. The appropriateness of the titles for both book production processes is explained, and the 'wealthy friends' are described as a source to borrow books from, and, thus, a way of expanding one's reading. The boy was 'very likely' taught to read as well which implies that he was probably educated.

Passage 33

Question 169: C

Nowhere does it state that all European countries have similar creatures (though certain types can be found in both British Isles and Norway) nor does it state that the array of animals is limited to this one nation. Sharing animals and birds does not necessitate sharing geographical features, but it is said that a country with forest and moorlands is likely to have a variety of birds and animals, so one can see the link between forests and creatures.

Question 170: A

There was a time when the English dreaded wolves and bears, but that indicates the past, or at least does not include the present. Norwegians being superior is not suggested here.

Question 171: C

Bears are called destroyers, which is sufficient to conclude that they cause damage.

Question 172: D

They are ruthlessly hunted by farmers in country districts, but numerous only in the forest tracts in the Far North.

Question 173: A

The word 'fortunately' implies that it is good that the wolves are no longer central. The children are under no threat, as the threat of wolves belongs to a bygone time, there is no mention of regret that such a time is gone and Norsemen are not presented in the above passage as having respect for Nature, but instead, they are said to interfere with it through hunting and driving wolves farther afield from their current homes.

Question 174: A

This is not based on evidence and can vary from person to person. It can't be tested as being true or false and, therefore, is an opinion.

Passage 34

Question 175: A

Most requires over half by definition, and "most" of the people living in this area were the descendants of immigrants who moved to the country a "full century ago".

Question 176: B

Hall only makes a claim for New England, not the entirety of America, being the descendants of 20,000 immigrants. The 'one million' figure comes from Franklin, not Hall. Less than 80,000 ("under" 80,000) people led to the population boom of one million. One million is over ten times more than 80,000, so "B" is correct.

Question 177: C

It is said to be "distinct" to older aristocracy "of the royal governor's courts". It is not similar to any European aristocracy. There is no specific reference to it not being a system based on lineage.

Question 178: B

"A", "C" and "D" are cited in the passage (the journey took 'the better part of the year', it was 'hazardous' and 'expensive'), whereas 'B' is not referenced at all.

Question 179: E

The word 'reference' best fits in as the author is referring to the lack of citation of Shakespeare and the Puritan poet Milton.

Passage 35

Question 180: C

It would be a massive assumption to state that just because two characters in a book are 'vicious', all of them will be, so 'A' is not necessarily correct. 'B' also believes in a despair that is described to belong to the Comedian, but not Rorschach. The argument of the passage is that 'D', which Moore may believe, is not the case - the beloved character is not simply worshipped for his violence, but for his belief in justice. 'C' is correct, as Moore describes how he did not wish for Rorschach to be a favourite character, but rather a warning.

Question 181: D

Rorschach is considered as a hero, despite using violence. Therefore, A and B are wrong. The author then goes on to highlight that the issue with the Comedian is that he has no purpose but at least Rorschach does have a purpose, even if he does use violence. Accordingly, D naturally follows from this. C and E do not follow from what is discussed.

Question 182: B

He does not mention madness ('D'), or invoke shame ('C'), or simply state it is good to be good ('A') - specifically, he states we must act as if the world is 'just', even when it is not, in order to attain dignity.

Question 183: D

No value judgment is made regarding violent actions or on the Comedian's jokes so 'A' and 'B' are false. The passage also acknowledges Rorschach's violence, showing 'C' is wrong but does state that his actions are due to the fact he believes that he is acting in the name of justice, which lends him an ethical justification for his actions.

Question 184: A

The lack of meaning in anything is what leads him to treat everything as a joke - it is not hatred, but the inability to see 'purpose' in himself or his fellow man.

Passage 36

Question 185: C

This is the most accurate definition of a fallacy. Indeed, the first paragraph discusses that the gambler's fallacy is indeed a falsehood.

Question 186: C

The Monte Carlo fallacy and the Gambler's fallacy are labels for the same phenomenon. Accordingly, the events in Monte Carlo are not the definition of the fallacy, but merely why it has its name. Answer C fits in best with the explanation provided in the first paragraphs. It is the fact that there have already been lots of (previous) black rolls that people believe that there will be a red due.

Question 187: D

People were betting before the 26th spin so C is incorrect. E is incorrect as the passage does not indicate that they knew of the gambler's fallacy. A is impossible as it's not possible to know with certainty that a future event will happen and there's no evidence to suggest that B was the case. As explained in the passage, the gambler's fallacy operated to engender a false belief that one's odds of getting an *independent* future result (i.e. a red) increased because of the fact that there were lots of blacks beforehand.

Question 188: D

This can be assumed as the gambler's fallacy involves at least two independent events – that because one (or more) previous events occurred in a given way means that it's likely that a future event will occur in a particular way.

Question 189: D

Since the experimental group did not differ from the control group, it is clear that teaching probability alone may not change things. It does not prove A as not all possible methods of alleviating the gambler's fallacy were used in the study. B and C don't follow, and E is explicitly mentioned in the passage.

Question 190: D

The gambler's fallacy appears to operate whether or not one has a belief in it (based on the study in the preceding paragraph) therefore, B and C are incorrect. While A is true, it doesn't explain _why_ those in the first group tended to predict tails more. Also, point A was the same for both groups. E is too vague. Thus, D has to be correct. It also follows from the author's explanation of the gambler's fallacy in the beginning of the passage that that is what was being tested here in the final paragraph.

Passage 37

Question 191: C

The essence of the libertarian argument is that people should do what they want so long as they're exercising 'free will' (according to the passage). Therefore, A, B, D, and E are all incorrect. It doesn't matter if there is some risk to the provider or if it is against nature, as long as people consent, it's fine. Accordingly, when one does not give informed consent, arguably one is not exercising free will. Therefore, the libertarian argument would not be relevant in those circumstances.

Question 192: D

It is explicitly stated in the passage that people in Wales are presumed to consent unless they have indicated otherwise, so it is thus clear that people in Wales still have a choice as to whether to donate their organs.

Question 193: A

The author throughout has based the arguments in the passage on the case for allowing the sale of human organs, so point A is correct.

Question 194: A

Whether the poor are exploited through a system for organ sales is open to debate and arguments on either side can be equally valid. This can't be tested as being true or false. Therefore, it is an opinion. All other options are assertions of fact.

Question 195: D

This is never suggested in the passage – while the author believes that there wouldn't be exploitation of the poor, it is never suggested or implied that the poor should donate their organs as opposed to the rich.

Passage 38

Question 196: E

This is the only option that has to be true in order to sustain the author's argument. The author based his point on the fact that the officer didn't want to speak with Tatchell. However, if the students unions would have been fine with Tatchell speaking, then the author would, in fact, have a platform to speak.

Question 197: D

It is clear that the author's assertions and arguments point towards Tatchell being allowed to speak but the question was asking what the author's views were as to freedom of speech generally. Option D, thus, better describes the author's main position as adopted throughout the passage.

Question 198: E

Whether such censoring is concerning or not is simply an opinion – some may find it wrong but others may legitimately find it OK because it might offend people. It cannot be tested as it can't be true or false. On the other hand, all other options are assertions of fact. In regard to B, while the question of 'offensiveness' in itself is an opinion, the statement that "Offensiveness has a number of different meanings' is a factual assertion – we can simply have a look at whether there's more than one meaning of offensiveness or not. Therefore, point B taken as a whole is an assertion of fact.

Question 199: E

This is the only point that logically follows from the passage. It is not suggested that any students are extremists (so A is incorrect), it is not suggested or implied that students generally are against free speech (so B is incorrect), it is not clear that many students are against it (so C is incorrect) and finally, the passage does not assume or imply that student unions express the views of the student body.

Passage 39

Question 200: D

While A is true, it is not the main reason for the US bringing the case against Great Britain. B isn't correct as it's not clear that the seals were the US' and C isn't correct as it's neither stated nor implied in the passage that the Canadian ships breached US waters. The passage explicitly states that the depletion of seal stocks led to the conflict.

Question 201: C

The passage explicitly says that Great Britain's counsel persuaded the tribunal. Therefore, it's clear that Great Britain won the case. Canada was not a party to proceedings as Great Britain was representing them (so B is incorrect).

Question 202: B

This explains why the US' argument was convenient because this means that Canada was acting illegally while the US was acting lawfully and would have supported the US' case. A does not provide an explanation. C is not true and D, in fact, makes the US' argument less convincing. The author did not use the word 'convenient' in the context of E.

Question 203: D

This is true as while both sides could seal in the exclusion zone, the tribunal held that they were both restricted in the extent of their sealing.

Question 204: C

The regulation didn't refer at all to the Pribilof Islands. Accordingly, the US would still be free to seal there as they always had done (so long as there were enough seals there).

Passage 40

Question 205: D

A is obviously incorrect as opponents and supporters disagree on the existing wage rates. B is incorrect – we don't know whether opponents and/or supporters believe that fairness has different meanings in different contexts. The author seems to imply this, but the author does not imply that opponents and supporters think that it can have different meanings. However, the author points out that the supporters have one view of fairness (market-based wage rate) and opponents have another view (based on differences in wages). Therefore, both sides consider fairness and so D is correct. E is obviously incorrect because opponents of footballers' wages do not believe that the current outcome is fair.

Question 206: A

This is a tricky question involving the use of two negatives ('not' and 'inconsistent') in the question. Simply put, which of the following responses doesn't contradict both the opponents and supporters arguments? Opponents and supporters all disagree on B, C, D, and E. Therefore, A has to be correct. Indeed, supporters of footballers' high wages never deny that luck played a part.

Question 207: A

This is because the passage never says that hard work is unimportant or not a factor. It's perfectly logical for hard work to be a factor alongside one's birth talents. B is wrong because it's not obvious – the author says that 'careful' political debate is required to determine whether there should even be reform. C is wrong because two different formulations of fairness are highlighted in the passage. D is wrong because the author states that luck has a role to play. E is incorrect.

Question 208: E

It is clear that this sentence is introducing a contrasting point because of the author's use of the word 'however'. Accordingly, having just brought up the possibility of a wage cap, it means that the author is suggesting some kind of limitation to that in this sentence. E is the only choice that fits in with this.

Question 209: D

The author does not argue that the footballers' wages are correct or incorrect but *considers* the arguments as to whether they are *fair*. Therefore, A and B are incorrect. C and E are incorrect as they are merely assertions in the passage as opposed to arguments. D is correct because that is an intermediate conclusion, backed up by reasons immediately preceding it.

Passage 41

Question 210: B

The author asks readers to 'consider how you might feel if your employment contract was suddenly changed without your consent. The author also states: I would further ask how you would feel if that contract was changed such that your working hours became significantly less convenient without additional remuneration. Both of these statements invite readers to empathise with the tube workers.

Question 211: A

The doctors' strikes are the only one of the answers mentioned in the passage. They are mentioned in a way which contrasts the position taken (by the public) on their strikes with the position taken towards tube workers. As such, the answer is A because the author is highlighting an inconsistency in the way in which the industrial action of both groups is received by the public.

Question 212: C

The author states we should hesitate before using social media as a useful poll of public opinion. This indicates that it may offer a skewed view of public opinion. The other answers are not mentioned in the passage.

Question 213: D

In the passage, 'conservative' is used to describe newspapers, rather than people. As such, A & C cannot be correct. B cannot be correct because the passage does not indicate a view as to how conservative individuals view or interact with other people. Therefore, D is the correct answer.

Question 214: C

The author states that many of us may have unwittingly taken on the views promulgated in such media and that we might benefit from rethinking such views'. As such, she is suggesting that we should re-evaluate some of our beliefs.

Passage 42

Question 215: D

The author does not state any political views. Although the Labour Party is used as an example, and the Conservative Party briefly mentioned, the author does not comment on whether or not their views align with them. The Liberal Democrats are not mentioned. As such, there is not enough information to tell.

Question 216: B

This is explicitly given as a reason why people should join political parties in the introduction. Although the other answers are mentioned in the passage, they are not used to demonstrate why people should join political parties.

Question 217: A

The whole piece assumes that it is a good thing to engage with politics. If that were not the case, the whole piece falls down. B and C are mentioned explicitly so they cannot be correct. D is not an assumption of the piece – most of the piece is not about young people.

Question 218: B

'Untenable' means not able to be maintained or defended against attack or objection. It is a word which especially applies to a position or viewpoint.

Question 219: D

The statement that a large proportion of the UK population are completely disengaged from politics is one of the three arguments which follows directly from the statistic given. Although the other points are mentioned in the passage, they are not connected by the author to the statistic given.

Passage 43

Question 220: A

To perpetuate something is to cause it to continue, perhaps indefinitely.

Question 221: C

The author says 'only' 7% and that it was not clear whether this increase was caused by the policy. The use of the word 'only' indicates the answer cannot be A, and we do not have enough information to determine whether it could be D.

Question 222: D

The author states that 'if such privileges are given only to women, the idea that only women should be responsible for childcare may be perpetuated'. Answers A and B are not mentioned.

Question 223: B

The author states that 'the aim of this policy was to allow women to continue working, rather than leaving their jobs after having a child'. This is distinct from C because the author uses the word 'allow' rather than 'make'.

Question 224: A

The author says 'emphasising women's differences and adjusting for them may actually have done more harm than good in this case', indicating that it may not have helped gender equality.

Passage 44

Question 225: A

This statement is made without any evidence to support it and does not follow from any other arguments made in the piece. The other statements have evidence to support them.

Question 226: D

Although A is mentioned in the passage, it is not in the first paragraph. Peel's Metropolitan Police is discussed in the first paragraph but there is no mention of it being old fashioned. Visible patrol is mentioned but there is no indication that it wasn't working. The idea that governments were keen to secure value for money is explicitly mentioned, however.

Question 227: B

This is the only option, which is actually given as evidence. C is not present at all; although D and A are present, they constitute merely a further explanation of the point. B is given to support the original statement and, as such, is being given as evidence for the claim.

Question 228: C

Although this might be useful evidence for the statement, it is not present in the passage. A, B, and D are all referred to in the final paragraph.

Question 229: B

The passage is not designed to persuade the reader of anything. Persuasive language is not used and, as such, A and D are wrong. Although C is mentioned, it is only part of the last paragraph and is not the main point of the piece. The idea expressed in B runs through the entire extract.

Passage 45

Question 230: B

Is the central point of the first paragraph. A and C are not paradoxes at all; they are merely pairs of facts. D is incorrect because there is no suggestion that people are not interested in the law.

Question 231: A

The key phrase to understand this answer is "although many treaties now", indicating that the situation was different in the past. The other answers are not mentioned in the passage.

Question 232: D

A and B are not mentioned in the passage. C is only briefly mentioned. D is frequently referred to and runs throughout the passage and is the central argument of the extract.

Question 233: C

This option is not mentioned in the passage. The other answers are all contained in the last paragraph.

Question 234: A

The sentence on 'rules on international postage and telephone calls' is followed by the statement that 'this shows that the legislative process does work reasonably well'. B is the opposite and is, therefore, wrong. D and C are entirely separate points and mentioned elsewhere in the passage.

Passage 46

Question 235: D

C is not given as a reason for why it's important to find an answer to the question of whether or not law is coercive. A and B are both mentioned but there is no indication given as to which is more important. As such, the question cannot be answered with the information available.

Question 236: A

This is the only mention of Hart in the passage so we cannot know what he thinks about the other statements. The passage says "The criminal law is the clearest example of law being like commands (Austin), and despite Hart's criticism of Austin, he too concedes this", indicating that Hart and Austin agree on this point.

Question 237: B

C and D are not correct statements at all and misinterpret the 'puzzled' idea in the extract. There is no indication that lawmakers actually enjoy the use of sanctions so A cannot be correct. B is stated in the first paragraph.

Question 238: B

Coercive means using force or threats to make someone do something. The other answers do not relate to this idea at all.

Question 239: B

C is not mentioned at all. D cannot be correct because the passage states "serious sanctions are needed". B is correct because the passage states "a substantial portion of society obeys the law because of fear of sanction." A is an unconnected statement.

Passage 47

Question 240: C

Paragraphs two and three provide different perspectives on the new management structures. One offers an advantage and one offers a disadvantage and the author does not come down on either side.

Question 241: A

This is the only idea which is presented as a positive consequence of the new management structures. B is not mentioned. C and D are presented as potential negatives of the new structures.

Question 242: B

The introduction tells you that consultancies have been advising big companies on hiring fewer people from outside their companies and the second paragraph provides statistics to show that this is what Sahara have been doing.

Question 243: C

A is mentioned in the last paragraph, B in the second paragraph and D in the introduction. Although staff is discussed, they are management level, not entry level.

Question 244: A

The author says avoiding another financial crisis would "mean that we all avoid a great deal of misery". This suggests that there was a great deal of misery this time round.

Passage 48

Question 245: B

The final line of the passage is "There is no place to argue they are treated too leniently". The author tries to correct the ideas expressed in A, and D. C is not expressed in the passage.

Question 246: C

This might be very useful information to have but it does not appear in the passage. A, B, and D are all mentioned at different points in the passage.

Question 247: D

The enduring belief is discussed but not explained. C is not mentioned. A and B are included in the passage but not in connection with the enduring belief.

Question 248: A

The author states that these programmes "emphasise dependency and traditional femininity, and most crucially, fail to aid rehabilitation".

Question 249: A

We know that 70% of women have addiction problems, which is double the amount for men.

Passage 49

Question 250: C

B is not mentioned at all. A and D are mentioned as minor pieces of evidence. C is a repeated theme throughout the passage.

Question 251: A

The author states "prison has in fact been found to have an overall negative effect", which rules out B, C, and D which are all positive outcomes.

Question 252: B

The author states "making people miserable is not the way to bring about reform", meaning both that people are sad and that rehabilitation is, therefore, not being achieved.

Question 253: D

A and C are explicitly stated so cannot be unstated assumptions. B is incorrect – the passage repeatedly explains that prisons do not have a positive effect. The passage criticises the prison system for not achieving rehabilitation without ever explicitly saying that rehabilitation should be a goal, making it an unstated assumption.

Question 254: A

This might be a valid criticism but it does not appear in the passage. All the other answers are contained within the passage.

Passage 50

Question 255: C

All the answers are mentioned in the passage. However, readers are warned to be careful so as not to knock the water over, so it is not an advantage.

Question 256: B

This is the only contradiction. It cannot actually be true that everyone is sleeping badly if the author herself is doing well with sleep.

Question 257: A

To say 'we all are these days' is clearly an exaggeration. The other statements are literally true and not hyperbolic.

Question 258: D

Although the author says that the third piece of advice is important, she does not say it is the most important. There is no way of knowing which she thinks is the best piece of advice.

Question 259: C

The author tells us explicitly she is not a doctor. The final line, "I hope this helps you all sleep much better. Leave your comments below with tips and tricks which work for you!" indicates the piece may well be from a blog.

Passage 51

Question 260: C

In the introduction, the author states "It is submitted that all four are useful and are relatively complete as a set, meaning the criteria do not need to be re-written." B and D are not stated at all.

Question 261: D

The author tells you that Taiwan lacks independence. Prior to that, she has explained that capacity to enter into legal relations is essentially the same as independence.

Question 262: D

The author merely states that "The government need not be 'legitimate' or 'democratic' – it just needs to be in effective control". She does not state what she thinks about this.

Question 263: A

In the third paragraph, the author writes that "States are necessarily legal entities made up of people." B and C are not present. D is mentioned but is an explanation, rather than an argument in favour of the permanent population requirement.

Question 264: B

The author states that "it is self-evident that we expect states to be territorial entities and it is clear that this is an appropriate criterion for statehood". The other criteria are explained much more thoroughly.

Passage 52

Question 265: B

Although A and C are mentioned, they are only briefly referred to. B runs throughout the passage. D is not contained in the passage.

Question 266: B

Although A is made as a concession, this is not in the third paragraph. C and D are not concessions. The phrase "if it's really necessary" indicates a concession is being made.

Question 267: A

This might be a good argument but the author does not make it. B, C, and D and E are all referred to in the passage.

Question 268: C

The author states "the whole system ought to be built around teaching healthy and happy relationships, with the biology section firmly as a secondary point." This shows that the author thinks that biology should be taught but it's less important than relationships.

Question 269: B

Although A, C, and D could lead to the problem highlighted, the author does not connect them to the issue. She states in the introduction "children are going to the Internet or their peers to glean more information and this is leading to unstable adolescents."

Passage 53

Question 270: A

We know this because she says it would be "no bad thing" if "all businesses (were) required to have this policy" in the final paragraph.

Question 271: C

We know this because, in the first paragraph, the author states that it is "especially useful for new parents, who want to return to work but not do as much as they had previously".

Question 272: D

When discussing this option, the author writes that "this is unlikely, however, because businesses which offer flexi-time have largely invested in the scheme wholesale."

Question 273: B

The author states that "one of the major benefits from the perspectives of businesses is that their employees become much more productive." C is discussed as a possibility but not something of which we can be certain.

Question 274: A

This piece is not aimed at persuading anyone of something. It is written as a news article, reporting on a changing business practice.

Passage 54

Question 275: C

We know this because the author states "More significant, however, is that they employ even more people in the planning and sales of such projects."

Question 276: B

The author states that "One wonders, then, why the government is choosing not to invest much further. The economic recovery will not happen by itself." The unstated assumption behind this is that it is the responsibility of the government to stimulate the economy.

Question 277: A

The author offers criticism of the other groups. She also states that "One wonders, then, why the government is choosing not to invest much further."

Question 278: B

A and D are mentioned but would not impact the economy. C is incorrect because the author repeatedly states that the reason we need big infrastructure projects is to stimulate the economy.

Question 279: D

The author says that this group has a "selfish attitude". Although she disagrees with the environmental groups, she acknowledges they are thinking of people's welfare.

Passage 55

Question 280: C

The author states that "We discovered that London is the world's most popular tourist destination!" This necessarily means that London receives more visitors than anywhere else in the UK.

Question 281: B

We know this because the author states "councils should not be investing in tourism. This is because it is such a high-risk industry."

Question 282: A

All the other ideas are mentioned in the passage. Although it might be interesting to know how many tourists go to the beach, the author does not discuss it.

Question 283: D

The author expresses a great deal of surprise that people want to go for Madame Tussauds and queue for it. She goes as far as to call it 'bizarre'.

Question 284: C

It might be reasonable to think the other answers could be true but they would be going beyond what the author tells us. All seasonal means here is that the income is dependent on the seasons.

Passage 56

Question 285: B

The author states "Tax, it seems, makes all the difference." She alludes to Americans tweeting that Europeans are alcoholics but does not give credence to this claim.

Question 286: C

We know this because the author states "There is more sugar in American fizzy drinks than their European equivalents. In fact, there is even less in Asian markets." This necessarily means the American market has more than both the European market and Asian market.

Question 287: A

We know this because earlier in the paragraph, we read that "American senators have had very little joy in trying to tackle this problem because drinks companies provide financial support to so many of those in politics in the US."

Question 288: D

We know this from the following sentence: "The French – very meritoriously – take children with them to dinner parties and slowly allow them to start drinking alcohol in a sensible way."

Question 289: D

We know this because when the author discusses the tax, she says "but this appears to have made little difference to those in their twenties."

Passage 57

Question 290: E

The other options could happen and are discussed in the passage. However, none of the options are presented as definite results of the ageing population. D would only occur if governments do indeed struggle to pay for their ageing populations; however, the passage tells us only that they are 'likely' to struggle.

Question 291: B

All the other answers are perfectly credible reasons, but the author does not, in fact, choose to explain why there is a difference.

Question 292: A

We know this because although the author expresses concern for all the countries mentioned, she uses the phrase "Most alarmingly, in countries such as Niger…" indicating that this is the place she is most concerned about.

Question 293: B

A and C are stated explicitly in the passage so cannot be unstated assumptions. D is not relevant at all. The paragraph discusses the limits on development as being problems in themselves, meaning B is the right answer.

Question 294: D

We know this is the answer because the paragraph discusses different ways of paying for an ageing population as well as how to save costs. Thus, we can infer that they cost a lot of money.

Passage 58

Question 295: A

We know this because she uses phrases such as "swan around thinking they are so on trend and yet most of them produce nothing of value whatsoever", indicating an attitude of derision towards them.

Question 296: C

We know this because the author criticises tech for being 'dull in practice' and states that jobs in arts are 'more interesting' – and therefore, better.

Question 297: A

The sentence "Jobs in finance may seem like good ideas for graduates but jobs in publishing actually offer more opportunities for the truly talented" is a direct comparison.

Question 298: D

The author states that "publishers are much more rigorous in their hiring procedures – and rightly so" indicating the hiring procedures are tough. She does not think they are too hard, however – as we know from the use of the word 'rightly'.

Question 299: C

This is only revealed in the final paragraph when the author says, "The point here is that although I'm sure you've been told that jobs in the industries that graduates typically go into have a lot to offer, you should not close your mind to other options. Think a little bigger!"

Passage 59

Question 300: E

The passage begins by speaking of 'social convulsions' and quickly switches to speaking of 'revolutions' without any indication that the author is changing the topic. 'Social convulsion' must, therefore, refer to revolutions. Option C is incorrect as although a revolution necessarily brings about change, the notion of change is not synonymous with that of revolution (given that change can come about otherwise than through a revolution).

Question 301: E

The author writes that 'This skin, or envelope, however, does not expand automatically.' This suggests that law and institutions (which are represented by 'skin') do not change by themselves or without much pain/suffering.

Question 302: D

The author argues that Washington was able, because of his genius and because of the men who supported him, to postpone revolution. If, therefore, one could prove that the men around his were not particularly remarkable, it would weaken the argument that they contributed to his success. Option B is incorrect because even if intelligent men opposed Washington, this has no effect on the author's argument that some intelligent men also supported Washington.

Question 303: E

All the other options appear in the second and third paragraphs in relation to Washington. Option E is used in relation to the men who supported Washington.

Question 304: C

The author writes that: 'Yet even Washington and his adherents could not alter the limitations of the human mind'. This immediately precedes the statement to the effect that Washington could postpone but could not prevent revolution, suggesting that the natural limitations of the human mind were to blame.

Passage 60

Question 305: A

The first line of the piece hints at option A as the correct options: 'the whole scope of the essay is to recommend culture as the great help out of our present difficulties.' Later on, the author makes clear the fact that the value he places on culture lies in its ability to give us 'a fresh and free play of the best thoughts'. Options B, C, and D are just used to support the main argument. Option E is an argument advanced by the author only to further elucidate the nature of culture (the fact that it is inward).

Question 306: D

The author emphasises the fact that his critics have lost sight of 'the essential inwardness of the operation of culture, and that 'it is just our worship of machinery, and of external doing' that led to the bringing of the charge. This suggests that the entire example is designed to illustrate our focus on external actions.

Question 307: C

The third paragraph suggests that the effect of culture is to bring about a change inwardly in our thoughts – hence, we need not follow a specific set of outward actions to obtain culture, if we can obtain these same effects without following that set of outward actions.

Question 308: A

The author emphasises that an academy constitutes 'outward machinery', whilst culture is 'inward', suggesting that an academy is unhelpful because it is outward.

Question 309: A

The lines which immediately follow that line suggest that people can obtain culture through different means and need not necessarily do so by reading. The author, however, does not suggest that the meaning of culture itself is open to individual interpretation – only the means by which one strives to obtain it.

Passage 61

Question 310: A

The question which follows the line in which 'basal' is used ('what common ground…') sheds light on the sense in which it is being used by the author. If a feature is common to many different situations in which laughter occurs, it suggests that the feature is the 'essential' element in the laughable.

Question 311: C

In the second paragraph, the author suggests that his exploration of the topic of laughter cannot fail to throw light on the way the human imagination works and that it must have something to tell us about life. Although A, B, D, and E are mentioned in the passage, nothing in the author's language suggests that these are the goals of his investigation – and they seem to relate to the method of the investigation A, background information about the topic E, and mere observations on the topic D and B.

Question 312: B

All the examples in the paragraph are geared towards demonstrating that, as stated in the penultimate line, 'the comic demands something like a momentary anaesthesia of the heart'. The use of 'then' in the penultimate line affords strong evidence that what is to follow is a summary of all that has been discussed earlier in the paragraph and is, therefore, the conclusion of the paragraph.

Question 313: A

A metaphor involves the representation of one thing with another. A is the correct answer as the other options are either not figurative, or even if they do employ figurative language, do not concern the representation of one thing with another.

Question 314: C

The second paragraph states that laughter demands an anaesthesia of the heart, suggesting that the absence of sentiment is a precondition for laughter. A can be found in the penultimate paragraph, B is derived from the statement in the final paragraph that laughter has a social function, D is derived from the statement in the first paragraph that laughter can throw light on human life/imagination and E is also stated in the first paragraph ('we regard it as a living thing'). Therefore, C is the only answer not used to characterise laughter.

Passage 62

Question 315: C

A premise of the author's argument that the patriarchal nature of a society influences their conceptions of God is that the Jewish God embodies masculine qualities and is to be found in a patriarchal society. However, for this premise to support her argument, she must assume that the Jewish conception of God stemmed from the people themselves (and, thus, from the patriarchal society), and not through divine revelation (which would have been completely unrelated to the people, and, thus, unrelated to the patriarchal society). Options A and B are incorrect because whether or not there is a god or whether or not the Jewish God exists has no direct bearing on whether the conceptions of God *endorsed by a society* are endorsed because of the structure of that society. Option D is incorrect because it is the writer's argument, rather than an assumption of her argument. Option E is an implication of her argument, not an assumption of it.

Question 316: A

This question requires the student to pick up on the word 'special' and to identify the subtle distinction between the motive for subjecting religious conceptions to examination, and the intended outcome of such an enquiry. The question, which asks 'why' the author believes that religious conceptions should be examined, relates only to the former. The author states: '*as* the forms and habits of thought connected with worship take a firmer hold on the mental constitution than do those belonging to any other department of human experience, *religious conceptions should be subjected...*' - suggesting that this is the motive for her subjection of religious conceptions to examination.

Question 317: D

The author's argument is that as society evolved into a patriarchal society, the religious ideas endorsed by society came to reflect a more masculine conception of God. The author, therefore, argues that society's conceptions of God follow changes in the structure of society. It is true, of course, that different societies have different conceptions of religion, but this doesn't come *closest* to describing the author's stated views on the relationship between religion and society – hence E is incorrect.

Question 318: B

The author offers no reasoning to support this view – it is merely stated – hence it is a mere assertion. D is supported by the reasoning in the first paragraph (that as society became more masculine, conceptions of God changed) and E is merely an application of D to a specific society – it is, therefore, supported by the same reasoning which supports D. C is supported by the statement that 'every branch of inquiry is being subjected to reasonable criticism', and A is supported by the reasoning in the second paragraph.

Question 319: E

The author mentions that with the rise of male dominance came the valuing of male qualities over female qualities. Both these factors were necessary for producing the contemporary (masculine) ideas of God. The answer is to be found in the final paragraph: 'After women began to leave their homes at marriage, and after *property, especially land, had fallen under the supervision and control of men*, the latter, as they manipulated all the necessaries of life and the means of supplying them, began to *regard themselves as superior beings, and later, to claim that as a factor in reproduction, or creation, the male was the more important*. With this change, the ideas of a Deity also began to undergo a modification.'

Passage 63

Question 320: E

Independence is not attributed to goodness, as the author states that the idea of goodness appears only as an '*element*' in other things, and we must ourselves strive to '*give goodness an independent meaning*'. The other options are to be found in the first and second paragraphs. Option D is incorrect because, at the start of paragraph two, goodness is described as being somewhat paradoxical in that it is familiar and yet remote and, therefore, curious.

Question 321: A

The author states: '*But while thus familiar and influential when mixed with action, and just because of that very fact, the notion of goodness is bewilderingly abstruse and remote.*' The other options are not stated as reasons for which the notion of goodness is difficult to understand. Option E is a restatement of the problem, not an explanation of its origin.

Question 322: B

A statement to this effect comes immediately after the author's declaration that '*familiarity obscures*' – this gives the student a strong hint that B is the answer and constitutes an explanation of the phrase '*familiarity obscures*'. The next closest option is A, but this is not stated as an explanation for the phrase '*familiarity obscures*'. It is presented, rather, as the false belief of most people (which the phrase '*familiarity obscures*' is intended to correct).

Question 323: D

Immediately after making this statement, the author uses another example: '*To determine beforehand just how polite we should be would not facilitate human intercourse*' which reiterates the idea that understanding the theory of a topic does not help in its practical everyday application.

Question 324: A

Although the author does mention that understanding goodness will not necessarily help and may hinder everyday life, that statement acts merely as a caveat to the central point – which is that understanding the idea of goodness is necessary for ethical students (but if we mean to be ethical students and to examine conduct scientifically, we must evidently at the outset come face to face with the meaning of goodness). The entire text concerns coming to grasp the idea of goodness, so it is also likely that the main conclusion of the final paragraph will relate to the importance of understanding the idea. Option E is incorrect because the statement that the goodness is the ethical writer's chief tool is a premise which is designed to support the idea that students need an understanding of goodness; in critical thinking, it cannot, therefore, be the ultimate conclusion.

Passage 64

Question 325: A
In the final paragraph, the author is speaking against the idea that women ought to have children only in wedlock. This suggests that when she encourages having children '*in freedom*', she means having children outside of marriage.

Question 326: A
The fifth paragraph speaks of the fact that women are taught to only consider money in determining who to marry. This suggests that they employ a flawed criteria in determining who to marry, and the author suggests at the beginning of the paragraph that this is in fact the reason why '*Nora leaves her husband*'. Option B is the next closest option – however, it is not the <u>ultimate</u> cause of unsuccessful marriages, as the use of a flawed criteria temporally and logically precedes the marriage to an unsuitable person.

Question 327: C
The author makes it clear at the beginning of the passage that '*Marriage and love have nothing in common*', suggesting that there is no relationship between the two. She goes on to mention that love doesn't spring from marriage (paragraph 3).

Question 328: A
If the parents of destitute children are unmarried, then the author's' argument in that paragraph (that marriage does not protect children) completely falls apart. The other options do nothing to damage the argument in that paragraph – e.g. option D, or, in fact, contradict the author's argument (option B), or are explicitly argued by the author – i.e. option C (see the final line of the paragraph).

Question 329: E
The examples and arguments advanced by the author all support the idea that marriage is a failure – i.e. it is a failure because it cannot protect children, and because it cannot bring about love and because divorce statistics are so high. The main argument does not concern children as only one paragraph is devoted to children, and that paragraph is itself designed to support the idea that marriage is a failure.

Passage 65

Question 330: A
The author mentions that '*the multitude, the mass spirit, dominates everywhere, destroying quality*'. She then goes on to give one example of how this has occurred, mentioning that mass goods are now being produced (quantity), which are generally injurious to us (reductions in quality). This example is given to support the idea that we prioritise quantity over quality nowadays, so the other options cannot be the main conclusion (as they support option A). Option E is not mentioned in the first paragraph.

Question 331: B
A metaphor is the representation of one thing with another. Only option B satisfies this criterion (a '*god*' is used to represent success).

Question 332: A
The main argument is that art is successful only if it appeals to the majority of people. The author then states that this is the cause of the ugliness of the art that can now be seen around the city. If, however, a minority are responsible for that art, it would undermine the author's argument that the prevalence of majority taste produces ugly art. Option E is irrelevant to the author's argument, as is option D. Option B is just a reiteration of what the author mentions already and supports her argument. Option C doesn't do much damage to the author's argument as the idea that for most of their lives artistic geniuses are unable to thrive as they don't appeal to the vulgar majority taste still stands.

Question 333: C

This statement supports the idea that the oft-repeated slogan is incorrect as it suggests that individuals are only successful due to the majority, and cannot obtain that success on their own. The other options do not give reasons for us to take seriously the author's rejection of the oft-repeated slogan and are mere assertions.

Question 334: A

The author suggests that even when politicians engage in disgraceful behaviour, they need only '*muster up their minions*' to ensure that they stay in power nonetheless. This suggests that democracy helps those with poor morals to hold on to power. Options E and B are incorrect because the author nowhere suggests that democracy actively encourages corruption. Option C is incorrect as it is the opposite of what the author is saying.

Passage 66

Question 335: D

That is the only option which is not supported by extra reasoning – e.g. the statement that the purpose-novel proposes to serve two masters is supported by the idea that the novel aims to entertain as well as to instruct.

Question 336: A

The author mentions that the bishop may be the best preacher there is, but the viewer is still entitled to get their money back because they came to watch a play, not a sermon. This idea is encapsulated by option A. Option D is incorrect as the purpose-novel is said to be a failure in that it fails to entertain and fails to instruct; however, the bishop example is an expression of the author's view even in the event that the purpose-novel is successful in providing excellent instruction for the reader.

Question 337: C

In paragraph 2 the author mentions that '*it is often said that the novel should instruct as well as afford amusement*', before branding this species of novel a '*purpose-novel*' and going on to state that such novels propose to '*serve two masters*'. The two masters are, therefore, likely to represent the two purposes of a purpose-novel. Option D is incorrect as the author says '*besides procuring…*' and this suggests that providing an income for the author is a separate function of the purpose-novel in addition to its service of '*two masters*'.

Question 338: C

A novel about politics is most likely to be aiming to instruct readers, and the author views instruction or education in serious topics as unsuitable objects of a novel. He would, therefore, believe that topics/themes which are more entertaining are more suitable (i.e. options A, B, D or E). Option D could be considered trivial, and, therefore, appropriate for a novel.

Question 339: A

This would strengthen the idea that the purpose-novel is a fraud because it tricks people. Option B would weaken the author's argument. Option C is irrelevant, as are option E and option D (as they do not relate to the topic of books which purport to be works of fiction but which, in reality, are not).

Passage 67

Question 340: B

This is the only option that can be directly implied from the line quoted. Option A is close but incorrect because it is expressed in absolute terms 'no one', whilst the quote is not ('no one who was older than a baby). Option C is incorrect because the statement itself doesn't tell us anything about how widespread the destruction was. Option D is incorrect as the opposite is implied by the quote.

Question 341: D

Option D provides a reason for abandoning Big Media. Options A, B, and E are merely factual statements with no normative content. Option C is an argument in favour of Big Media.

Question 342: E

In the second paragraph, the author mentions that the changes were inevitable because of *'new publishing tools available on the internet'*, suggesting that this was the principal trigger for the changes. The changes were manifested after 9/11 – but 9/11 didn't itself trigger the changes, and option D constitutes a result of the changes, not a trigger for the changes.

Question 343: B

The author mentions that the evolution *'will force the various communities of interests to adapt'*. Option C is incorrect as it constitutes merely a repetition of the idea that there has been an evolution. Options D and A are not mentioned.

Question 344: C

The meaning of the phrase *'we can't afford'* is that we cannot continue in our present state – this suggests that the author repeats the phrase to reinforce the idea that change is necessary. Option B is incorrect as repetition of an idea does not make it any more convincing; it is, therefore, unlikely that the author employs repetition to this effect. Option D is incorrect as finance is only mentioned towards the end of the paragraph in question. Option A is vague – so option C is the better option.

Passage 68

Question 345: D

This option is merely an assertion of fact, not an argument (it has no normative content), although it later constitutes a part of the author's argument that intellectual property rights shouldn't be tradable. The other options are all arguments (premises) found in the text. Option E can be found in the final paragraph where the author suggests that knowledge is power and that the abolition of copyright laws would enable the education of more people.

Question 346: B

This can be found in the following line: *'There is no cost for materials and the same labour that went into making what I bought also went into what everyone else bought'*. The idea that intellectual property is not physical merely sets the stage for the invocation of the argument found in option B (i.e. it is because intellectual property is not physical that there is no extra cost involved in the sales of the produce of intellectual property).

Question 347: B

The protectionism example states that we try to improve transportation but immediately offset any gains by using tax laws. The author is, therefore, trying to make the point that *'we have advanced methods of storing and promulgating information, we replace any advantage gained in that respect with legislation that restricts the flow of information'*.

Question 348: E

An assertion is a statement that is unsupported by evidence or reasoning. Option E is the only statement which fits that description; all the other statements are supported by reasons given by the author – e.g. option C is supported by the idea (in the first paragraph) that we can use *'information as a framework for future advances'*. Option B is supported by the statement that *'all the knowledge of the world could be given away freely in a digital format'* (i.e. it would be easy to educate people.)

Question 349: A

The author argues, effectively, that intellectual property rights should not be tradable for value because, for instance, it doesn't cost the seller anything extra to sell one more e-book and no extra labour goes into it. It would follow that whenever the provision of a good or service doesn't cost the seller anything extra, and no extra labour goes into it, they should not be able to charge for it. Only option A fits this criteria, as it doesn't cost anything extra (beyond the initial cost of building the house) to rent out a house, and no extra labour is involved.

Passage 69

Question 350: C

The tenor of the author's argument is that humans are all similar such that we may '*see [our] own vices without heat in the distant persons of Solomon, Alcibiades, and Catiline*'. Options E and A are incorrect as the author focuses on learning not just on seeing what historical figures have done – e.g. '*...we shall learn nothing rightly*'.

Question 351: B

Most of what comes after this statement consists of examples of the way in which humans have shared characteristics – e.g. '*What Plato has thought, he (any other man) may think; what a saint has felt, he may feel*'; '*All that Shakespeare says of the king, yonder slip of a boy that reads in the corner feels to be true of himself*'. Option D is merely another example of this broader idea.

Question 352: E

The author uses an encyclopaedia to represent a man. He, therefore, employs a metaphor. None of the other options involve the representation of one thing with another.

Question 353: D

The author dedicates a large part of the passage to showing that humans all have shared characteristics. However, for him, the significance of having shared characteristics lies in the fact that we can then study history and, through the study of men in the past, learn about ourselves in the present. The title – '*History*' – also hints at option D as the correct answer.

Question 354: A

The use of '*we*' seems to suggest that it is indeed possible to make generalised statements, yet it doesn't go as far as to suggest that humans are '*identical*' in every respect other than in the respects alluded to in the statement. Option B is also a step too far and cannot be directly inferred from the use of 'we' alone.

Passage 70

Question 355: B

In the first paragraph, the author writes: '*it is easy in solitude to live after our own; but the great man is he who in the midst of the crowd keeps with perfect sweetness the independence of solitude*' – i.e. it is harder to live amongst other people and '*live after*' one's own opinions. Option D is incorrect as not all forms of contempt are necessarily difficult to handle, according to the author (e.g. '*the sour faces of the multitude, like their sweet faces, have no deep cause, but are put on and off as the wind blows and a newspaper directs*').

Question 356: A

The answer is to be found in paragraph three, where the author suggests that the rage of the '*multitude*' is more fearsome than that of the upper classes. Options C and D are incorrect as the author argues the opposite. Option E is not mentioned.

Question 357: A

This is the argument advanced by the author against reverence for past acts; he states: 'suppose *you should contradict yourself; what then?*' The next closest option is option E, but a close reading of the final paragraph will reveal that the author is not saying that reflecting on the past is in itself <u>always</u> unhelpful.

Question 358: C

In the passage, the author argues that a person must follow their own opinion. It follows from this that social conventions should only be followed if a person personally considers them worth following. It does not follow that they should never be followed since the point is that an individual should think for themselves, and not that social conventions should be shunned per se. The same reasoning applies for option D.

Question 359: A

The use of the word 'corpse' emphasises the fact that the events which are the subject of memories have already passed and are of little relevance to life in the present. Nothing else in the passage seems to suggest that memories are unpleasant – so option D is incorrect. The phrase '*corpse.*' doesn't itself suggest that memories are burdensome – hence options B and C are incorrect. Portraying memory as '*the ghost of past events*' does little to add to the author's argument or support that argument – therefore, option E is incorrect.

Passage 71

Question 360: A

The author states: '*The student receives at one glance the principles which many artists have spent their whole lives in ascertaining*', suggesting that the student is able to gain an insight into the principles of art without going through the same lifelong process as the great masters. This is the main advantage – although the implication may be that students need not think for themselves; this is not stated by the author as an advantage. Option C, D, and E are also incorrect, as they are not stated by the author.

Question 361: C

The author states: '*he who begins by presuming on his own sense has ended his studies as soon as he has commenced them*', suggesting that no sufficient progress can be made by those who already think they are enlightened enough to criticize the great masters. Option A is incorrect, as the author is not necessarily arguing that the works are themselves infallible – just that they should be '*treated*' as being infallible. Options B and E are not mentioned. Option D is also a departure from what the author says (which is that principles only constitute a fetter to students who have no talent – not that such students focus on criticism).

Question 362: A

The author's argument is that men of great natural abilities can improve through observing the works of the great masters. He uses Rafaelle as an example of this process. However, if Rafaelle produced his best work before gaining access to the work of great masters, it would suggest that observing those works did not improve him. Option B is incorrect as it is irrelevant to the main argument. Option D is incorrect because the author is already speaking of talented artists, and never disputes that artists may be talented before viewing the works of great masters. Option E is incorrect because the author argues only that exposure to such works improves the artist and it may not be necessary for the artist to value the art highly (it may suffice that they view the art as a repository of art principles/rules) for the improvement to occur.

Question 363: A

The author writes that armour is '*an ornament and a defence*' for the strong, but is a '*load*' for the weak. The illustration, therefore, highlights the point that rules give additional strength or help to those with talent, whilst constituting a '*load*' – i.e. a hindrance for the students who lack artistic talent.

Question 364: B

The argument is that students can and should learn from the works of the great masters in order to produce good work and progress. The author must, therefore, assume that art has an objective value and is not purely subjective, otherwise, it would make little sense to speak of improvement in one's art. Options C and E are incorrect as they are more or less explicitly argued by the author, and cannot, therefore, constitute assumptions. Options A and D are not strictly relevant to the argument that studying the works of the great masters is a <u>means</u> to improvement.

Passage 72

Question 365: C

In the first paragraph, the author declares that the value of art is to be measured by 'the mental labour employed in it, or the mental pleasure produced by it'. The most mental labour involved in the painting of the scene is likely to be the most, especially given the amount of time it takes. It would, therefore, be of most value.

Question 366: A

The author insists that only general topics (topics to which people can relate generally) are appropriate subjects of art. He implies, therefore, that the recording of a subjective experience through art would be inappropriate. Option D is incorrect as it does not necessarily follow that the scope of choice for artists is very narrow (i.e. there may be many topics which fall under the requirements laid out by the author). Options B and E are explicitly stated. Option C is incorrect because the majority of people can relate to the subject without enjoying the piece of art.

Question 367: B

The author does not stipulate that art must embody a grand style to constitute valuable art – he just mentions *'grand style'* as one form of art. All the other options are described as features of good art – e.g. in speaking about an appropriate topic for works of art, the author writes: *'there must be something either in the action or in the object in which men are universally concerned'*.

Question 368: B

The author argues that works of art should focus on the broad picture and not on tiny details because when people picture a story in their minds, they see only the broad picture. The author must, therefore, assume that the purpose of art is not to show tiny details (and, therefore, not to record reality as it really is). Option D is irrelevant to the author's argument. Option A is stated. Option C is not really relevant to the argument – it only offers to explain why people only remember the big picture in their minds. Option E is irrelevant because the paragraph focuses on situations in which we are concerned with stories, and doesn't speak of what should prevail in other situations; the author, therefore, doesn't assume that the situations of which he speaks (where a story is involved) are the only situations in which art is produced.

Question 369: D

The author states: *'The power of representing this mental picture in canvas is what we call invention in a painter'*, making it clear that the answer is D.

Passage 73

Question 370: D

The author writes that unlike the *'superstition of religion'* which resulted organically from man's experience with unexplained natural phenomena, the superstition of patriotism is man-made. Options C and E are not used to distinguish between religion and patriotism. Option A is not used to characterise the *'superstition of religion'*. Option B is only used to characterise patriotism.

Question 371: B

Gustave Herve, and not the author, refers to patriotism as a superstition.

Question 372: B

The author says that the rich are perfectly at home in every land. This is a matter of opinion as there are no objective means of proving this statement true or false. The other statements are statements of fact, not of opinion, given that they can be proven to be either true or false.

Question 373: D

The author mentions that the money used to fund the army and war is *'taken from the produce of the people'*, suggesting that the people bear the financial burden of patriotism. The author does mention factory children and cotton slaves but this is in relation to the wealth that the rich are able to accumulate, not in relation to the cost of patriotism, and, in any case, it is not argued that only those two groups bear the burden of patriotism.

Question 374: C

The rich are presented as hypocrites in that, whilst they endorse patriotism '*theirs is the patriotism*', they are still able to send messages to foreigners and live abroad (i.e. they do not act in accordance with patriotism because they do not promote the interests of their country above that of other countries). Hypocrisy consists in endorsing or promoting an idea but not acting in accordance with it – hence, option C comes the *closest* to the author's description of the rich.

Passage 74

Question 375: D

The context in which the phrase is found sheds light on its meaning. The author states afterward that '*We doubt that we bestow on our hero the virtues in which he shines*', suggesting that we almost worship our friends and attribute many '*virtues*' to him. This suggests that we have an extraordinarily positive outlook on our friends. Options C and E are incorrect as the focus of the first paragraph is not on our <u>expectations</u> of our friends or our <u>expectations</u> of friendship – its focus is on how we currently view our friends.

Question 376: C

The author compares friendship to the immortality of the soul by saying that they are both too good to be believed. None of the other options involve the comparison of one thing (or of a feature of one thing) to another. Option E is a metaphor.

Question 377: B

'*The slowest fruit*' is a reference to fruit which takes a long time to grow. The fact that we '*snatch*' at it signifies that we are impatient and don't want to wait for the fruit to ripen. Similarly, the author is arguing that we don't take the time to explore and cultivate deep friendships, and want a '*swift and petty benefit*'.

Question 378: A

The benefit at which people '*snatch*' is described as being both '*swift and petty*'. These two are not, therefore, contrasted. The other contrasting pairs are to be found in the text. Eternal laws and sudden sweetness are contrasted in the lines: '*The laws of friendship are great, austere, and eternal, of one web with the laws of nature and of morals. But we have aimed at a swift and petty benefit, to suck a sudden sweetness.*'

Question 379: E

This is the author's main argument because all the other points raised in the paragraph are designed to support the idea that we need to explore our friend's true characteristics. For example, options B and D attest to the need to explore our friends' true characteristics. Moreover, option A is incorrect as the statement '*friendship is too good to be believed*' merely suggests that we do not view our friends as they really are – which itself lends support to the idea that we need to explore their true characteristics. Therefore, option E, and not A, is the main argument.

Passage 75

Question 380: B

The author states that receivers of gifts wish to assault givers and that receiving a gift is a very '*onerous business*', suggesting that givers of gifts ought not to expect gratitude on the part of recipients. Option E is not really stated by the author. The other options support the idea encapsulated in option B, and cannot, therefore, constitute the main argument. Option A is incorrect as it is not stated by the author who is likely to use the term '*mean*' to mean '*poor*' or '*mean-spirited*' – rather than unkind.

Question 381: D

'*generous*' is the meaning of the term '*magnanimous*' which best fits in this context; a magnanimous person is said to put us in his debt – suggesting that he himself gives to us, such that we feel that we owe him.

Question 382: B

The author states in the final paragraph that: '*our action on each other, good as well as evil, is so incidental and at random, that we can seldom hear the acknowledgments of any person who would thank us for a benefit, without some shame and humiliation*'. The third paragraph doesn't really consider our <u>entitlement</u> to thanks and is more focused on reasons why we cannot <u>expect</u> thanks from beneficiaries – hence option E (which could possibly be derived from the third paragraph) is incorrect.

Question 383: A

The meaning of the expression is that receiving a gift well is a particularly rare quality. This suggests that most people do not possess it – hence option A is the correct answer. Option B is incorrect as '*good*' is not being used in the sense of moral status – as evidenced by the author's suggestion that we may be wronged by gift giving (hence we would incur no moral obligation to receive a gift well in the first place). Option E is incorrect as the author speaks only of one particular instance in which we could be gracious – and not generally. Option D is incorrect as it cannot be inferred directly from what the author says.

Question 384: C

This answer can be derived from the explanation by the author of why gifts are offensive. He suggests that we are offended because the giver aims to '*bestow*' on us and we don't like the idea being provided for by another person: '*We sometimes hate the meat which we eat, because there seems something of degrading dependence in living by it*'. It would follow that gifts out of love, which are acceptable, lacks this feature which makes other gifts undesirable.

Passage 76

Question 385: E

Irony is the expression of one's meaning through saying the opposite. Options A, B, C, and D are incorrect because the author really means what he explicitly says – e.g. he says '*We write from aspiration and antagonism, as well as from experience. We paint those qualities which we do not possess*', to suggest that he really means that he has no prudence, and is only writing about it because of his '*aspiration*' towards prudence.

Question 386: A

This illustration is immediately preceded by: '*We paint those qualities which we do not possess*', suggesting that the illustration is designed to support the idea that we esteem qualities which we do not possess. Option B is incorrect because even though the illustration might suggest that sometimes we want our children to have different careers to us, that isn't the <u>purpose</u> of the illustration.

Question 387: E

The author states: '*A third class live above the beauty of the symbol to the beauty of the thing signified*'. He then describes that class as having '*spiritual perception*' and being able to reverence: '*reverencing the splendour of the God which [they] see bursting through each chink and cranny.*' These two statements suggest that to live above the beauty of the symbol means to focus on the spiritual significance of physical things. It involves an appreciation of nature, however, option E is the fuller description of what the statement means.

Question 388: C

Most of the author's reasoning is designed to support the idea that prudence involves spirituality. The second paragraph speaks of prudent people as being spiritual, and the third paragraph criticises a view of prudence as relating to physical matters. Although option B is implied in the introduction, it doesn't feature anywhere else in the argument. Option E does not feature in the piece. Option A is not mentioned by the author either (he speaks of a different form of prudence in the last paragraph, but it is clear that he does not view it as being really prudence at all).

Question 389: A

The author refers to the false idea of prudence as being '*a disease like a thickening of the skin until the vital organs are destroyed*', clearly manifesting his disapproval. The other terms aren't themselves used to convey disapproval, even if they appear in disapproving statements.

Passage 77

Question 390: B

The author states that *'these sacrifices are by far not the only standard of measurement of Russia's participation in this gigantic struggle'* before going on to the other means of gauging Russia's role in the war. All the reasoning in the paragraph is, therefore, designed to support the conclusion that the losses sustained by the Russian Army should not be the primary means of gauging their role in the war.

Question 391: A

The answer is to be found in the following quote: *'The final catastrophe of the Central Powers was the direct consequence of the offensive of the Allies in 1918, but Russia made possible this collapse to a considerable degree, having effected, in common with the others, the weakening of Germany'*. Option E is not stated in the text. Options C and D are incorrect because they do not constitute descriptions of the significance of Russia's contributions to the defeat of Germany. Option B is not stated by the author.

Question 392: D

The fact that Russia acted in self-sacrifice tells us nothing about the effect of her self-sacrifice on the course of the war. That statement, therefore, unlike the other options, does nothing to advance the case that Russia played a significant role in the defeat of Germany.

Question 393: A

If the German's lost very few men in the conflict with Russia, it might suggest that Russia did not do very much to weaken the German army. Options C and D are incorrect because the author does not deny the relevance of the other allies to the defeat of Germany. Options B and E are irrelevant to the argument that Russia weakened the German army.

Question 394: C

The author's statement suggests that it is only the recognition of those facts that is necessary for forming the conclusion that Russia played an enormous role in the victory against Germany. It is implied, therefore, that the facts alone are very compelling. Option D cannot be inferred directly from the author's statement – especially as historians may fail to concede Russia's enormous role due to ignorance of the facts, not only due to bias. Option E cannot be inferred from anything in the statement, and neither can option B. Option A is incorrect as it constitutes a mere repetition of the explicit statement.

Passage 78

Question 395: D

The answer is to be found in the author's statement to the effect that: *'finding, however, that the alterations this would have involved would have been incompatible with a clear and connected view of the author's statements, he preferred giving the theory itself entire'*.

Question 396: A

This is added by the writer as a reason for the unfavourable treatment of Goethe's work – it is not stated by Goethe himself. The other options, which are stated by Goethe, can be found in the second paragraph.

Question 397: B

The author's main argument is that Goethe's ideas were not accepted because of the way in which he presented them. This argument would be strengthened by option B because the option suggests that in Goethe's days, the manner in which ideas were presented had an effect on how they were received. Options A, C, D, and E tell us nothing in relation to the main argument – e.g. it may well be the case that all of Goethe's work was expressed in uncompromising language, but we would need to know of the success of those pieces of work to form any sort of judgment in relation to the main argument.

Question 398: B

Immediately after using this phrase, the author states *'the translator begs to state once for all, that in advocating the neglected merits of the "Doctrine of Colours," he is far from undertaking to defend its imputed errors'*. This suggests that he doesn't want to accept Goethe's views just because opinions about them are changing, unless there is reason to accept the views – hence, he refuses to deny the errors in Goethe's doctrine. All this suggests that *'fashion in science'* refers to changes in opinion which are not based on reason.

Question 399: B

The author speaks in favour of *'a due acknowledgment of the acuteness of [Goethe's] views'*. In this context, 'due' signifies approval as the author, in using it, portrays the *'acknowledgement'* as being fitting or appropriate. Option E is not used to portray approval or disapproval. Option D is simply a description of an alternative to the harsh treatment Goethe received, but the author expresses neither approval nor disapproval. Options A and C are also neutral.

Passage 79

Question 400: E

The author mentions servitude to the aristocracy but that is preceded by the line *'With the loss of their municipal independence went the loss of their political authority'*, which suggests that the loss of municipal independence is the trigger for everything else that happened.

Question 401: B

A principality is a state. The phrase *'little principality'* is likely to be used in this sense here as the towns which are described as little principalities are also described as having political/governmental independence – e.g. they *'elected their own rulers and officials'*. A little principality differs from local government in that local government is often still subject to central control – however, the text doesn't suggest that any central control was imposed on the people.

Question 402: C

The phrase *'this free and vigorous life'* refers back to the three paragraphs which precede it. The life described in those paragraphs comes closest to what is described by option C. Option D is incorrect as the text does not say that the treaties established were peace treaties. Option A is incorrect as there is no suggestion that the reform bill itself negatively affected the towns. Option B is incorrect as the mayors are stated as defining the boundaries, not the citizens.

Question 403: A

The author's main argument is that the loss of municipal independence led to political apathy in towns. If A is correct, this severely damages the author's argument. Option C is incorrect as it is already conceded by the author. Options D and E do not really relate to the author's main argument, neither does option B.

Question 404: B

Given that the author speaks of ordinary townspeople in a generally positive manner, he is likely to use quotation marks around *'mean people'* – a derogative term – in order to make it clear that he is not endorsing the term.

Passage 80

Question 405: B

The author states that there was a shift from 'attention on the punishment of terrorists and the criminalisation of new offences following their occurrence' to greater emphasis on the *'management of anticipatory risk'*. Option C is wrong as it relates only to how the Terrorism Act 2000 differs from its predecessors, not to legislation post 9/11.

Question 406: A

This is the only option that can be directly derived from the statement quoted. No extra information is given to support B or D. Options C and E do not necessarily follow from the statement.

Question 407: E

The author mentions that the Terrorism Act 2000 set aside article 5 of the HRA; however, the author doesn't really contrast the two acts with each other. In contrast, all the other options involve each part of the pair being presented as an alternative.

Question 408: D

The author's main argument is that terrorism laws, in general, are discriminatory. However, if only one Act has been found to have been discriminatory, this would erode the author's main argument. Option A is incorrect as the fact that one specific group has not been affected doesn't make terrorism legislation any less discriminatory. Option C supports the author's argument. Options E and B in no way affect the main argument.

Question 409: B

This is the only option which is stated in the text ('*the government felt that in order to prevent mayhem their actions needed to be swift and decisive*'.

Mock Paper Answers

Paper A		Paper B		Paper C		Paper D	
1	C	1	B	1	C	1	D
2	D	2	E	2	D	2	C
3	B	3	C	3	A	3	A
4	C	4	D	4	A	4	E
5	A	5	A	5	E	5	A
6	D	6	B	6	D	6	E
7	E	7	B	7	B	7	E
8	A	8	D	8	C	8	C
9	A	9	E	9	E	9	C
10	B	10	B	10	C	10	A
11	E	11	C	11	C	11	A
12	C	12	D	12	A	12	B
13	B	13	A	13	B	13	C
14	D	14	D	14	A	14	E
15	E	15	B	15	A	15	C
16	A	16	C	16	B	16	A
17	C	17	A	17	A	17	D
18	C	18	D	18	C	18	A
19	B	19	B	19	B	19	D
20	D	20	C	20	C	20	B
21	E	21	A	21	E	21	A
22	D	22	B	22	A	22	A
23	A	23	D	23	A	23	D
24	E	24	D	24	E	24	D
25	E	25	E	25	C	25	C
26	E	26	C	26	B	26	C
27	A	27	B	27	D	27	C
28	C	28	A	28	C	28	B
29	B	29	B	29	C	29	C
30	D	30	A	30	E	30	E
31	B	31	E	31	A	31	D
32	A	32	D	32	A	32	C
33	B	33	C	33	E	33	E
34	E	34	A	34	C	34	D
35	C	35	C	35	D	35	C
36	A	36	D	36	A	36	D
37	E	37	B	37	D	37	C
38	D	38	A	38	E	38	A
39	C	39	B	39	A	39	A
40	E	40	E	40	C	40	D
41	B	41	C	41	D	41	E
42	C	42	A	42	C	42	B

Mock Paper A: Section A

Extract 1

1. C is the correct answer. A is not incorrect, but is not the purpose of the comparison. B is about the importance of universality, not efficiency. D is not particularly relevant. E is correct, but it is more of a *conclusion* of the author, not the purpose behind the comparison

2. D is the correct answer. A is overtly mentioned in the first sentence. B is not really relevant to the piece, there is no implication that the younger generation will change their gaming habits on the basis of efficiency. C is explicitly mentioned in the sentence beginning 'Now, one…'. D is almost certainly correct, but is still an assumption which leads to the assertion in the first paragraph. E is irrelevant.

3. B is the correct answer. A is incorrect as it appears to be a rhetorical question. B Is most likely to be correct, as it draws parallels between the choice that the younger generation make on a day to day basis, and the choice that they often do not think about. By bringing them into the same sentence, he makes them seem equal. C is not correct, as although he does mention that younger people have lots of choices to make, he does not imply that this is overwhelming. D is correct, but the question is placed there in the context of the paragraph, which is clearly to do with other choices to be made. E, again, might be correct, but is not the reason he concludes the third paragraph in that way.

Extract 2

4. C is the correct answer. Although the statement says that the researchers 'propose', it is a statement of fact that this is what the researchers propose.

5. A is the correct answer. The author mentions that even though it may not help academically, it can help create a human being with good habits. This, along with the author's tendency to consider homework in a positive light implies that good habits and personality is as important as academics. B is stated outright. C, although may be inferred, is not necessarily implied by the author in the first sentence. D seems a little too far-fetched. It could be implied that busy work is not as important as other work, but it's too much of a leap to assume that she meant to imply that these teachers are *failing* their students. E, again is a step too far. It is clear form the piece that she does not agree with Kohn's work, but to conclude that it is untrustworthy would be incorrect.

6. D is the correct answer. This question refers to the second sentence of the last paragraph. A assumes that earning money is the equivalent of success. B refers to the child's mindset, which is not the same as 'methods'. C is a direct quotation from earlier in the extract, but this is not asserted as something that a child *needs*, instead it is stated as an inevitable result of children doing homework. E is irrelevant.

7. E is the correct answer. This question will be between C and E. C, however is too black and white, saying that homework is *not* helpful for academic achievement, where the article is less direct on that issue.

Extract 3

8. A is the correct answer. Although B is stated explicitly in the text, it is not the reason *why* the author uses the example, instead it is a question that is raised as a result of the use of the example. C cannot be correct as the author cannot emphasise a point that he has not yet made. D is incorrect, not only for the purpose of the example, but the piece altogether, as is E.

9. A is the correct answer. This is simply a question of close textual interpretation. A is identical to the first sentence of the last paragraph. B states that they are entirely different concepts, therefore not linked, which is not true. C is not true, as although the last sentence states that they must be *viewed* in the same way, the identities themselves are clearly not identical. D is not really dealt with at all, and E is not dealt with in the final paragraph (and is an incorrect interpretation of what was said previously)

10. B is the correct answer. B is taking a statement too far. The extract only suggests that the theory of tolerance is 'yet to be tested', not that tolerance has failed. A and C are stated in the piece (with different words). D is implied by the phrase beginning 'whatever else 'Britishness' might be'. E is implied with the brackets '(as we might be of 'Englishness' if we looked close enough)'.

Extract 4

11. 11. E is the correct answer. All of the statements could be a valid reason why the author has included the reference to certain airlines.

12. 12. C is the correct answer. This may well be a method used by retailers, but is not mentioned outright in the article. A is mentioned in relation to furniture shops. B is 'drip pricing'. D is the 'BOGOF' offers. E is in relation to the paragraph about 'price discrimination over time'

13. B is the correct answer. This question is about very close interpretation of the structure of the question. There is a difference between 'Why is it suggested that "people are not as rational as the standard economic model implies"?' and 'Why is it suggested that people are not as rational as the standard economic model implies?'. The use of a quotation in the first example implies that the question is asking why the author suggests something, or why the author uses a particular piece of evidence. However, the second structure is really asking 'Why are people not as rational as the standard economic market implies'. The inclusion of the 'is it suggested that' is only to remind the reader that the following part of the sentence is only a suggestion by Heuristics, it is not fact. Thus, the students may get confused between A and B. C is another example of the incorrect analysis of the question. D is incorrect as it is too strong a conclusion to draw. E is simply incorrect.

14. D is the correct answer. This will likely be a choice between A and D. Although the last paragraph lists some advice for consumers, this seems almost an odd addition to end the extract with. The purpose of the main extract, excluding this section, seems to be to inform the reader about the different methods used. Thus, overall, the extract seems to be informative, rather than advisory. The last paragraph is simply an addition on how to best make use of the information gained from the piece. C is also focusing on the 'advisory' last paragraph instead of the meaning of the piece as a whole. E is a homage to the title of the piece, but is not really relevant to the content.

Extract 5

15. E is the correct answer. A is stated. B is not an assumption. It may be implied by the author, or inferred by the reader, but an assumption is something that must be true in order for the author's argument to make sense. B does not fulfil this criterion. C, again might be implied by the author, especially for subjects that he considered irrelevant (namely homeopathy in this instance), but it is not rightly an 'assumption'. D is not relevant to the discussion of university as an investment. E is the correct answer, as, for the author's argument about university being an   investment to be correct, he must admit that there are no other reasons why someone could want to go. If he admits that they are other reasons, then his argument is fatally flawed.

16. A is the correct answer. A is explicitly stated at the beginning of the penultimate paragraph. B is only correct in relation to non-Russell Group universities. C is incorrect, as the author only implies that university is too expensive for certain degrees. For D, the author never goes so far as to suggest that the degree is worthless, he just suggests that it is not worth the fees. E is incorrect, as the author suggests only that Russell Group universities are worth attending. This extends further than Oxbridge.

17. C is the correct answer. The author describes how university is beneficial for some people but not for others, his approach to this suggests that he believes the system is flawed. A is clearly an accurate description of the author's views of some parts of the university system, but not others. B is the option which may cause the most confusion for students. Whilst the passage does not seem to be supportive of the university system, it does make clear that there are certain universities that are beneficial to their attendees. Thus it is a step too far to argue that the author is not supportive of all aspects of the university system. D, again is too far in the opposite direction. The author, despite arguing that some aspects of the system are overpriced and unhelpful, does not conclude that the system as a whole is completely useless. E is clearly wrong.

Extract 6

18. C is the correct answer. A is pretty much irrelevant, but for the fact that money for arts could be going on emergency services, but this is a weak argument. B would strengthen the author's argument. For E, correlation does not necessarily mean causation and thus is not as strong as it would   seem. C and D could both potentially weaken the author's argument, but C is more damaging to the author's perspective.

19. B is the correct answer. The use of the word 'desperate' indicates the author's aversion to Ms Miller. A is not the author's opinion, it is a direct quote from Richard Morrison. C is not about the author's distaste for Ms Miller, but a description of the quote that he has just used. D is clearly positive. E does not show any opinion either way, it is merely a statement of fact.

20. D is the correct answer. A is stated at the beginning of paragraph 4. B is stated in the sentence 'The knives appear to be out for her in government'. C is stated in the first paragraph. E is clear. D is the answer as, even though the article states that 'much of the effect of the cut might be ameliorated through the National Lottery', it does not go so far as to state that it will make up the whole of the 5% cut in the arts.

21. E is the correct answer. A is incorrect as the question is not whether they are currently widely accepted. B is the option that the students are most likely to get confused with. This is very slightly wrong, as the question is not whether the arts are more worthy than they once were (although this is mentioned earlier), the question is whether they should be get 'special treatment'. Furthermore, it is not about being 'widely accepted' it is about whether local authorities will accept the idea. C is not really relevant - there is no mention of different opinions. The 4th answer is a moral question -should it be considered a special case, instead of whether it will be considered a special case.

Extract 7

22. D is the correct answer. A is mentioned in the observation that 'almost all respondents grew up in a male-dominated household'. B is explicitly mentioned in that 96% of respondents were male. C is mentioned in the penultimate paragraph.

23. A is the correct answer. B is causally related, as shown in the penultimate paragraph. C is mentioned in the exploration of the respondents' male-dominated households. D a relatively obvious point shown in the first sentence of the second paragraph. E is more subtle, with reference to people being 'pushed over the edge' in response to 'all Muslims [being] treated as terrorists'. A is the only one that is not causally related, there is no mention that the domination of extremism in the media has led to more extremism.

24. E is the correct answer. A and B are both mentioned in the extract, but neither is mentioned as being 'the definitive reason'. C is not even mentioned in the extract (though it may be inferred). D seems to be correct, as none of the above are 'the definitive reason'. However, choosing this option would admit that there is, in fact, one definitive reason. This is incorrect, as the author does not, at any point state that one correlation is more important than any of the others.

25. E is the correct answer. A is mentioned in the main body of the text, but is not the conclusion of the final two paragraphs. B is not mentioned at all, so is clearly wrong. C and D are mentioned in the penultimate paragraph, but are only part of the conclusion, not the main conclusion.

Extract 8

26. E is the correct answer. A is incorrect, but is likely to be picked by students. The growth of online pharmacies is not the reason why the misuse of medicine is becoming more of a topic. If anything, the growth of online pharmacies has led to medicine becoming more readily available which has in turn made it more of an issue. However, there is a missing link in the analysis to say that the growth of online pharmacies had led to a large issue of misuse of medicine. For example, the growth of online pharmacies could conceivably make it less of an issue, as it could make it more difficult for people to get medicine - perhaps because it's more difficult to prove to a computer that you deserve the medicine. B, too, is not directly responsible for the rise in relevance of misuse of medicines, it is only a factor. C and D may be inferred from the first paragraph, but it is not suggested. E is directly mentioned by the Chairman.

27. A is the correct answer. A is implied by Prof. Iversen in his 'Cat and mouse' analogy, where he states that the cat 'should not withdraw in defeat'. B and D are stated explicitly and thus cannot be said to have been implied. C is implied, but not by Prof. Iversen. Though both academics discuss pregabalin, it is only Professor Hill who mentions anything to do with the drug's widespread use. E is simply incorrect.

28. C is the correct answer. This question has been added to see if the students can determine whether a new person is being referred to, or simply more information is given about an already present character in the story. For example, the first paragraph refers to 'the government's senior drug adviser' and the second to 'Professor Les Iversen', where these are in fact the same person. The correct answer is C, the three people being Professor Iversen, the Home Secretary and Professor Hill.

Extract 9

29. B is the correct answer. A is incorrect - the argument relies on the premise that late-term abortions are legal, not that they have recently become legalised. B is correct, as the author simply assumes that mothers have the choice to terminate the pregnancy - this statement relies on the assumption that mothers alone are legally able to make this decision. C is incorrect - the author suggests that it may be more positive to wait until a child is born to decide whether or not to terminate its life, as there will be certainty that the child has no chance of survival. This would lead to more cases of infant euthanasia, but the reason they have increased in frequency is development in ultrasound scans such that spina bifida can be detected before birth. D is explicitly stated in the description of the illness as 'lethal'. E is a difficult one - the author may be implying this, by stating that it is 'not necessarily the best course of action', or indeed the reader might have inferred it, but it is not quite correct to state that the author assumed it in his argument - an assumption is an assertion of a fact or set on facts upon which an argument is based . This is not quite the case here.

30. D is the correct answer. In this case, incorrect students will most likely pick E, as it is the least controversial statement and thus most likely to be right. But the author does more than simply state that more discussion should be had. Throughout the article he mainly suggests positive things about a narrow legalisation of the euthanasia of children, both explicitly and implicitly, by favouring the arguments and countries that have accepted it. Further, in the conclusory paragraph he states 'parents, physicians, hospitals and nation need to confront this issue'. This is one step further than simply 'discussing': the use of the word 'confront' implies that they should decide positively on the benefits of legalisation and make changes.

31. B is the correct answer. Proof of parents supporting child euthanasia suggests that even parents, whose natural instincts are to care for their children, agree that in certain circumstances child euthanasia is beneficial. Both A and D suggest that spina bifida is a bad disease, which would support the view that late-term abortion is preferential to child euthanasia and thus sit against the author's stance. A significant rise in child mortality rate, as suggested with C is also detrimental to the author's cause. At first glance E may seem to be positive, as it looks like the President of the EU supports the author's plight, but upon further inspection, the President does not actually support the legalisation of child euthanasia, but merely stated that it should be explored more. This, in itself, does not support the author's argument.

32. A is the correct answer. A is the most likely of the correct answers. B may well be true, but it does not explain why the author began with that particular sentence. C may well be true of the first paragraph, as a push (though it seems more likely that the paragraph is included to shock the reader), but not of the first sentence, as this does not inform of the methods used, merely that they are easy. D is incorrect as both A and B could be possible reasons: E is incorrect as C is not a possibility.

Extract 10

33. B is the correct answer. This is a 'process of elimination' type question - A is not the 'ultimate aim', it is simply the method used. C is merely an intermediate step towards meeting the ultimate aim. D is incorrect, as the article does not suggest that this method can be used to stop any other disease except for Dengue Fever. E is irrelevant. Therefore, it must be B.

34. E is the correct answer. This question is likely to be a toss-up between A and E. Both B and C basically conclude that the project will not work, but this is not terrible, as it doesn't have any negative impact, it only reduces the potential positive impact. D shows that the evidence might be wrong, but this is not as bad as it definitely being wrong. The problem that arises between A and E is that, whilst one (A) is only a potential problem, if it were to arise the consequence would be worse, whereas (E) is a definite problem, but the consequence is not as bad as would arise if A came to fruition. On balance, though, the most damning statement should be E, as the fact that it is definitely transmittable is more damning, especially given that the consequence (infertility) remains quite serious. If for example, E were 'The discovery that Wolbachia can be transmitted to, and causes the sniffles in, humans', then on balance A would be more damning.

35. C is the correct answer. A is addressed in the reference to 'transparency and proper information for the households'. This implies that without such information, people may panic. B and E are directly addressed in the phrase ending 'concerns at the time that Wolbachia could infect humans and domestic animals'. D is less obvious, but arises from the fact that the author reassures the reader that, using this process, researchers will not have to 'constantly release more contaminated insects'. This suggests that, were the process not as adequate, this may have been a problem. C is the correct answer because, though the 10 week period is mentioned in relation to the experiment in Australia, there is no suggestion that taking longer than this would be problematic.

Extract 11

36. A is the correct answer. For A, the 'word seriously' is mean as a descriptive word about how affected the women's liberation movement is. B brings with it images of people putting their two cents into an argument- 'weighing in' usually carries with it negative connotations. C is pretty clearly negative, with the use of the word 'sacrifice' immediately bringing negativity with it. For D, the use of 'the sort of men' again carries with it less than friendly implications. E's use of the description 'hot pink', which tends to be associated with air-headedness for the word 'equivocation' suggests that she does not believe that it is.

37. E is the correct answer. Although some of the reasons identified are more prominent than others (for example, explaining what the fonts are like is probably not the main reason that Laurie Penny uses the word) all of them are possible reasons.

38. D is the correct answer. The answer most likely to be considered sarcastic is C. However, this is not sarcasm in the true sense of the word. Sarcasm occurs when someone says the opposite of what they mean in order to draw attention to it. The archetypal example being a grumpy teenager moaning 'oh great' in response to being made to do something that they don't want to do. C is not an example of sarcasm. It is factitious and full of attitude, but is not rightly categorised as sarcastic. More accurately it would be described as litotes – understatement for comic effect.

39. C is the correct answer. A is found in the idea that feminists want to change the may people think - they 'ask men to embrace a world where they do not get extra treats. B is mentioned in the discussion of why you cannot simply 'rebrand' feminism. D is implied by the paragraph beginning 'It is not "young women today" who need to be convinced that feminism remains necessary and "relevant". E is explicitly stated: 'the ugly, unfuckable feminist'. C, on the other hand is used to describe women, specifically how the magazines want women to be - it is not directly related to feminism as a concept

Extract 12

40. E is the correct answer. A is clearly wrong. B, C and D are all possible interpretations of what he's said, but his main opinion is E, as he mentions the implications for further study three times.

41. B is the correct answer. A is incorrect as the dust is simply the 'petri dish' of the experiment, the surface upon which the chemical reactions take place. C is correct in that it is radioactive, but there is no mention of it being an atom. D is clearly incorrect as the first paragraph notes that Iso-propyl cyanide is 'closer to' complex organic molecules. This by its very nature means that it is not the same thing. E is incorrect as it is only 'a step closer to discovering [...] the precursors ... of amino acids.'

42. C is the correct answer. A and D are incorrect as they are stated explicitly. B is incorrect as it is not about amino acids. E is incorrect as it is never mentioned that amino acids are essential, but only a possible way to explore life outside of earth. C can be inferred by joining together the ideas that Iso-propyl cyanide is radioactive, and that Iso-propyl cyanide could be the 'origin of more complex branched molecules, such as amino acids'

END OF SECTION

Mock Paper A: Section B

1. **"The government should legalise the sale of human organs" Discuss.**

For
- Bring it out of the black market
- Make it safer
- Could ensure that there are certain forms, stamps etc. to make sure that a good price is being paid, the person is consenting etc.
- Stop the need for human trafficking
- Bodily autonomy
  - People should be allowed to do whatever they want with their body
  - We allow this when we are *donating* organs - why not allow it into the capitalist system - it is proven to deal best with supply and demand issues
- Will allow desperate people a final option
  - Is it not better to sell organs than sell your body or descend into a life of crime

Against
- Why should we allow rich people with the means to pay to be able to buy organs to prolong their life, where other people may not be able to
- Is it *right* for life itself to be part of a capitalised system
- Perhaps we already do this? Allow rich people access to more experimental and expensive medicine - perhaps the question should be more about whether all humans should have access to the same medical treatment?
- Is it *right* to sell organs - should it not remain a donation based system - is making it a commodity the best way to deal with the issue of not having enough organs? Perhaps an 'opt-out' rather than 'opt in' system would create the supply required without needed to address the moral issue of whether one should be able to sell bits of themselves.
- Do we want desperate people to have to resort to selling their own organs?
  - Legalising it may make it an option for people, where it may not have been before

2. **"Developed countries have a higher obligation to combat climate change than developing countries" Discuss the extent to which you agree with this statement.**

For
- Higher GDP - more 'disposable income'
- Moral obligation - MEDCs have benefitted from the earth's resources more than other countries, so have more of an obligation to make it better
- MEDCs have the technology and capacity to develop new systems to prevent/slow down climate change
- LEDC's have more important things on their mind
  - See 'pyramid of needs' below - those in MEDCs have already dealt with the bottom layer, so they can deal with the second layer.
  - Those in LEDC's cannot deal with the red layer until the yellow layer is secured

Against
- Do they have a higher *moral* obligation? Perhaps it is true that practically the MEDC's should be targeting climate change more aggressively than LEDC's, but maybe this is because they have more resources to do it, it is not a matter of morality
- We know that one of the main causes of methane in the atmosphere is agriculture - cows mainly. These are more likely to be found in less developed countries. Thus, logic would suggest that these countries are equally responsible, if not more so than more developed countries
- It is not disputed that MEDCs should do more, but it is not because of *morals*
- It's a universal problem - it faces everyone, so should be deal with by everyone

3. **"Putin is a serious threat to global stability" Discuss.**

For

➤ It is a rarity in this day and age for a country to simply decide that they want land back, and just go in to take it

➤ His intention is ultimately to undermine democracy - he is a traditional 'baddie' - power hungry

➤ He cannot be defeated by manpower. He is not a 'extremist organisation which can be bombed. He is in a position of power, and thus has a unique position to undermine *global* security. He is a permanent threat

➤ He cannot be defeated by manpower. He is not a 'extremist organisation which can be bombed. He is in a position of power, and thus has a unique position to undermine *global* security. He is a permanent threat.

➤ Having such a strong, power hungry person in charge of such an extensive nuclear arsenal could result in a new cold war.

Against

➤ Need to explore whether this means Putin as a person, or Russia. How much power does Putin as a person really have, in the grand scheme of things?

➤ People have been scared of Russia as a concept since the cold war

➤ Its perceived character makes Putin seem more intimidating than other, similar people

➤ Is it Putin himself that is causing a threat to global stability, or the world's reaction to his actions

➤ We need an enemy, and if it weren't Putin, it would be someone else

➤ The media needs a target - and politicians follow - e.g. - The Ukraine problem got 'boring', so the media moved onto Islamic state

➤ How often do we hear about the Ukraine now? And has the problem actually ended?

4. "Sufferers of anorexia nervosa should be force fed" Do you agree with this statement? If so, evaluate at what point of an individual's disease this measure should be taken.

For

➤ This is a question about consent to medical treatment - As a result of thumb we do not treat people unless they consent to the treatment. Sufferers of anorexia nervosa who do not want to eat clearly do not consent to being fed. However, there are situations when we let this happen, if someone is considered too mentally ill to be able to give a free and informed consent.

➤ Anorexia is a disease whose physical symptoms all stem from the person not eating. Often the only way to save someone's life is to give them food. If they do not consent, it is indicative of the fact that they are incapable of logical thought, and the state should step in to help them, despite the intrusion on their bodily autonomy.

➤ There is no other option - they need food, they won't eat it.

Against

➤ Bodily autonomy - anorexia is not a mental illness like other mental illnesses - they still have their faculties and should be able to refuse treatment - even if this results in death

➤ Force-feeding is a humiliating, horrendous thing to have to go through (it has been used as a form of torture). There must be better ways of keeping someone alive until the disease releases its grip

➤ Could the humiliation of being force-fed aggravate the mental disorder?

END OF PAPER

Mock Paper B: Section A

Extract 1

1. B is the correct answer. Firstly we can discount E. A cannot be correct as the government cannot have implied anything specific about Egypt and Sudan, as these countries are not mentioned. C is too strong a sentiment to have been implied from what is a fairly neutral statement. D is most likely to be correct, but the statement cannot stretch so far as to have implied this. They only admit to having 'redressed' the balance. This cannot be taken to mean 'retaliation'. B is correct as the government states outright that Ethiopia was left bereft of access to the Nile, which suggests that they were overlooked in the past.

2. E is the correct answer. A can be discounted outright. Both C and D show passion, but do not go so far as show disdain. B is merely a statement of fact and does not suggest how the people of Ethiopia feel. E is correct as the use of the strong 'nefarious' suggests an underlying feeling of distaste.

3. C is the correct answer. The question asks why the particular phrase is used. All of the other answers refer to the context in which the phrase is used (the building of the Dam) and do not focus on the phrase itself. B and D might confuse some students, but it must be remembered that it is not asked what Mr Hossam's argument is, nor why he chose to explain that sentiment, but why he chose to use exaggeration within that particular phrase.

Extract 2

4. D is the correct answer. C can be discounted immediately. A and B are both conclusions reached in the article but are not the *main* conclusion of the article. E is a conclusion that might be inferred from what the author has said throughout, but is not stated within the article and thus it would be a stretch to say that it is the main conclusion. The article keeps returning to the idea of the healthy robbing the ill of their choice to i.e. - this must factor in the main conclusion of the article and so the answer must be D.

5. A is the correct answer. C and D are expressly stated in the article so there is no room to infer them. B is not even alluded to, so to infer something so extreme would be illogical. E is touched upon in the piece, however, the author states that his work regarding those who wish to die with dignity should be commended, so clearly he is rightly placed to discuss *some* aspects of the laws. The correct inference here comes from the statement 'the fact remains that he is a rich, powerful man and it is highly unlikely that his wishes would be ignored'. A logical inference here would be that others who aren't in his position might not be afforded the same luxury.

6. B is the correct answer. A will not support the author's argument as she clearly says that some people's opinion of what they think they will want if they become ill may change if and when they find themselves in that situation. C may rectify the problem that the author identifies with regards to the choices of the rich and famous being respected above the choices of the vulnerable, but this is not the main problem that the author identifies in relation to euthanasia and thus does not sufficiently answer the question posed. D is irrelevant. E might rectify a problem relating to euthanasia - the cost of keeping people alive, but definitely not the main problem that the author identifies in the extract. B would likely rectify the problem of a seriously ill person not being able to choose - since this analysis is as close to choice as is possible given the extenuating circumstances.

7. B is the correct answer. A is too far removed from the term to suggest that it is the reason she used it. C is the alternative answer most likely to be considered correct, but does not quite fit. The paragraph about how 'it's what they would have wanted' is irrelevant for those who have not yet passed. It would be illogical for the author to start the paragraph with a discussion that people should think further about what they say and then go on to argue something completely different. D cannot be correct as it suggests that the argument that follows is useless, but this is incorrect, the author in fact acknowledges that the statement is clearly true. E is incorrect as it cannot follow simply from using the term 'knee jerk', that the argument that follows is either good or bad, the author goes onto explain this herself in as many words, so cannot expect the term 'knee jerk' to make that implication. B is correct as the adjective 'knee jerk' casts doubt upon the value of the following statement, suggesting that it has not been sufficiently analysed or evaluated.

Extract 3

8. D is the correct answer. A is impossible to know for sure, B is an opinion, C required knowledge of how the poll was expected to go (which is not given in this piece). D is correct as 8/10 (i.e. *most*) of people who admitted to illegally downloading music also said that they bought music.

9. E is the correct answer. A is irrelevant. B, C and D make the evidence questionable. E makes the evidence simply false.

10. B is the correct answer. This is the only statement of opinion: 'need'. The others are clearly all fact. The students might get confused about E. This is still a statement of fact as there is a condition in there - not that they *will* rise, but that they can *expect* to, based on the data.

Extract 4

11. C is the correct answer. C is too draconian. B is incorrect as the schools are obliged to hand over the information. Both D and E are specifically referred to in the piece as not being correct, as Fairport central school district was found to be non-compliant when they only gave over the information once they had received consent. C is the correct interpretation of the law.

12. D is the correct answer. D is the only opinion that is not specifically stated in the extract. Although Tina Weishaus alludes to the idea that the strong tactics used by the recruiters are inappropriate for children, this is not *stated*. Thus, D, if anything, is inferred by the reader from Ms Weishaus' statement.

13. A is the correct answer. C is clearly incorrect. B may at first glance seem correct but the question asks what is the *purpose* of the inclusion of the *phrase*. The use of language may well add something to the piece, but we are not discussing why the author has used that particular literary device. We are discussing the reason the *phrase* is included, and as such it follows that the phrase is included to add something to the argument, namely to draw attention to it.

14. D is the correct answer. All of the answers can be inferred from the figures given. Most students clearly don't want to be approached, as 95% of them didn't consent. Given that recruitment among seniors hit 2% it is fair to assume that most of the 5% who consented must have come from the senior class, thereby suggesting that they are more likely to consent than younger students. Finally, it would be fair to assume that, given that recruitment figures were fairly high among seniors, that the current recruitment strategies work best amongst this group of students.

Extract 5

15. B is the correct answer. The students will probably end up with a choice between A and B. The correct answer is B because in the first paragraph the author clearly states 'our best theories of physics imply we shouldn't be here'. It cannot be said to be 'inexplicable' as clearly, given the piece, there could be an explanation. Thus, A is incorrect. C is incorrect as it is too strong a suggestion - that it 'can' be explained instead of it 'might' be explained. D is simply incorrect as our best theories *do not* explain our existence, and E is too strong a conclusion - it is never stated that no answer will ever be found.

16. C is the correct answer. This is described in the paragraph beginning 'So Vachaspati and his colleagues...'. Only C is identified and all others are clearly incorrect – this should be a fairly simple question as long as the candidates are looking in the right place.

17. A is the correct answer. The section in the final paragraph in the hyphens relates to the definition of the term 'new science', so only the description 'stuff beyond the standard model of particle physics' is the definition and as such the answer must be A. Students might be caught out by the section after the hyphens, but this refers to what is required for CP-violation to provide enough matter in the universe, not to the 'new science' definition.

Extract 6

18. D is the correct answer. B and C are both wrong because the author states that it is impossible to keep addressing the new approach as it arises, the issue must be addressed from the source. E is the conclusion of the first counterargument, but it would be incorrect to say that it is the conclusion of the whole article. A was a point that was made, but again was not the *main* conclusion of the article. The answer is D as the final paragraph concludes that the government must adopt a ground-up approach to tackling extremism.

19. B is the correct answer. A is addressed in the first paragraph of the author's counter argument, C is addressed in the second, D in the penultimate paragraph about over-stretched resources in the security and police departments, and E in the discussion of TPIMS. Although those on the 'fringes' of groups are mentioned as a point of concern, the author does not suggest that the measures will encourage people to move out to the fringes of groups, and as such B must be the correct answer.

20. C is the correct answer. A and B are clearly wrong. E is far too vague and cannot be the reason why the example is used. The choice will come down to C or D. D might seem like it might be correct, however, the question asks why the specific example is used. D would have been correct had the question been, 'Why does the author use an example in this instance'. However, the question asks why he uses that *specific* example, so the answer must have some reference to the example used.

21. A is the correct answer. This is the only statement of fact. B suggests that it is 'plain to see', but this is subjective. C suggests that 'we need...'. This too must be an opinion. D is identified as an opinion by the inclusion of the word 'likely'. E is probably the most obvious opinion as the word 'should' is used.

Extract 7

22. B is the correct answer. A would not be in support of the hypothesis. C and E would all be one stage of extraction further away than B. C shows that murders are premeditated, which gives the murdered the opportunity to contemplate the consequences of their actions. E is not particularly compelling as it only questions free citizens, who might and probably do think very differently to murderers. D would eliminate some evidence in favour of the opposition, but does not put forward any evidence of its own. Thus, B is the only option that gives conclusive results.

23. D is the correct answer. A is explicitly mentioned and thus inference is not required. B is impossible to infer as the likelihood of change is not touched upon. C and E are simply incorrect. It can be inferred from the words which the authors choose to use that they have some level of disdain for those who oppose the death penalty. It almost suggests that they oppose it for the sake of opposing it, and that they will not take kindly (as they must 'come to terms with' it) to new information disproving them.

24. D is the correct answer. A is simply incorrect which in turn means that it cannot be E. B is too vague and does not really address why he uses that specific word. In this question, most students will be choosing between C and D. C is not quite right in that it does not address the word itself - it gives the author's motivation but it does not explain why he chooses to use the word 'admittedly'. The reason he chooses this word is to portray reluctance in showing that his argument may be flawed. D addresses this directly.

25. E is the correct answer. D is irrelevant and not touched upon in the piece at all. A, B and C are all discussed in the article as flaws in the research, but on balance, the main reason that the author suggests the studies have attracted criticism is due to the invalidity of the conclusion as a result of insufficient data. This is shown clearly as the author keeps returning to this point through the piece.

Extract 8

26. C is the correct answer. A reader of the piece could infer from what Saadia Zahidi has said that too much attention is spent on youth unemployment, but it is certainly not an assumption made by the authors, and thus the correct answer cannot be A. B is simply incorrect and not suggested in the piece at all. D could be inferred from the fact that they top the ranking, but is definitely not assumed by the authors. E is an assumption made by Klaus Schwab, but the authors make clear that it is not their opinion and thus not their assumption.

27. B is the correct answer. A is not really a description of the index, but rather an opinion of how useful the index will be for employers. C describes the focus of the index, but not the main purpose of the index itself. D, once again, shows how useful the index may be and not what it actually is. E describes only the part of the index that is discussed in detail throughout the extract, but is not the only information contained within it. B is correct as it accurately describes the whole index, and basically is a paraphrasing of the description given in the second paragraph

28. A is the correct answer. B and D are both explicitly stated, so there is no room for implication. C takes the idea a little too far, and it cannot be said that Schwab intended to imply that people are naive based on what he said. E again is far too extreme to have been implied. A is the correct answer as he suggests that one thing is important and then goes on the state 'we must look beyond campaign cycles and quarterly reports'. This wording implies that he considers that these are not worth spending time on.

Extract 9

29. B is the correct answer. B is the most relevant emotion. Kolby clearly states that the situation is fairly positive at the moment, but has the potential to get worse. B is the only option that takes into account both of these feelings.

30. A is the correct answer. B is mentioned as something that may become a problem in the future, it would not be right to infer from this that chytrid is always supervirulent, in fact it is shown that in some cases (such as this) it is not. C and D are both stated in the article and thus there is no scope for them to have been inferred. E cannot be correct as it would be too far to assume that chytrid is not harmful to any frog – the article shows that in this specific case, for some unknown reason, it is not currently harmful to the frogs in Madagascar. A is the correct answer as the statement 'It could mean we just caught it very early' coupled with the knowledge that the frogs had chytrid on their skin but are not sick would logically lead to an inference that it takes a while for chytrid to make the frogs sick.

31. E is the correct answer. A and C are explicitly mentioned in the final paragraph, whereas the reproduction centres mentioned are not restricted to only Madagascan frogs.

Extract 10

32. D is the correct answer. This is the only answer that conveys sadness, the rest all have an element of disapproval or anger.

33. C is the correct answer. A is incorrect as she clearly goes on to say that these experiences are normal for a child going through a parental separation. B is incorrect as she mentions she had trouble deciding whether to tell her friends about her new familial situation. D is incorrect as she demonises Baynes for making 'gross generalisations'. E is not even touched upon in the piece. C is stated in the paragraph beginning 'I had some really horrible times...'

34. A is the correct answer. This is in reference to where the author states that it is 'gut-wrenching, whatever the circumstances, to watch your dad pack his things into sports bags and leave the family home'. In order to make this statement, she has assumed that there is no possible scenario in which a child would be happy to see their father leave, or else she could not have used the phrase 'whatever the circumstances'. B is incorrect as the author makes no assumptions about how logical resentment is as a response, merely states that it was experienced within her family. C is incorrect as she only refers to Dolce and Gabbana as an example of how Baynes' article made her feel and does not suggest that they believe anything other than what they have said that they think. D is incorrect as she does not compare how she was after the break up to how she was before the break up. She notes how unhappy the new family dynamics made her but does not suggest that she would have been happier had the break up not occurred. E is the opposite of what she suggests in her discussion of Dolce and Gabbana's comments on IVF children.

35. C is the correct answer. The author uses the term 'poison dart' after discussing that Baynes' may well be justified in speaking about her family in a negative way. This suggests that the term relates to only the extension of the argument to include all gay parents. Both B and D must be incorrect, as it has been shown that the author does not dislike the whole of Baynes' argument. A is incorrect as there is no question of what is the most important part of her argument, merely the most damning part.

Extract 11

36. D is the correct answer. All of the answers are potential reasons that the author started the piece in this way

37. B is the correct answer. A is discussed in the first section. C and D are mentioned explicitly in Dr Pardo-Guerra's discussion of AI in the financial sector. E is discussed in relation to Netflix. B is the correct answer as the article discusses how there are certain elements of human behaviour that computers simply cannot predict.

38. A is the correct answer. C is purely incorrect. E is irrelevant. D is true but cannot be said to be the main reason that AI beings are inferior. A and B are both identified in the piece as being restrictions that AI beings have, however, A is discussed first and at further length than B, and so it must be concluded that it is the main reason that the author considers that AI beings are inferior.

39. B is the correct answer. B is the only thing that is actually *implied* by the author in the piece. She notes that 'computers are becoming king', and not just in the tech sector. This is an explicit explanation of A. C is described in her rhetorical question that it may become 'dull' and D is discussed in the final section, with 'And the good news is there is still room for the human touch - at least for now'

Extract 12

40. E is the correct answer. The US position is given in two halves, the first in the very first sentence, and the second at the end of the paragraph. Thus, the answer must be in two parts, so either D or E. E is the correct answer as C is more similar to the wording given in the article, and is marginally different to the meaning in A so both cannot be correct.

41. C is the correct answer. The author uses the phrase 'The FDA and USDA actually had the audacity to include in the draft position…'. The use of the word audacity is the inspiration for this question. Although all of A through to D could be correct the *best* description of the author's position must be C, given that C is a synonym of the word 'audacious' as explicitly mentioned in the piece.

42. A is the correct answer. C is clearly incorrect as it argues against the statement given. D is an 'ad hominem' argument and only undermines the person, not the person's argument. B proves that there is support for the US position but does not disprove the author's position. An article may well undermine the position if backed up by evidence, but that is unknown. The publishing on an article may not address the issues the author raises and as such would not be as damning as the disproval of the studies that that author identifies.

END OF SECTION

Mock Paper B: Section B

1. **Tennessee currently protects teachers who wish to teach children to explore the potential value of following creationism. Do you think that this is correct? Identify and analyse any legal problems that may arise in discussion of this law.**

For - Point
➢ Freedom of speech should apply to teachers as much as anyone else
➢ Students should be given the opportunity to choose between evolution and creationism.
➢ It is possible to teach creationism in a critical and analytical way so as to allow children to decide themselves on what they should believe.
➢ This in turn will help ensure that science class is about critical discussion rather than just imbibing facts - education instead of indoctrination
➢ The bill doesn't exclude evolution from being taught alongside creationism - it merely allows room for other theories to be taught.
➢ Evolution will still be taught in school, but the bill creates more options an opens up academic enquiry

For – Counterpoint
➢ This is not a simple freedom of speech issue.
➢ Teachers have freedom of speech in their own time, but during their working hours they have a responsibility to only teach what is relevant to their pupils.
➢ For example, teachers should not tell their student their own opinions on certain issues, nor should they divulge personal information - their job is to teach fact and curriculum
➢ Creationism can be taught in a religious context, but should be kept out of schools. Evolution should remain in science classes
➢ Allowing room for other 'theories' may not be an accurate portrayal of the situation. Impartiality and objectivity is something to strive towards, but creationism is not just another 'theory', it is active denial of scientific fact, and as such should not be fit within an academic environment
➢ Teaching purely evolution does not mean that students cannot discuss and analyse how well evolution fits with the facts - it can still give rise to a valid discursive environment.
➢ Furthermore, whilst we want children to be critical of what they are being taught, a large part of education is about the transferral of information and fact - it is widely accepted now that creationism does not have its roots in fact, and as such the teaching of it could confuse children.

Against – Point
➢ Teachers do not and should not have the freedom to teach whatever they want as fact
➢ Teachers should have to stick to a syllabus
➢ Children should be given a rounded education that includes knowledge and how to critically analyse that knowledge. However, including too many options in early education could confuse a child. In primary school children should be taught what science is, and then critical analysis could be incorporated at a later date.
➢ Children should have the right not to be misled by their teachers.
➢ Telling children that evolution and climate change are scientifically controversial is misleading, as there is no controversy amongst scientists
➢ Creationism is not science and should not be taught as such
➢ Science requires testability and falsifiability, this is satisfied by the theory of evolution, and not creationism

Against – Counterpoint
➢ It is never too early to teach students the value of questioning ideas and theories.
➢ Just taking the consensus view because it is the consensus view does not encourage students to question what they are told. We do not want students to just believe everything they are told, we want them to be critical and decide for themselves.
➢ Teaching children only one viewpoint could mislead them into thinking that the issue is fact and settled.
➢ Arguably, evolution cannot be tested either

2. **"There is a time and a place for censorship of the internet." Discuss with particular reference to the right of freedom of expression.**

For - Point

- It is the purpose of the government to protect its citizens from harmful sites
- Certain social media sites can be used to harm others, via cyber bullying resulting in psychological and even physical damage. More recently sinister terrorism recruitment sites have become more proactive and successful in recruiting young people. These results can and should be protected against, and the best way to do so is with censorship
- The government has a right to restrict free speech in certain circumstances
- People can express their beliefs and opinions but only where it does not impact other people's human rights. For example we restrict the expression of racial hatred.
- Protection of people from cyber harm would be a logical extension to the valid restrictions on the freedom of speech
- Other forms of media are already regulated
- Newspapers and books are subject to censorship, as are television and film. The internet is arguably more dangerous by virtue of its accessibility, so should be censored in the same way.
- Even sites that everyone thought were innocent have been used in devastating ways
- For example - rioters organising themselves via Facebook and instant messenger. London riots resulted in numerous acts of criminal damage, theft and violence towards others.
- In order to protect people the government must be able to censor sites that can be used disruptively

For – Counterpoint

- Social networking sites can be used malevolently, but they can also be used as a force for good. Surely we should not censor a site which has the potential to do good just because it also has the potential to have a negative impact.
- At what point do we decide that something's negative impact outweighs the potential value of its impact?
- Perhaps censorship *can* be justified but only where the site is objectively and completely harmful and thus the justification of protecting citizens can be used.
- Banning prejudice does not actually address the problem. Perhaps a better way to deal with the issues that arise on the internet, particularly those that arise on social media sites is to engage with it publicly, rather than trying to avoid it.
- The internet is different from other forms of media, in that it is a forum for free information and expression for individuals rather than corporations. Its value is in its freedom and to censor it is to change it irrevocably.
- As a proportion of the number of people who actually use these sites, those who misuse them are very small.
- The site was merely the platform and not the reason behind the disenchantment. The government will be continually chasing these people around the internet from site to site - censoring the site will not address the problem.

Against – Point

- Censorship is incompatible with the notion of free speech.
- It is hypocritical for a government to give people free speech and them ban certain areas of the internet.
- The internet is a free domain and should not be controlled by the government
- The internet is also international, so it is harder to identify which sites would fall under a particular country's laws and regulations
- Censorship may do more harm than good, particularly in terms of losing respect of the government
- For example, people are so dissatisfied with the dictatorial government in China that there has been growing public outrage and disenchantment with the authority of the government

Against – Counterpoint

➢ No rights are complete, there can always be a reason for restricting a right if it protects people.

➢ The relevant question is whether the government should restrict freedom of expression, but about how far into a person's autonomy the paternalistic state can venture under a justification of protectionism.

➢ Arguably however, if the information within a particular site has the capacity to harm people form a certain state, there is a reason and a justification for governmental intervention

➢ The government can weigh up whether the potential detriment caused by disenfranchisement is worth the benefit of protecting people online.

➢ For example, people don't like to pay taxes or for taxes to be raised, but the government has the power and the right to collect and raise taxes if they believe it to be for the good of the people.

3. **"The UK should codify its Constitution' Discuss.**

For - Point

➢ England will integrate better with Europe if we have a similar legal foundation.

➢ For protection. A codified Constitution would help prevent tyrannical leaders coming to power.

➢ For example, the German Constitution allows the German federal government to declare parties unconstitutional and dissolve them.

➢ A codified, written Constitution provides framework for a successful separation of powers. This is essential for a system of checks and balances.

➢ We already have a number of written document outlining different sections of our uncodified Constitution. It wouldn't take a lot to codify everything into one manageable document.

For – Counterpoint

➢ That suggests that we want to stay within Europe. There's a growing national movement of distrust of Europe and to change our constitution to suit Europe would exacerbate the problem.

➢ The likelihood of that happening is very small - we still have rules and regulations, they just aren't written down in one document.

➢ Does it though? We wouldn't necessarily be changing the way that our political system works - our executive would still be within our legislature.

➢ We shouldn't do something just because it might be easy - there needs to be a valid reason for doing it.

➢ Furthermore it may not be as easy as it seems - there are a number of elements of our constitution that are not enshrined in writing

Against – Point

➢ Codification of the Constitution eradicates the flexibility of our current uncodified Constitution.

➢ Sometimes extreme action can be necessary to protect people in new and changing circumstances (e.g. infringing on people's privacy rights in order to protect about new forms of terrorism). Having a codified constitution inhibits quick and efficient action.

➢ Too much emphasis on the letter of the law leads to an obsession with the text itself and can result in laws that don't quite fit with current popular opinion

➢ One danger of codifying a constitution is to give the judiciary too much power.

➢ One of the great things about Britain's common law system is that judges can adapt the law to suit real life situations. Codification could interfere with this and allow unelected officials to 'legislate from the bench'

➢ Codified constitutions are difficult to change

➢ A decision to codify our constitution would grant too much power to the government who is able to write it

Against – Counterpoint

➢ Part of the point of a constitution is to protect people's rights. It is important for a government to have to jump through certain hoops to legitimise their actions. If we don't have sufficient checks and balances the government could be free to abuse their powers.

➢ Codifying the constitution wouldn't necessarily mean more power for the judiciary. It can be written to protect the system that we already have.

➢ Furthermore the Constitution can always be changed if it isn't right

➢ Being resistant to change is the point of a constitution, it provides security.

4. **"The general trend towards the liberalisation of marriage undermines its religious basis." Discuss this comment with reference to the idea of abolishing marriage as a legal concept.**

For - Point

➢ Marriage from a religious perspective is between a man and a woman and the liberalisation of this - the rise of divorce and the legalisation of same sex marriage DOES undermine its religious basis, in that it's a union between a man and a woman in the eyes of god.

➢ This is not, however, necessarily a bad thing - one idea would be to abolish marriage as a legal concept - making a joint union for the purposes of taxation that would be between whoever wants to create that union. Marriage then would be left as a separate union in the eyes of god alone and not in the eyes of the law. Whatever happens to the legal union will be separated from the religious one.

➢ Religion and the law should be separated, especially given our aim to be a multi-cultural and multi-religious society.

➢ It is to be unfairly preferential to one group of people to integrate one religion with the law above all others

For – Counterpoint

➢ Why do we support relationships in the first place - partially mutual support that a long term commitment gives to someone, but also procreation, which is lost by extending it beyond heterosexual couples

➢ This may be the case if we were creating the law now, but Christianity is inherently connected to the English legal system by virtue of its history and fused past.

Against – Point

➢ Religious basis is changing - lots of Christians believe that the concept of marriage should be extended to fit in one with modern perceptions

Against – Counterpoint

➢ But this is only as a response to the law changing what the understanding of marriage is.

➢ Marriage is traditionally a religious concept and its being integrated with the law means that the religious community has lost control of what marriage is.

END OF PAPER

Mock Paper C: Section A

Extract 1

1. C is the correct answer as the writer mentions in the last paragraph that 'the time will come, and that the present legal conditions of wedlock will be altered in some way or other'. A is wrong because the writer says 'the time has not yet come for any such revolutionary change'. B is wrong as the writer says a change will occur in the future. D is wrong as the writer does not mention the fact that marriage will become 'temporary', this was suggested by Tolstoy in the second last paragraph. E was not mentioned by the writer in the passage.

2. D is the correct answer. Tolstoy mentions that 'the relations between the sexes are searching for a new form', meaning that marriage as a concept is evolving. Whilst A is not entirely wrong, it does not show that marriage is evolving as opposed to being outdated. Tolstoy does not mention B, C or E.

3. A is the correct answer as the writer mentions that 'people are always interested in matrimony' (B), 'marriage has been the hardy perennial of newspaper correspondence' (C), 'whether it be a serious dissertation…or a banal discussion' (D) and 'well-worn' (E). Only A is not mentioned by the writer.

4. A is the correct answer as hardy is not being used as a criticism – rather it is being used to show how marriage has been an 'unfailing resource'. Silly, banal, superficial and distasteful all connote a form of criticism.

Extract 2

5. E is the correct answer as the writer mentions that nature and nurture are 'so widely current'. B is wrong as Shakespeare was the one that created the terms, Galton merely popularised it. C is wrong as the writer is criticising Galton's use of the terms. D is also wrong as the writer says only environmental influences do not undergo much change.

6. D is the correct answer as the writer is referring to humans as a whole. A is wrong as the writer does not mention that 'Man' is the official name of a species, B is wrong as the writer is not trying to emphasise the importance, and E is wrong as it is the opposite of what the writer is trying to refer to.

7. B is the correct answer as the writer mentions that 'nature vs. nurture cannot be solved in general terms' and 'can be understood only be examining one trait at a time'. A is wrong as it is the opposite of B, and C, D and E do not reflect precisely what the author said in the passage.

8. C is the correct answer. Soil and climate are not being used as a contrasting pair, rather they are being referred to as related examples. The rest of the pairs are all used as contrasting pairs by the writer.

Extract 3

9. E is the correct answer as the writer states that biology 'is a generic term applied to a large group of biological sciences all of which alike are concerned with the phenomena of life'. A-D are not mentioned by the writer.

10. C is the correct answer. C was presented as a fact by the writer in the second paragraph, whereas A, B, D and E were all opinions of the writer and the writer did not claim that these were facts.

11. C is the correct answer as the writer states in the second paragraph that the biologist 'may not hope to solve the ultimate problems of life…'. A, B, D and E were all mentioned by the writer as issues that a biologist may resolve.

12. A is the correct answer as A is the only option that does not refer to the metaphor of a 'machine', whereas B-E all refer or relate to the 'machine' metaphor.

Extract 4

13. B is the correct answer as B is given only as an example by the writer, whereas A, C, D and E all introduce a new argument.

14. A is the correct answer as only A was used as a complete definition of fossils by the writer, B-D were not exhaustive definitions of fossils and E is also wrong as a result.

15. A is the correct answer as 'animals' and 'plants' were referred to as a collective, not as a direct comparison, whereas B-E were all used as direct comparisons.

16. B is the correct answer as the writer explains in the last paragraph that 'unless the conditions were such as to preserve at least the hard parts of any creature from immediate decay, there was small probability of it becoming fossilised'. A, C and D were not mentioned by the writer, and hence E is wrong as well.

Extract 5

17. A is the correct answer as the writer alludes to this in the first paragraph by stating that 'Sanskrit was the eldest sister of them all'. Hence all the other options are wrong.

18. C is the correct answer as 'human mind' is being used literally in this passage, whereas the other options are being used metaphorically.

19. B is the correct answer as the writer alludes to this by stating that 'as Sanskrit stepped into the midst of these languages, there came light and warmth and mutual recognition'. A, C and D were not mentioned by the writer, and hence E is wrong as well.

20. C is the correct answer as the writer alludes to this by stating that 'take the words which occur in the same form and with the same meaning…'. A, B and D were mentioned by the writer but were not suggested as precise ways of understanding a language but merely examples, and hence E is wrong as well.

Extract 6

21. E is the correct answer. The writer suggests this by stating that 'we must teach nature…but we must not, in so doing, wean still more from, but perpetually incite to visit, field, forest, hill…'. A-D do not precisely describe the writer's argument put forward in the last paragraph.

22. A is the correct answer as the writer states in the first paragraph that 8-12 is where the acute stage of teething is passing, not when it is starting. B-E were all mentioned by the writer as happening from 8-12.

23. The correct answer is A as Rosseau states that we should 'leave prepubescent years to nature', and this is agreed on by the writer in the last paragraph. B-D are wrong as Rosseau argues that children should grow up in the wild independently, whereas the writer argues that we should be supplementing reading with outdoor activity. Hence E is wrong as well.

24. E is the correct answer as E provides the best summary of the author's argument in the last paragraph, whereas A-D do not reflect the writer's argument.

Extract 7

25. C is the correct answer as the writer mentions that 'we must still be permitted to doubt if the time has even yet arrived…can be appreciated in accordance with his just value'. This suggests that the writer thinks Chopin's work may be underrated, and hence the other options are wrong.

26. B is the correct answer as this was alluded to by the writer in the second paragraph where he states 'is it not equally true that…are never recognised as prophets in their own times?'. Hence, the other options are wrong.

27. D is the correct answer, as 'composition' is being used in a more literal sense as compared to the other options.

28. C is the correct answer as the writer alludes to this when he says that 'musicians who do not restrict themselves within the limits of conventional routine…more need than other artists of the aid of time'. The other options are a misreading of the writer's argument.

Extract 8

29. C is the correct answer as the writer comes off as being critical about the argument that morality and religion are related. The other options do not fit the tone the writer has adopted.

30. E is the correct answer as all the philosophers mentioned agree about the lack of connection between morality and religion.

31. A is the correct answer as only A argues that morality and religion are related, whereas the other options disagree.

32. A is the correct answer as 'backward' is not used negatively in the first paragraph, it simply refers to the past, whereas the other options are used negatively.

Extract 9

33. E is the correct answer as the writer states this in the last paragraph by saying 'Australia was the last part of the world to be thus visited and explored'. Hence the other options are wrong.

34. C is the correct answer as this was stated in the last paragraph when the writer mentioned 'scarcely any one cared to run the risk of exploring it'. The other options were not cited as reasons for not exploring Australia.

35. D is the correct answer as this was cited in the first paragraph where the writer states that 'they believed the man who could penetrate far enough would find countries where inexhaustible riches were to be gathered'. The other options were not cited as reasons for venturing into the unknown.

36. A Is the correct answer as only A was not mentioned as a factor behind the ability to travel in the passage. A was cited as a motivation instead.

Extract 10

37. D is the correct answer, as the writer alludes to this in the last paragraph by saying that 'but the explanation of the phenomenon and the name that is given to it matters little…it constitutes…the principal material of which we shape our dreams'. The rest were mentioned by the author as periphery matters instead of the heart of the question.

38. E is the correct answer as the writer merely states some examples of opinions in the last paragraph, but does not show that there is a consensus. A-D were all cited as examples of differences in opinions.

39. A is the correct answer, as the writer only agrees with the fact that we see colours when we close our eyes, but does not attempt to agree on what are mentioned in B-D. Hence E is wrong as well.

Extract 11

40. C is the correct answer, as this was stated in the last paragraph when the writer mentioned that 'pathological lying is very rarely the single offense of the pathological liar'. The other options were not mentioned by the writer.

41. D is the correct answer as this was mentioned in the third paragraph when the writer states that 'where the individual by the virtue of language ability endeavours to maintain a place in the world which his abilities do not otherwise justify'. The other options are a misreading of what the write explains using Case 12.

42. C is the correct answer, as the writer is providing an academic analysis of the condition of pathological liars, as opposed to adopting a different tone suggested by the other options.

END OF SECTION

Mock Paper C: Section B

1. Is social media damaging for teenagers?

Argument (For)
- Social media is damaging to teenagers because it lowers their self-esteem.
- Social media is flooded with pictures of good-looking people and unrealistic body images that perpetuates a need for teenagers to pursue that image.
- Excessive consumption of social media has been linked to lower self-confidence and a higher rate of depression amongst teenagers, especially amongst girls who are more likely to be affected by a lack of confidence in their body and looks.

Counter-point
- Social media does not always necessarily lead to lower self-esteem and perpetuate an unhealthy body image.
- Social media can be used to share uplifting posts, encouraging messages, and it is also a good platform to spread awareness of certain issues.
- Social media is a double-edged sword and can be used for good or bad, and if utilised correctly it can be useful in increasing the awareness of important issues such as fighting against eating disorders and mental disorders.

Argument (Against)
- Social media does more good than harm to teenagers as it allows teenagers to express themselves more and provides an outlet for them to showcase their creativity.
- For example, some teenagers may showcase their photography skills through Instagram, or video-editing and producing skills through YouTube, and they can receive support easily from their peers through such social media website.

Counter-point
- Even though such social media platforms provide an opportunity for teenagers to showcase their creativity, they are more often than not used to showcase frivolous material and perpetuate low-level humour.
- For example, YouTube is flooded with distasteful videos and the latest drama involving Logan Paul shows how the content available in YouTube usually leaves much to be desired.

Argument (For)
- Social media is damaging to teenagers as it leads to an obsession with popularity and creates an unhealthy culture of emphasising the need for validation.
- Teenagers are increasingly obsessed with having more followers on Instagram and having more likes for their pictures, and this creates a situation where many young girls are posting revealing pictures in order to garner more followers and likes, when they are too young to understand the consequences and repercussions of doing so.

Counter-point
- Not all teenagers see social media as a platform for increasing their popularity and receiving validation from strangers.
- Many teenagers use social media as an efficient platform for connecting with and keeping in touch with their friends and families, especially when they are geographically separated.
- Social media platforms such as Facebook provide a good method for teenagers to easily stay in touch with their friends and families and be updated about their lives.

Argument (Against)

➢ Social media is potentially very dangerous for teenagers due to a rise in the number of sexual predators utilising social media to take advantage of teenagers.

➢ There have been many reported cases of sexual predators posing as someone on Facebook in order to establish contact with a teenager, and later on coerce them into doing sexual acts or sexually grooming them.

➢ The lack of adequate safeguards and parental control over a teenager's use of social media makes it potentially very dangerous.

Counter-point

➢ The use of social media platforms will not be dangerous as long as adequate safeguards are put in place.

➢ For example, parental control can be implemented on social media sites, and social media companies are taking down false accounts or suspicious activities and posts in order to reduce incidents of sexual predators searching for victims on social media sites.

2. To what extent should journalists be responsible for 'fake news'?

Argument (For)

➢ Journalists have a huge influence on the public and should be responsible for ensuring that whatever they publish is accurate and not misleading.

➢ It has been argued that misleading news have led to more people voting in favour of Brexit than there would be had certain news outlets been more responsible in portraying accurate facts that were not sensationalist and misleading.

➢ Journalists should be held responsible should they fail to fulfil their duties and lead to the public being misinformed.

Counter-point

➢ Journalists are merely responsible for conveying facts and opinions, it is up to the public to be discerning in reading the news and forming an opinion themselves.

➢ Different news outlets are allowed to have different political opinions – for example, The Guardian is known to be left-wing in general whilst The Daily Mail tends to be more right-wing.

➢ Journalists should not be held responsible if someone fails to read critically and takes what the newspapers say as the gospel.

Argument (Against)

➢ Making journalists responsible for 'fake news' will be a serious impediment to the freedom of speech and the need for news outlets to be independent and allowed to voice their opinions and independent thoughts.

➢ Trump's allegations that journalists are irresponsible and often write 'fake news' in order to incite hate fails to take into account the fact that freedom of speech is a fundamental right behind journalism and without such freedom the public will not be able to learn about both sides of the argument.

Counter-point

➢ As important as freedom of speech is, such freedom should be exercised responsibly and journalists should be held responsible if they fail to verify facts and only focus on increasing readership by posting sensationalist and controversial news.

➢ The general public may not always be very discerning or have the time or initiative to read widely from different sources, and they may easily believe the first thing they read and be affected as a result.

➢ Hence, journalists have an inherent duty to ensure that what they are reporting are factually accurate as much as possible.

Argument (For)

➢ Fake news is potentially highly dangerous to society as journalists are capable of inciting hate in the public and this may lead to a division amongst different religious and racial groups, which leads to social disintegration.

➢ Journalists need to exercise a level of responsibility in ensuring that they are not publishing news that denounce a certain racial or religious group, and ensure that they take into account the sensitivity and delicacy of the situation before posting any news article that might incite hate.

Counter-point

➢ The journalists job is simply to report on facts and opinions and they should not be treated as politicians and be a moral arbiter in deciding what news should be published and what should not be in the interest of the public.

➢ It is important for journalists to publish any news or facts that the public deserves to know instead of being limited by cultural sensitivities or controversial topics, as without such publications the public will be deceived about what is going on and will not be properly informed about current affairs.

Argument (Against)

➢ It is not appropriate to make journalists liable for 'fake news' as it is ultimately the readers fault for not verifying the source of the article or cross-referring to other publications.

➢ Journalists are not public figures and they are not expected to be held to a standard and scrutinised the same way that we would do with elected politicians or academics with an established standing.

➢ The job scope of a journalist does not entail making them responsible for 'fake news' – rather, it is up to the public to be discerning about what they read and choose to believe.

Counter-point

➢ Due to the important nature of news and the increasing importance of journalists in shaping public opinion especially during revolutionary political occasions, they should likewise be held to a high standard and be scrutinised if they fail to fulfil their duties and let down the public as a result.

➢ We should be elevating the role of a journalist and ensure that they are equally held accountable for misleading the public as would a politician be.

3.   Limitations should be put in place for scientific discovery. Discuss.

Argument (For)
- Certain limitations should be placed on scientific discovery due to the potential ethical and moral issues that may arise if scientific discovery was unlimited.
- For example, human cloning poses thorny ethical issues and despite the perceived benefits such discovery may bring to mankind, it carries potentially unethical outcomes such as flouting human rights and creating an unprecedented morality issue.
- Genetic engineering is another area where even though the potential benefits are great, but it may lead to unethical outcomes such as 'designer babies' or only allowing the rich to alter their genes beneficially.

Counter-point
- Scientific discovery is limitless and if limitations are sought against scientific discovery, mankind will not be able to progress quickly and we will not be able to resolve many problems plaguing society nowadays.
- For example, research on GMO food can potentially resolve nutritional deficiencies in developing countries, but due to religious backlash, such research has not received adequate funding and regulatory approval.
- Genetic engineering is also an area that can possibly lead to cures for certain genetic diseases such as Parkinson's disease, but limitations put in place will severely slow down any progress that can be made.

Argument (Against)
- We should not place limitations on scientific discovery as there can never be a consensus on certain ethical, morality and religious issues, and the progress of society and technology should not be impeded by the opposite of certain religious or fundamentalist groups.
- For example, stem cell research has proven to contain immense potential in helping to find cures for genetic diseases, yet certain religious groups are against such research as they deem an embryo to be a human life and thus extracting stem cells from an embryo is deemed against their religious values.
- Such limitations are damaging and can slow down the discovery of important solutions to debilitating diseases that plague many individuals.

Counter-point
- Even though the need to find a cure for genetic diseases is pressing, we cannot ignore morality or ethical issues that can arise from such research.
- Certain controversial research such as stem cell research, if left unlimited, may lead to a slippery slope where scientists are not afraid to do anything in the name of advancing science.
- This can lead to extreme situations where human testing is allowable in the name of progressing science.
- Not all methods of advancing science are ethical, and we need to draw a line between what is permissible and what is not.

Argument (For)
- Limitations on scientific research is needed because there are absolute moral rights that should not be infringed and science cannot be used as a reason for flouting such rights.
- For example, torturing a human or killing a human for the sake of scientific discovery will never be permissible, even if it is in the name of medical discovery, as the right to live is an absolute human right that should never be infringed.
- In less clear cut areas, such as stem cell research or genetic engineering, the pros and cons need to be weighed carefully in deciding whether certain research should be allowed in order to improve technologies, and scientists need to always ensure that they have adequately considered the ethical issues arising before conducting research.

Counter-point
➢ Scientific discovery can be used to save and improve millions of lives and if limitations are put in place to hinder such discovery, we are essentially allowing millions of people to die from a lack of cure to their sickness or exposure to dangerous elements.
➢ Scientific discovery needs to advance quickly in order to provide a solution to numerous fatal diseases that have no cure as of now due to the limited nature of scientific discovery, and mankind will benefit as a whole with scientific discovery being made unlimited.

Argument (Against)
➢ Scientific research is inherently expensive, and scientists need to be incentivised in order to produce quality research and unravel the complexities of science.
➢ Instead of placing further limits and hindering scientific research further, we should be doing more to catalyse scientific research instead, such as having more funding opportunities and initiatives for scientists to engage in their research.
➢ More people should also be willing to sacrifice in the name of benefiting science and discovery, such as opting to donate their organs when they die so that such organs can either be used to save another life or be used for important scientific research.

Counter-point
➢ Not all scientists may be engaging in research and development for altruistic purposes, and some may harbour bad intentions or may be purely looking at making a profit.
➢ If we leave scientific discovery unlimited, this creates a huge risk of scientists abusing their powers and benefiting from their research more than what the public stands to gain as a result.
➢ They may create a cure for HIV, but acquire rights over the cure and sell the drug at such a high price in order to make a profit that only the rich can afford such a cure.
➢ Hence, limits need to be put in place in order to ensure that scientific discovery will benefit mankind.

4. Too many students are going to University. Discuss.

Argument (For)

➢ Too many students are going to universities, and many of these students end up unemployed, or being over-qualified for jobs that place a greater emphasis on skills and know-how.

➢ Having too many university graduates causes a mismatch between the demand and supply of jobs, and students become increasingly disenfranchised after spending many years in university and paying a substantial amount of tuition fees, only to find out that they cannot get the jobs that they want.

➢ Many degrees are also irrelevant to the job market, and students end up being ill-suited for the kind of jobs that require employees.

Counter-point

➢ Going to university is more than just preparing students for work.

➢ University also provides a good opportunity for students to mature and develop, and acquire an important network of friends.

➢ Furthermore, university not only teaches academic content but also teaches crucial soft skills that are highly demanded in the workforce such as teamwork, leadership skills, communication skills, writing skills and negotiating skills.

➢ University provides a good opportunity in terms of training a student's mind and a more highly-educated populace is desirable because it allows the populace to be more discerning and well-informed.

Argument (Against)

➢ University education is important as it provides a good opportunity for students to discover what they enjoy doing and what they are good at, and gives them a safe ground for developing their potential and capabilities.

➢ Even though not all students go on to work in a field that is relevant to their degree, university education is important nonetheless as it trains students to think critically and be able to evaluate information and convey it in a clear manner – something that is highly emphasised on in the workforce.

Counter-point

➢ University education is becoming increasingly unaffordable and despite the perceived benefits of university education, too many students do not take university seriously and end up wasting their time in university and racking up a huge amount of debt as a result without any tangible benefits upon graduating.

➢ This problem is particularly acute in lower-ranked universities that have less stringent academic requirements and rigour, which results in many students not learning much during their university education and incurring unnecessary debt.

Argument (For)

➢ It has come to a point where universities are producing degrees en masse and there is too much emphasis on quantity over quality.

➢ Too many degrees being handed out results in the value of a degree being depressed as a result, and no one will take a degree seriously if everyone has a degree.

➢ Universities are exploiting the demand for degrees by having lax standards and low entry requirements, which results in students who are not academically-inclined being able to enrol into a university and get a degree that is not of much use in the workforce.

➢ There should be a greater emphasis on quality over quantity and there should be greater focus on ensuring that the education being given is of a high standard, as opposed to being mass-produced.

Counter-point

➤ Everyone has a basic right to education, and university education should not be limited to the 'elites'.

➤ The problem with reserving university education only for the top few students is that this results in greater inequality, where poorer students may fail to get an opportunity to have a university education just because they could not afford better teaching or resources in order to do well for their exams.

➤ Education is the great leveller of society and we should not deprive students of education by making university education reserved only for the elites, hence university education should be more accessible.

Argument (Against)

➤ University places should not be restricted to the top few universities as even though some universities may not be as highly-regarded or well-ranked as others, they may be able to provide good education in terms of achieving high student satisfaction and providing greater attention between teachers and students.

➤ University education should be made available for students who have a curiosity to learn and improve themselves, and we should not deprive such students of university education just because they are not as academically-inclined as the 'elite' students'.

Counter-point

➤ Society should place less emphasis on the need for a degree in order to survive in society.

➤ There are many countries with far fewer university graduates but high employment rates, such as Germany where many students opt to go into apprenticeships instead in order to acquire skills that cannot be taught in universities.

➤ We should not place so much emphasis on ensuring that everyone gets to go university, as not everyone is cut out for academic and there are many people that are talented in other areas.

➤ Alternative education and training should be given more emphasis in order to diversify our talent pool and ensure that students are getting the skills required for jobs and not just going to university for the sake of it.

END OF PAPER

Mock Paper D: Section A

Extract 1

1. D is the correct answer as the writer mentions throughout the passage that the laws only benefited certain classes at the expense of the public. A-C were not mentioned in the passage; hence E is wrong as well.

2. C is the correct answer as this was alluded to in the sentence 'to render the articles which such classes deal in or produce dearer than they would otherwise be if the public was left at liberty to supply itself with such commodities'. The other options are not alluded to in the first paragraph.

3. A is the correct answer as the writer states that patents were initially meant to result in 'exclusive exercise of certain trades or occupations in particular places'. The other options were mentioned as the unintended side-effects of patents.

4. E is the correct answer as the writer does not agree with any of the 'excuses' put forward on behalf of imposition of high duties.

Extract 2

5. A is the correct answer as the writer mentioned at the start of the passage that camouflage is used in 'modern warfare' which refers to humans, as opposed to animals. The other options all refer to animals.

6. E is the correct answer as all the options were mentioned by the writer to describe the two different forms of camouflage developed in nature.

7. E is the correct answer as all the animals mentioned adopt a similar method of camouflage – blending in with their surroundings.

8. C is the correct answer as the writer is providing an analytical review of camouflage in this passage, as opposed to the other tones mentioned.

Extract 3

9. C is the correct answer as the writer capitalises the word 'Nature' as he is referring to it as a distinct entity, as opposed to what the options suggest.

10. A is the correct answer as 'Universe' is used in the literal sense in this passage, whereas the other options are all used metaphorically.

11. A is the correct answer as it is presented as a fact in the first paragraph, whereas the other options are all presented as opinions.

12. B is the correct answer as the writer alludes to this in the last paragraph by mentioning that 'our widest and safest generalisations are simply statements of the highest degree of probability'. The other options were not alluded to in the passage.

Extract 4

13. C is the correct answer as the writer mentions in the second paragraph that selection for breeding differs between artificial selection and natural selection.

14. E is the correct answer as all of the examples were mentioned for examples of 'struggle for existence' behind the natural selection theory.

15. C is the correct answer as only 'avoiding accidents' was not mentioned by the author as a factor behind the selection of the best for reproduction.

16. A is the correct answer as only A was used as an example of artificial selection, not natural selection.

Extract 5

17. D is the correct answer as 'dropping out of one joint' was stated as the main effect of brachydactyly, whereas the other effects mentioned were all side-effects.

18. A is the correct answer as the writer explains in the first paragraph that one affected parent is enough to produce an offspring with brachydactyly, hence only two normal parents will not produce an offspring with the condition.

19. D is the correct answer as the writer explains in the third paragraph that 'in each individual there is a different set of modifying factors or else a variation in the factor'. The other reasons were not stated by the writer as a reason for the difference in seriousness.

20. B is the correct answer as the writer alludes to this in the last paragraph by concluding that 'it must be recognised that every visible character of an individual is the result of numerous factors'. The other options were all listed as examples to support his argument or are wrong.

Extract 6

21. A is the correct answer as the writer states it originated from Gautama of India in the first paragraph.

22. A is the correct answer as the writer states in the first paragraph that 'no further steps' were taken after the emperor set up the image of the Buddha, whereas the other options all contributed to the rise of Buddhism.

23. D is the correct answer as the other options were not given as reasons for the decline of Confucianism.

24. D is the correct answer as this was explained by the writer in the third paragraph where he says that 'this dream story is worth repeating because it goes to show that Buddhism was not only known at an early date…'.

Extract 7

25. C is the correct answer as only C is being used as a matter of fact, the other options are all presented as the writer's opinion.

26. C is the correct answer as this passage is being addressed to universities in general to encourage them to make students study more sciences.

27. C is the correct answer as the writer argues throughout the passage that the sciences are being neglected by students.

28. B is the correct answer as the writer is critical about the lack of education in science in England.

Extract 8

29. C is the correct answer as only C is presented as a fact, whereas the other options were all presented as opinions.

30. E is the correct answer as all the options are being used with a negative connotation in the passage.

31. D is the correct answer as the writer comes across as highly opinionated in the passage as opposed to the other options suggested.

32. C is the correct answer as the writer alludes to this in the last paragraph by stating that 'women's steady march onward, and her growing desire for a broader outlook, prove that she has not reached her normal condition'.

Extract 9

33. E is the correct answer as all the examples mentioned were not the writer's own opinion, rather they were viewpoints of other people.

34. D is the correct answer as the writer states in the second paragraph that analysing Muhammedanism helps us to be able to critically evaluate Christianity.

35. C is the correct answer as only C would be a neutral party who is merely analysing the religions to understand them better as opposed to strengthening their conviction in a particular religion.

36. D Is the correct answer as the writer mentions in the second paragraph that 'the evolution of such a system of belief is best understood by examining a religion to which we have not been bound'.

Extract 10

37. C is the correct answer as the writer is claiming the opposite of C as the reason why he chose yeast as a topic.

38. A is the correct answer as 'succulent' is not being used to describe the formation of yeast, but rather the fruits that are used to form yeast.

39. A is the correct answer as facts and phenomena are being used as a collective and not as a form of comparison.

Extract 11

40. D is the correct answer as both A and B are being illustrated by the examples used by the writer.

41. E is the correct answer as all the examples given are raised as differences between love in mankind and the animal kingdom.

42. B is the correct answer as the writer explains in the last paragraph that there is a scientific reason why good-looking people tend to be favoured.

END OF SECTION

Mock Paper D: Section B

1. Should the voting age be reduced?

Argument (For)
➢ The voting age should be reduced because policies that are being decided on now will affect young people the most in the future.
➢ If young people do not get the chance to vote, they do not have a say regarding who should be elected and what policies will be implemented as a result.

Counter-point
➢ The voting age should not be reduced as young people are not always sufficiently politically-aware in order to make an informed decision and the fact that voter turnout is usually low amongst the youngest age group shows that it is unnecessary for the voting age to be reduced as young people do not know how to exercise their voting rights responsibly.

Argument (Against)
➢ The voting age should not be reduced as young people are easily influenced and swayed by external opinion and tend to favour populist policies that may well harm their long-term interests.
➢ For example, young people may be overly-myopic and only focus on issues that provide a tangible benefit to them in the short term, such as lower tuition fees, without considering issues that might affect them in the long run.

Counter-point
➢ The voting age should be reduced as there is a greater emphasis on political education at a young age these days and many policies being proposed by the political parties have a direct and actual impact on the younger generation.
➢ Young people are well-equipped and well-informed to exercise their votes responsibly, and should have a say regarding which policies will benefit them the most.

Argument (For)
➢ The voting age should be reduced as it is unfair that the older generation gets to vote on certain policies that will benefit them at the expense of the younger generation, even though the younger generation will be the ones who have to suffer the consequences of such policies in the long term.
➢ For example, the older generation may prefer policies such as increasing pension funds, which will mean higher taxes on the younger generation, or bringing back compulsory national service, which will only be applicable for the younger generation.

Counter-point
➢ The voting age should not be reduced as every generation faces the same problem of not getting a say in the policies being implemented until they are of sufficient age and maturity to make an informed-decision.
➢ The older generation also once did not get the chance to vote on policies to benefit them, and they are fully-qualified to vote on policies that they think are beneficial for the country with their life experiences and maturity.
➢ Many older generations will also have children and it is not always true that they will vote on policies that will benefit them at the expense of the younger generation.

Argument (Against)

➢ The voting age should not be reduced as young people simply lack the maturity and political thoughtfulness to be able to decide which political party will serve the nation's interests.

➢ Surveys overwhelmingly suggest that young people are ignorant when it comes to politics and have an overly-simplified or non-existent knowledge of current affairs and the nuances in views between different political parties.

➢ They are too easily swayed by sensationalist and populist arguments made by politicians and they tend to vote without putting in much thought and analysis.

Counter-point

➢ The voting age should be reduced as young people get increasingly-educated nowadays and many young people are engaging in active discussion about politics online and are discerning when it comes to analysis political news from multiple sources and forming an independent opinion based on the information available.

➢ The fact that the political inclinations of young people tend to show a marked difference with the older generation simply shows that times have changed and the priorities of young people going forward are vastly different from the older generation.

2. Should tuition fees be reduced?

Argument (For)

➢ Tuition fees should be reduced as high tuition fees are hindering low-income students from considering higher education, resulting in a lack of social mobility and a denial of the right to education based on income-levels.

➢ Tuition fees have become increasingly unaffordable in recent years, and this has become a huge deterrent for students who are not from well-to-do families from pursuing higher education, even though they may be academically-capable of doing so.

Counter-point

➢ Tuition fees should not be reduced as tuition fees are needed for universities to remain competitive and hire the best professors and have the best resources for students in order to ensure quality teaching.

➢ If tuition fees were reduced, even if more students will be enticed to enrol in university as a result, it will mean that all students will end up receiving sub-standard education with the lack of resources and financial-backing needed.

Argument (Against)

➢ Tuition fees should not be reduced as there is already the student loan scheme in place which ensures that students only need to start re-paying their loan upon graduation if they earn a certain amount of income.

➢ This helps to ensure that low-income students will still have easy access to universities, and they will only need to pay off the loans if they manage to secure a job that pays enough for them to repay the loan.

Counter-point

➢ Tuition fees should be reduced as the high level of tuition fees being charged causes many students to be heavily-indebted upon graduation, causing an immense financial burden to them and provides a disincentive for many students to consider university in the first place, even if they qualified for university academically.

Argument (For)

➢ Tuition fees should be reduced as universities already have the benefit of huge donations and grants being given by alumni and research organisations in order for them to survive and provide quality education, it is unfair to charge students exorbitant tuition fees and create a high barrier to entry based on financial means.

➢ The argument that tuition fees are needed to sustain a university is weak in this day and age when the bulk of a university's revenue comes from research grants and legacy donations.

Counter-point

➢ Tuition fees should not be reduced as not all universities have the benefit of large grants and legacy donations.

➢ Only the top-ranked universities and the most prestigious universities will be able to attract sufficient funding and donations from successful alumni and be able to survive even without charging high rates of universities.

➢ Many other universities will struggle to survive without charging sufficiently high tuition fees, and reducing tuition fees might be counter-intuitive and result in less university places being available.

Argument (Against)

➢ Tuition fees should not be reduced as too many students are going to university for the sake of it and do not take their degree seriously.

➢ It is well-known that many students in less rigorous courses and universities only treat university as an extra 3-4 years of socialising and partying.

➢ Taxpayers should not have to subsidise these students when they are not doing something of value and tuition fees should remain as it is in order to act as a deterrent for students who are not naturally inclined for university education in the first place.

Counter-point

➢ Tuition fees should be reduced as university not only provides academic teaching, it also provides important soft-skills and allows students to figure out what they are good at and what they want to do in life.

➢ Studies have shown that university education is highly beneficial in terms of a person's success later on in life as well as their earning capacity.

➢ Hence, we should not deny this opportunity to many students who might otherwise be put off by the high tuition fees.

3.   **Tourism does more harm than good. Discuss.**

Argument (For)
➢ Tourism does more harm than good as it leads to issues such as gentrification of towns that are heavily visited by tourists.
➢ Popular destinations such as Paris and Bali have experienced this phenomenon whereby due to the heavy inflow of tourists every year, more and more shops are opened for the benefit of tourists, placing increasing competition on local shops and causing them to close down.
➢ This leads to an erosion of culture and to a process of gentrification whereby towns with significant culture start to become more standardised with similar shops to cater to the taste of tourists.

Counter-point
➢ Tourism provides more benefit than harm as tourism is a significant contributor to economic growth in many countries, and has led to the creation of many jobs.
➢ For example, destinations such as Hawaii are heavily-dependent on tourism for economic growth, and tourists provide a lot of potential economic benefits for the locals, such as sustaining certain trades that might die off without the demand from tourists, as well as sustaining the hospitality industry that is highly lucrative for the home country.

Argument (Against)
➢ Tourism provides more benefit than harm as tourism allows a greater awareness of the cultural differences between countries and acts as a form of diplomacy between different countries by increasing the level of interaction between people of different nationality.
➢ Gap years are an example of students being able to experience the culture of different countries if done responsibly and tourism helps open up certain countries to the rest of the world and allow people from all over the world to appreciate the beauty of a country.

Counter-point
➢ Tourism is more harmful than good as many tourists do not visit a country with an open-mind and there are many instances of tourists being an annoyance to their host country and causing discontent amongst the locals.
➢ For example, there are stories of tourists vandalising cultural artefacts in the host country, or tourists disrespecting the local culture of a country by dressing inappropriately or taking inappropriate photos. Students on gap years also often do so irresponsibly by partaking in activities that exploit the host country and are unethical without realising it.

Argument (For)
➢ Tourism causes more harm than good as it leads to environmental degradation, erosion of cultural sites and also contributes to a larger carbon footprint.
➢ Many tourists visit cultural sites without being responsible and helping to conserve such sites, and a high human traffic often leads to an acceleration of the degradation of heritage sites and an erosion of the natural beauty of a cultural site.
➢ Tourism also encourages a large amount of travelling by airplanes and trains which contribute significantly to the rising global carbon footprint.

Counter-point
➢ Tourism does more good than harm as tourism is becoming increasingly more sophisticated these days and many cultural and heritage sites are protected in order to preserve them, allowing tourists to fully enjoy the beauty of such places without destroying them.
➢ For example, there are plenty of UNESCO heritage sites that are being actively protected and works are often done in order to restore such sites and maintain them. Technology has also increasingly allowed more environmentally-friendly travel options to be made available, such as electric cars and bullet trains.

Argument (Against)

➢ Tourism is more beneficial than harmful as it closes the distance between countries and allows more individuals to be exposed to different cultures and practices that they would not be able to experience back at home.

➢ This fosters a more liberal and understanding mindset and well-travelled individuals will tend to be more well-informed, culturally-aware and accepting of differences due to their interactions.

➢ If tourism was reduced, more people would be insular and fail to understand differences between cultures due to their limited exposure and interactions with different people.

Counter-point

➢ Tourism causes more harm than good as it provides an artificial image of a country to tourists and merely scratches the surface, without allowing tourists to fully understand the way of living and the difficulties faced by a country.

➢ For example, a tourist will not be expected to have fully-immersed in the local culture if they spend a week in a luxury resort, without any interactions with the locals, and only visit tourist attractions that are catered and commercialised for tourists.

➢ This leads to great misconceptions about countries and does not allow a greater cultural awareness.

4. **Banning the wearing of a headscarf in the public sector is discriminatory. Discuss.**

Argument (For)

➤ Banning the wearing of a headscarf in the public sector is discriminatory as this causes a natural barrier for religious Muslim women who wish to enter the public sector, but are not allowed to exercise their freedom of expression of religion accordingly.

➤ This ban hence deters Muslim women from applying for public sector jobs especially if their religion is a fundamental part of their identity and they strongly believe in the requirement to wear a headscarf in public.

Counter-point

➤ Banning the wearing of a headscarf in the public sector is not discriminatory if it is done for a legitimate reason, and the ban extends to other forms of religious expression as well.

➤ For example, a legitimate reason will be that the public sector has to appear as secular and must not be affiliated with any religious belief.

➤ Hence, if headscarves are banned because of their affiliation with Islam, along with other forms of religious clothing such as wearing a cross for Christians or wearing a turban for Sikhs, such a ban will be justified.

Argument (Against)

➤ Such a ban is not discriminatory as it is within the control of applicants to decide whether or not they are suitable for a job and are able to adhere to the dress code.

➤ Many jobs require specific dress codes, such as the military and uniformed services, and we should not relax our standards just to cater to one particular group especially if there is a strong reason for the dress code, such as uniformity and symbolism.

➤ Hence, as long as the ban is not specifically directed towards a particular religion, but rather a general requirement of specific clothing that should be worn, it is not discriminatory.

Counter-point

➤ Such a ban is discriminatory as even though the ban is not a direct ban on Muslim women from serving in the public sector, they are indirectly discriminated against and suffer a disproportionate effect as they as a collective group are more likely to be required to wear a headscarf as per their religious norms.

➤ Having such a ban will result in much fewer Muslim women being able to apply for the job as it will go against their religious convictions, and this constitutes indirect discrimination against them.

Argument (For)

➤ Such a ban is discriminatory as there is no legitimate reason why the wearing of a headscarf will interfere with the job of working in the public sector.

➤ If wearing a headscarf will not interfere with the ability of the worker to complete her job, such a ban will not be justified and will be discriminatory.

➤ This is in contrast to jobs where eye-contact may be important, such as a teaching job for example, where in that case a burka might be unsuitable as it reduces the amount of eye contact between the teacher and student and interferes with the student's learning abilities.

Counter-point

➤ Such a ban is not discriminatory as there is an expectation in the public sector that the workers are representing the state, and there is a legitimate reason for expecting that a headscarf should not be worn, especially if it might cause discomfort amongst certain members of the public or if it will give rise to the wrong idea about the public sector.

➤ These are precautionary measures that have to be taken for a sensitive sector such as the public sector and candidates should be aware of these additional requirements before applying for such jobs.

Argument (Against)

> The ban is not discriminatory as such a ban can be justified if it is restricted to the more public-facing and sensitive roles within the public sector.
> For example, it is legitimate for an expectation that public sector workers facing applicants everyday should appear neutral and not be affiliated with any religion, so that there is no danger of the applicants thinking that the public sector worker is biased towards or against a certain religion.

Counter-point

> The ban is discriminatory as it can never be justified to impose a restriction that heavily impacts a specific religion.
> Many Muslim women will not be able to change their convictions and not wear their headscarves just because they have to work in the public sector as the headscarf forms an integral part of their culture and identity, and in many cases they will feel insecure and vulnerable without the headscarf.
> Such a ban is discriminatory towards a particular religion and should not be allowed.

END OF PAPER

Final Advice

Arrive well rested, well fed and well hydrated

The LNAT is an intensive test, so make sure you're ready for it. Ensure you get a good night's sleep before the exam (there is little point cramming) and don't miss breakfast. If you're taking water into the exam then make sure you've been to the toilet before so you don't have to leave during the exam. Make sure you're well rested and fed in order to be at your best!

Move on

If you're struggling, move on. Every question has equal weighting and there is no negative marking. In the time it takes to answer on hard question, you could gain three times the marks by answering the easier ones.

Afterword

Remember that the route to a high score is your approach and practice. Don't fall into the trap that *"you can't prepare for the LNAT"*– this could not be further from the truth. With knowledge of the test, some useful time-saving techniques and plenty of practice you can dramatically boost your score.

Work hard, never give up and do yourself justice.

Good luck!

About Us

Infinity Books is the publishing division of *Infinity Education*. We currently publish over 85 titles across a range of subject areas – covering specialised admissions tests, examination techniques, personal statement guides, plus everything else you need to improve your chances of getting on to competitive courses such as medicine and law, as well as into universities such as Oxford and Cambridge.

Outside of publishing we also operate a highly successful tuition division, called UniAdmissions. This company was founded in 2013 by Dr Rohan Agarwal and Dr David Salt, both Cambridge Medical graduates with several years of tutoring experience. Since then, every year, hundreds of applicants and schools work with us on our programmes. Through the programmes we offer, we deliver expert tuition, exclusive course places, online courses, best-selling textbooks and much more.

With a team of over 1,000 Oxbridge tutors and a proven track record, UniAdmissions have quickly become the UK's number one admissions company.

Visit and engage with us at:
Website (Infinity Books): www.infinitybooks.co.uk
Website (UniAdmissions): www.uniadmissions.co.uk
Facebook: www.facebook.com/uniadmissionsuk
Twitter: @infinitybooks7

Your Free Book

Thanks for purchasing this Ultimate Book. Readers like you have the power to make or break a book –hopefully you found this one useful and informative. *UniAdmissions* would love to hear about your experiences with this book. As thanks for your time we'll send you another ebook from our Ultimate Guide series absolutely <u>FREE</u>!

How to Redeem Your Free Ebook

1) Either scan the QR code or find the book you have on your Amazon
purchase history or your email receipt to help find the book on Amazon.

2) On the product page at the Customer Reviews area, click 'Write a customer review'. Write your review and post it! Copy the review page or take a screen shot of the review you have left.

3) Head over to www.uniadmissions.co.uk/free-book and select your chosen free ebook!

Your ebook will then be emailed to you – it's as simple as that!
Alternatively, you can buy all the titles at

<u>www.infinitybooks.co.uk</u>

Oxbridge
LAW PROGRAMME

UNIADMISSIONS

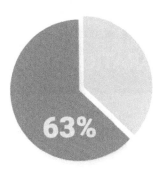

63%

UNIADMISSIONS 2019 Oxbridge Law Programme Success Rate

13%

The Average Oxford & Cambridge Law Success Rate

WHY DO OUR STUDENTS SEE SUCH HIGH SUCCESS RATES?

1 **30 HOURS OF EXPERT TUITION.**
UniAdmissions will guide you through a comprehensive, tried & tested syllabus that covers all aspects of the application - you are never alone.

2 **UNPARALLED RESOURCES.**
UniAdmissions' resources are the best available for the LNAT. You will get access to all of our resources, including the Online Academy, books and ongoing tutor support.

3 **WEEKLY ENRICHMENT SEMINARS.**
You'll get access to weekly enrichment seminars which will help you think like and become the ideal candidate that admissions tutors are looking for.

4 **INTENSIVE COURSE PLACES.**
By enrolling onto our Oxbridge Programme you will get reserved places for all of the Intensive Courses relevant to your application, such as the LNAT Intensive Course.

300+
Students successfully placed at Oxbridge in the last 3 years

50
Places available on our Oxbridge Law Programme in 2020

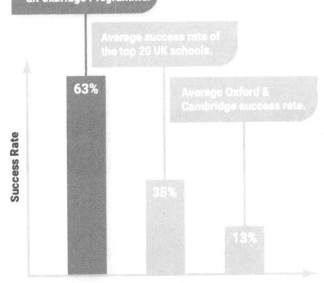

Students enrolled on an Oxbridge Programme.

Average success rate of the top 20 UK schools.

Average Oxford & Cambridge success rate.

63%

35%

13%

Success Rate

UNIADMISSIONS Oxbridge Programme Average Success Rate

58392431R00225